Air Power

Air Power
The World's Air Forces

Foreword by
Air Vice-Marshal Stewart Menaul

Edited by Anthony Robinson

Ziff-Davis Publishing Company
New York

First U.S. Edition
Ziff–Davis Publishing Company
One Park Avenue
New York, N.Y. 10016

First published in Great Britain by Orbis
Publishing Limited, London 1980
© by Orbis Publishing Limited, London 1980

Printed in Hong Kong by Toppan Printing
Company (H.K.) Limited

First Printing
ISBN 0–87165–080–0
Library of Congress Catalog Card
Number 80–80445

The text for this book was written by
J. M. Andrade (Chapter 8), Simon Clay
(Chapter 1), John Fricker (Chapter 11), John
Fricker and Paul Jackson (Chapters 6, 7, 12, 13,
15 and 16), Paul Jackson (Chapter 5), Peter
Kilduff (Chapter 2), John Lloyd (Chapter 4),
Group Captain R. A. Mason M. A., Director of
Defence Studies for the Royal Air Force
(Chapter 10), Kenneth Munson (Chapters 9
and 14) and L. Peacock (Chapter 3), and was
first published in the encyclopedia of
aviation, *Wings.*

*Title page: A Phoenix air-to-air missile is
released by a Grumman F-14 Tomcat carrier-
borne fleet defence fighter. (Photograph by
courtesy of the US Navy)*

Acknowledgments

The publishers would like to thank the following organizations for their permission to
reproduce the pictures shown on the pages listed:
W. D. Askham: 80T, 88/9, 92/3, 96. Aviation Photographs International; 1, 12C, 13B,
14/15C, 19CT, 24, 26/7B, 37C, 38, 39T, 47B, 50C, 52/3B, 53T, 54BR, 66/7B, 75T, 75B, 79B,
80C, 86/7TR, 93TL, 94T, 101, 104/5B, 106/7B, 107T, 116B, 136T, 136B, 140T, 140CT, 141L,
141R, 150/1B, 156T, 158T, 158C, 158B, 159, 160T, 160C, 160B, 162/3B, 176C, 177B, 180T,
180C, 184T, 185T, 186/7B, 189C, 190TL, 191T, 194/5B, 196/7T, 196/7C, 199, 204B, 216/7T,
218, 243, 248CT, 248B, 262C, 264CB, 268/9, 268T, 271T, 272/3, 273T, 273B, 274T, 274C,
275T, 275C, 282T, 283C, 283B, 284/5, 285T, 286T, 298TL, 300T, 300B. A. Balch: 164B,
172BL, 176T, 176B, 177T, 233T, 236, 240C, 299T, 300C. British Aerospace: 40, 263B, 281B.
C. Brooks: 189B, 222T, 261T, 261B, 298C. H. Cowin: 15T, 37T, 201T, 201BL. Dassault-
Breguet: 73R. Defence HQ of Canada: 113T. de Havilland Canada: 254C, 263R, 279TR,
280T, 295C. P. Endsleigh Castle: 22B, 27T, 32T, 34B, 76, 83, 110T, 111T, 114T, 147, 162T,
163B, 186T, 188T, 223B, 224T, 265T, 267T, 276T. Fokker VFW: 298/9. B. Freeman: 198L.
J. Fricker: 206TL, 206TR, 208T, 208/9B, 209T, 265CT, 265CB, 266T. J. Goulding: 4/5T,
12/13T, 16/17T, 19T, 54T, 55B, 59C, 63B, 90, 95B, 97T, 128T, 128B, 137T, 137B, 165T, 170T,
171T, 172T, 173T, 174T, 202T, 203T, 221, 255T, 260T, 260B, 289T, 295T, 296B. M. Hooks:
148T, 148C, 161C, 166/7TL, 167TR, 173BR, 175B, 179T, 179C, 210/1B, 215T, 252/3, 253T,
254B, 264T, 264CT, 278/9, 287, 287B, 292C, 297T, 297C. P. Jackson: 78T, 79T, 94B, 102,
188B, 190/1B, 192/3T, 192B, 206/7B, 249T, 249C, 249B, 250B, 251B, 254T, 268R,
270/1, 282R. M. Jerram: 252T. H. Kido: 222B, 225, 226, 226/7, 228/9, 232, 233B, 234/5.
D. Kingston: 205B, 244T, 244B, 246TL, 246TR, 266C. R. Kirkpatrick: 108CL, 115T, 123,
126TL, 133L, 137CT, 137CB, 140BL, 144L, 149T, 153BR, 155, 158TR, 161B. J. Lloyd: 3,
5C, 6/7, 8T, 8B, 10/11, 11TL, 11CR, 11BR, 12B, 14TR, 20TR, 51B, 60T, 63CT, 64, 65, 73B,
79C, 85B, 91, 94CB, 105T, 106T, 108T, 108CR, 108B, 110/11B, 116T, 119, 121T, 241, 242BR.
Lockheed: 242T. P. March: 7TR, 14B, 22T, 23, 28B, 28/9T, 34/5T, 36C, 45T, 46TL, 50B,
57T, 59T, 63CB, 68, 69B, 70B, 71T, 71B, 82B, 84/5T, 86/7B, 97C, 97B, 99R, 103B, 106B,
113C, 118B, 120C, 120B, 122/3B, 127B, 140B, 142/3C, 145, 146C, 146B, 150/1T, 152T,
292T, 293T. F. Mason: 47T, 73TL, 88TL, 120BR, 146TL, 164TL, 168CL, 169CR,
172C, 173C, 178TL, 179TR, 184BL, 187TL, 190C, 191C, 191BR, 193BL, 195TL, 197B,
198R, 201BR, 203B, 206C, 213TR, 213CR, 213BR, 215BR, 222TR, 223T, 231TL, 240T,
240B, 243T, 249TL, 252TR, 255B, 255C, 256T, 260C, 262T, 264B, 265B, 266B, 267L, 276B,
278L, 279L, 281T, 288TL, 288C, 295B, 296T, 297B, 298TR. MAP: 75C, 77, 81, 93TR, 94CT,
95TR, 100, 118T, 118C, 121B, 122T, 131B, 132C, 142T, 142/3B, 144R, 152C, 152B, 153T, 156CT,
156CB, 161T, 168T, 177C, 189T, 194T, 202CR, 203C, 204/5T, 205C, 207T, 210C, 211T,
212/3, 213T, 214C, 214/5B, 215CR, 217B, 219, 220, 262B, 276C, 277B. Ministry of Defence:
vi, 49T, 50T, 51T, 55C, 57C, 57B, 58B, 58T, 59B, 60B, 61, 63T, 66T, 70T, 71C, 72T, 72B,
166/7B, 169T. John Moore: 224B. Novosti: 165B, 175T, 179B. L. Peacock: 2, 4, 5B, 7CR, 9,
12T, 13C, 14CL, 15C, 17B, 19B, 20C, 20B, 21, 25T, 25C, 25B, 26, 30, 30/1, 32C, 32B, 33T,
33C, 34/5C, 35B, 36T, 39C, 39B, 42C, 43B, 44C, 44B, 45L, 46TR, 46B, 78C, 78B, 82T, 84B,
86TL, 98/9, 103T, 109, 112T, 112C, 112B, 113B, 114/5B, 117, 120T, 138, 139, 146/7T, 148/9B,
154TR, 156B, 205T, 216B, 230/1C, 237B, 238, 246/7, 248TL, 248CB, 250/1, 251T. S. Peltz:
184B, 185C, 185B, 289B, 293C, 294C. Pilot Press: 170/1B, 172BR, 173L, 178TR, 181T, 181C,
183T, 183C. C. Pocock: 256/7, 257T, 257C, 258T, 258CL, 258CR, 259TL, 259TR, 259BL,
259BR. G. Rhodes: 237T, 237CT, 239. Rolls Royce: 283T. Saab: 154TL, 154/5, 157.
H. Sixma (IAAP): 126/7, 127T, 128CT, 128CB, 129, 130T, 130B, 130/1C, 131T, 132/3T,
132B, 133R, 134/5, 136CT, 136CB. M. Turner: 48/9B, 122C, 288TR, 288B, 290/1T, 290/1B,
294T. US Navy: 41, 42T, 43T, 44T, 174B. Westland: 69T, 124. D. Wilton: 228T, 230BL,
231B, 237CB, 239T. M. Young: 210T, 245.

Contents

Foreword

The 1950s was a period of some uncertainty and indecision in military aviation. The ballistic missile had come of age with the deployment of medium-range weapons (MRBM) by the United States and the Soviet Union. Furthermore, it was widely accepted that, with the advent of inter-continental ballistic missiles (ICBM), manned aircraft would cease to be the primary instrument of strategic nuclear deterrence – a role it had performed since 1945 – and that the strategic bomber would be faded out of operational service. The famous 1957 British Defence White Paper went even further and suggested that missiles would eventually replace manned aircraft in both offensive and defensive roles. Thus began a debate on the future of manned aircraft which was to last for more than a decade, during which many advanced aircraft designs were either cancelled or placed in abeyance.

But, not for the first time, the pundits were proved wrong. Those who were most voluble in relegating aircraft to some subsidiary role in support of land forces, despite the lessons of World War II, Malaya and Korea, merely demonstrated their total lack of appreciation of the role of air power in war. Far from sinking into decline and giving way to missiles, manned aircraft today represent one of the most potent and powerful weapons in the armed forces of all industrialized countries and are much sought-after by the emerging Third World countries which recognize the speed and flexibility with which manned aircraft can bring immense fire-power to bear on a variety of targets.

Progress in the design and performance of modern aircraft in recent years, particularly in the strike, interceptor and close-support roles, has been rapid and dramatic. Developments in navigation, target acquisition and other systems have given the manned aircraft the ability to find and destroy targets in any weather by day or night. It is still the most flexible delivery system and the most accurate and reliable against moving targets.

There is a well-known law, never far from the mind of the military strategist, which says that for every measure there is a counter-measure and for every counter-measure there is a counter-counter-measure. Nowhere is this more pronounced than in the field of electronics, which plays such an important part in the operational efficiency of modern aircraft and the weapons they carry. The battle for superiority in the air see-saws between defence and offence and the competition to produce more sophisticated aircraft and weapons is unending. Missiles, however, can never totally replace manned aircraft, no matter how effectively they may perform in the variety of roles in which they are now employed. The cruise missile, with its terrain correlation guidance system, will take over some of the tasks assigned, until recently, to manned aircraft, but the computer responsible for the automatic control and guidance of the

Left: A Phantom FGR Mk 2 is pictured with a variety of weapons used for ground attack. The type performs equally effectively as an interceptor and air superiority fighter

missile to its target cannot be made to think and act in the rapidly changing environments likely to be encountered in modern war. Only air power can provide for swift and accurate fire-power in conventional and nuclear assaults on and beyond the battlefield.

The development of increasingly sophisticated aircraft has brought its problems, not least the vulnerability of the elaborate bases required to operate them. Despite protective measures such as hardening, camouflage and surface-to-air defences, the modern concrete air base remains vulnerable to attack by missiles and manned aircraft armed with conventional, nuclear or chemical warheads. Yet, strange as it may seem, the obvious answer to the problem was provided by Britain as long ago as the early 1960s with the introduction of the revolutionary vertical take-off and landing (VTOL) Harrier aircraft. Now operating in the more effective vertical/short take-off and landing mode (V/STOL) and by ski-jump, the Harrier is in operational service with the RAF, the United States Marine Corps and a number of other air and naval forces. Its most important theatre of operations, however, is Europe. Deployed on dispersed sites away from major air bases, the Harrier would survive attack by missiles or aircraft where other fixed-wing aircraft would be destroyed on the ground or rendered inoperable by the destruction of the air base itself.

The momentum in the development of military aviation, which resumed after the initial euphoria generated by the arrival of missiles of various kinds, has been maintained over the years. Both the major alliances have deployed new, more effective and more expensive aircraft and there has been no shortage of demand for them from developed and developing countries. It becomes increasingly difficult, therefore, for even the professional military strategist to keep abreast of air forces world-wide. Yet it is increasingly important for him to do so in an age when global strategy is the accepted discipline and no part of the world can be considered in isolation. Books such as this one, beautifully illustrated, should provide professional military strategists and laymen alike with a valuable reference source and at the same time an enjoyable read.

Air Vice-Marshal Stewart Menaul CB, CBE, DFC, AFC

Right: Vulcan B Mk 2 bombers of RAF Strike Command are parked on their operational readiness platform

CHAPTER ONE

United States Air Force

Below: the USAF's subsonic A-7D attack fighter has an advanced navigation and attack system, which facilitates very accurate weapons delivery. The pictured aircraft, from the 355th TFW was on detachment in Hawaii in August 1973. It carries 500 lb Mk 82 bombs on underwing racks. The unit has since re-equipped with the A-10, but A-7s continue to serve with TAC. Below right: the General Dynamics F-16 is Tactical Air Command's latest fighter. A two-seat F-16B is pictured

The world's most powerful air force can trace its origins back to the formation of an aeronautical division in the US Army's Signal Corps on 1 August 1907. For the next 40 years military aviation remained wedded to the US Army despite many inter-war attempts by air leaders to attain some form of autonomy. The decisive part played by air power during World War II was the major factor in the formation of the United States Air Force on 18 September 1947. Paramount in this development was the fact that the air arm held the supreme deterrent of those days, the atomic bomb. During World War II, the semi-autonomous United States Army Air Forces had a strength of over two and a quarter million men and some 20,000 aircraft, but postwar demobilisation saw these vast forces hurriedly reduced so that at the time the USAF was created only 300,000 personnel and a handful of combat-ready units remained. Soviet attitudes and ambitions provoked a gradual improvement in strength with emphasis on Strategic Air Command, which had the responsibility of nuclear delivery.

The outbreak of war in Korea in 1950 and the possibility of subsequent world-wide Communist military action, brought rapid expansion. The USAF maintained very large strategic and tactical forces for the next decade when the United States' support for South Vietnam brought another build-up, predominantly in Tactical Air Command at the expense of Strategic Air Command. With the termination of operations in South-East Asia there has been a gradual decline in personnel strength; the number of flying squadrons has been reduced from 427 in 1968 to 258 in 1979, while personnel has dropped from 904,000 to some 559,000 in the same period. The effective strike power, however, has not decreased due to new and greatly-improved weapons systems.

The USAF is composed of several commands and services, each devoted to a particular type of activity. The major flying commands are Tactical Air Command (TAC), Strategic Air Command (SAC), Aerospace Defense Command (ADCOM), Military Airlift Command (MAC) and Air Training

Command (ATC). There are also overseas forces which rate as major flying commands, notably the United States Air Forces in Europe (USAFE), Pacific Air Forces (PACAF) and Alaskan Air Command (AAC). Major back-up forces are Logistics and Systems Commands and Communications and Security Services. Aerospace Defense Command, Alaskan Air Command, Pacific Air Forces and the United States Air Forces in Europe are all linked with joint-service or multinational defence commands in their area of operations.

The structure of a command varies considerably, but the basic flying unit is the squadron. However, it has long been the practice to group a few squadrons with service and support units under a headquarters to form a more practical administrative and tactical organisation known as a wing. If flying squadrons are under a headquarters without full service or support units this is termed a group. The number of flying squadrons assigned to a wing or group varies from one to six but is usually two, three or four. A group or wing usually controls the same numbered squadrons which traditionally have always been assigned to it. Many wings and groups have only non-flying squadrons permanently assigned and merely function as holding units for flying squadrons from other wings or groups temporarily based at their airfield. There are also independent flying squadrons not assigned to a wing or group, but directly responsible to a higher command. The air division is such an organisation which normally acts in an administrative capacity between the wings and command.

The number of aircraft assigned to a squadron is

TACTICAL AIR COMMAND Headquarters: Langley AFB, Virginia, USA			
9th Air Force HQ: Shaw AFB, South Carolina			
12th Air Force HQ: Bergstrom AFB, Texas			
1st TFW	F-15A	Langley AFB, Va	9AF
1st SOW	AC-130A & H, CH-3E, UH-1N, OV-10	Hurlburt AFB, Fla	9AF
4th TFW	F-4E	Seymour-Johnson AFB, NC	9AF
23rd TFW	A-7D	England AFB, La	9AF
24th CW	O-2A, UH-1N	Howard AFB, Panama	
27th TFW	F-111D	Cannon AFB, NM	12AF
31st TFW	F-4E	Homestead AFB, Fla	9AF
33rd TFW	F-4E	Eglin AFB, Fla	9AF
35th TFW	F-4C, E & G	George AFB, Cal	12AF
49th TFW	F-15A	Holloman AFB, NM	12AF
56th TFW	F-4D & E	MacDill AFB, Fla	9AF
57th TTW	A-7D, A-10, F-4E, F-5E F-15A & B	Nellis AFB, Nev	
58th TTW	F-4C, F-15A & B, F-104G	Luke AFB, Ariz	12AF
67th TRW	RF-4C	Bergstrom AFB, Tex	12AF
347th TFW	F-4E	Moody AFB, Ga	9AF
345th TFW	A-10A	Myrtle Beach AFB, SC	9AF
355th TFW	A-10A	Davis-Monthan AFB, Ariz	12AF
363rd TRW	RF-4C	Shaw AFB, SC	9AF
366th TFW	F-111A	Mountain Home, Idaho	12AF
388th TFW	F-16	Hill AFB, Utah	12AF
474th TFW	F-4D	Nellis AFB, Nev	12AF
479th TTW	T-38	Holloman AFB, NM	12AF
507th TACW	CH-3E, O-2A	Shaw AFB, SC	9AF
552nd AWCW	E-3A, EC-135C	Tinker AFB, Okla	—
602nd TACW	O-2A, OV-10A, CH-53C	Bergstrom AFB, Tex	12AF

Abbreviations: AWCW, Airborne Warning and Control Wing; CW, Composite Wing; SOW, Special Operations Wing; TACW, Tactical Air Control Wing; TFW, Tactical Fighter Wing; TRW, Tactical Reconnaissance Wing; TTW, Tactical Training Wing.

dependent on its role. Tactical Air Command fighter squadrons have an establishment of 24 and tactical reconnaissance squadrons 18. Strategic Air Command bombers and tankers total 14 or 15 to a squadron, Military Airlift Command heavy transports 17 or 18 and tactical airlift squadrons are usually composed of 16 aircraft. Many specialised squadrons operate with much smaller complements.

The largest flying command in the USAF, Tactical Air Command currently has a personnel strength of 98,000 and some 2,000 aircraft. The Command is divided into two numbered air forces, the Ninth and Twelfth, with control of 11 flying tactical wings. There are also a number of specialised flying and training units directly under TAC Headquarters, which is located at Langley Air Force Base in Virginia. A wing normally consists of three or four squadrons, and a squadron aircraft establishment varies according to the mission of the unit.

Tactical Air Command has two major responsibilities; the first is to maintain a variety of first-line units for quick reinforcement of overseas areas in the event of an emergency. The Command also supplies the largely tactically-based air forces in Europe and the Pacific areas, providing units, equipment and trained personnel. Although PACAF and USAFE are independent commands, they are basically overseas extensions of TAC. In the United States TAC also oversees the equipping and operations of nearly 100 units of the 'part-time' Air National Guard (ANG) and Air Force Reserve (AFR). State territorial units make up the

Air National Guard of which more than half have a commitment to TAC. In the past ANG units have largely been equipped with aircraft retired from the regular Air Force. Policy changes now place more reliance on the ANG and many of the squadrons have received aircraft straight from the factory. The TAC-committed element of the Air Force Reserve is much smaller, numbering 10 squadrons in 1979, but with a variety of equipment. Both AFR and ANG squadrons take part in TAC-organised exercises, including overseas movements to locations in Europe and the Pacific.

TAC fighters have multi-role capability although some types are more suited to particular missions. The McDonnell Douglas F-15 Eagle was originally designed and developed as an air-superiority fighter to meet the ascendancy of Soviet fighters in this category which became very evident in the late 1960s. The USAF has a planned procurement of over 700, with production continuing until 1985.

The sophistication of this large twin-jet fighter has made the F-15 a very expensive project and led the USAF to order the General Dynamics F-16, which has its origin in a programme initiated in 1970 to study a relatively simple, lightweight

The Fairchild A-10 is specifically designed as a ground-attack aircraft and it has a 30mm cannon which can fire up to 4,200 rounds per minute. This weapon has proved to be extremely effective against heavily-armoured vehicles and the aircraft's role is primarily as a tank killer. Additionally the aircraft can carry a 6,800 kg (15,000 lb) load of bombs and missiles on 11 underwing and fuselage attachment points. The A-10 is intended for operating from forward bases so that its endurance can be best used for patrolling a battle area. It has also been designed to survive enemy ground fire through a combination of exceptional manoeuvrability and design features which enable it to keep flying even if badly damaged. A total of 733 A-10s has been ordered and will equip two TAC wings and three in Europe.

The largest overseas contingent of the USAF is located in Europe with a NATO commitment. The locations of the main combat units are in the United Kingdom and Germany but USAFE also has bases in Spain, Italy, Turkey, Greece and the Netherlands. Eight tactical fighter wings constitute the major strike force and are backed up by two tactical reconnaissance wings and a number of specialised flying units. Recent modernisation of the USAFE

Above: Nellis Air Force Base, Nevada, is the centre of the USAF's extensive tactical fighter training programme. An F-4E of the Base's 57th Tactical Training Wing is illustrated.
Left: Northrop F-5B two-seaters of the 58th Tactical Training Wing's 425th Tactical Fighter Training Squadron are used to train F-5 pilots of foreign air forces at Williams AFB, Arizona. The rest of the Wing is at Luke AFB.
Right: the 355th TFW's tank-busting Fairchild A-10A Thunderbolt IIs are based at Davis-Monthan AFB, Arizona.
Below right: the Cessna 0-2A undertakes forward air control duties with the 602nd TACW, marking targets for attack by tactical fighters

fighter design. With general escalation of aircraft costs, the programme assumed more importance and the F-16 was eventually chosen for production, receiving an additional boost from the decision by four NATO countries to re-equip their fighter squadrons with the type. Highly manoeuvrable and with a Mach 2 performance the F-16 offered exceptional value for money at the time of order and the USAF is scheduled to receive 1,388 of these fighters. However, adapting the aircraft to undertake all-weather and ground-attack roles has proved costly and its effectiveness in these missions has been questioned.

Although the F-15 and F-16 are scheduled to dominate the TAC inventory, the F-4 Phantom will play a major role in this and associated overseas commands for some years to come. The most important strike aircraft in TAC is the 'swing-wing' General Dynamics F-111 of which 427 were produced between 1967 and 1976. With its good payload, advanced avionics and high speed, the F-111 is likely to remain with TAC for many years and will be subject to regular updating.

inventory has included the establishment of an F-15 Eagle wing in Germany and a single squadron with this type in the Netherlands, primarily charged with an air superiority role. These and other units obtain realistic aerial combat training in exercises with the 527th 'Aggressor' Squadron operating from Alconbury in the UK with Northrop F-5E aircraft which simulate the interception tactics of a potential enemy.

The tactical fighter role is still performed by the faithful F-4 Phantom, units equipped with this type being located in Germany and at a base in Spain. The photographic reconnaissance version of the Phantom equips a wing in Germany and one in the UK. Latest addition to the inventory is the Fairchild A-10, the first USAFE wing to equip with the type being the 81st TFW. Based at Bentwaters in the UK, it is unique in that it is scheduled to have six squadrons instead of the usual three. However, Bentwaters is planned as a rear base and at any one time one third of the A-10 force will be at advanced locations in Germany.

A major element of the USAFE strike force is the

180 F-111 long-range, all-weather tactical fighter-bombers based at Upper Heyford and Lakenheath in the UK. With nuclear capability and terrain-following guidance systems, these aircraft are the most important factor in NATO's deterrent posture. There are some Military Airlift Command and Strategic Air Command units under USAFE to handle rotational aircraft from the United States.

The United States' involvement in the South-East Asia conflict of the 1960s saw a major consignment of USAF strength to the Pacific Air Forces. In contrast, a decade later PACAF is a very small force, although maintaining a high state of readiness. The Command has responsibility for US interests in a vast area approximating to half the world's surface. Major locations for the flying elements are Hawaii, Korea, Okinawa and the Philippines. The F-4 Phantom is the principal strike aircraft of the three tactical fighter wings located in Korea, Okinawa and the Philippines. Other specialised units are attached to these wings and operate a variety of different aircraft types. In a modernisation plan, the F-15 Eagle is due to replace the F-4s of the 18th Tactical Fighter Wing at Kadena, Okinawa and E-3A Sentry aircraft are also to be located at this base to give airborne early warning and control capability for the area.

The most famous of all USAF commands, Strategic Air Command maintained the western world's principal nuclear deterrent force during the height of the 'cold war' in the 1950s. Its then-enormous force of manned strategic bomber aircraft was eventually supplemented by intercontinental ballistic missiles. The bomber element has since waned, particularly with the advent of the nuclear missile-launching submarine and it was expected that the manned bomber would eventually disappear from the inventory. However, the decision not to put all the deterrent eggs in one basket found a resurgence of interest in the manned bomber, particularly with the advent of terrain-following radar, which provides a higher degree of surviv-

UNITED STATES AIR FORCES IN EUROPE Headquarters: Ramstein AB, Germany
3rd Air Force HQ: Mildenhall, England
16th Air Force HQ: Torrejon AB, Spain
17th Air Force HQ: Sembach AB, Germany

Wing	Aircraft	Base	AF
10th TRW	RF-4C & F-5E	Alconbury, UK	3AF
20th TFW	F-111E	Upper Heyford, UK	3AF
26th TRW	RF-4C	Zweibrucken, Germany	17AF
36th TFW	F-15A	Bitburg, Germany	17AF
48th TFW	F-111F	Lakenheath, UK	3AF
50th TFW	F-4E	Hahn, Germany	17AF
52nd TFW	F-4D	Spangdahlem, Germany	17AF
81st TFW	A-10A	Bentwaters/Woodbridge, UK	3AF
86th TFW	F-4E	Ramstein, Germany	17AF
401st TFW	F-4C	Torrejon, Spain	16AF
601st TCW	O-2A, OV-10 & CH-53	Sembach, Germany	17AF
32nd TFS	F-15A	Soesterberg, Neth	17AF

Holding wings for rotational units

Wing	Aircraft	Base	AF
406th TFT	F-4, A-10, F-15	Zaragoza, Spain	16AF
513th TAW	C-130 & EC-135	Mildenhall, UK	3AF

Abbreviations: TAW, Tactical Airlift Wing; TCW, Tactical Control Wing; TFS, Tactical Fighter Squadron; TFTW, Tactical Fighter Training Wing; TFW, Tactical Fighter Wing; TRW, Tactical Reconnaissance Wing.

ability for large aircraft in a hostile environment. In consequence, the number of manned bomber squadrons in SAC has been fairly constant for the past five years although efforts to obtain a replacement for the ageing B-52 Stratofortress have not met with success.

Reliance on the B-52 – which first entered service in 1955 – continues, with 17 bomb wings operating the type. Eighty B-52D models were rebuilt in 1975–1977 to prolong their useful service life. New bombing and navigational systems and defensive electronics made these aircraft superior in these respects to the later B-52G and B-52H models which have been retained by SAC. Both the B-52G and B-52H can carry the AGM-69 short-range attack missiles (a maximum of 20 per aircraft) and are equipped with an electro-optical viewing system employing low-light level television sensors to give the aircraft terrain-following flight capability. The G and H models are expected to be completely refurbished when the air-launch cruise missile, currently under development, is introduced. A maximum load of 12 is planned for each B-52.

Supplementing the 300 Stratofortresses are 60 General Dynamics FB-111As. Operated by two

wings (with two squadrons each) based in the north-eastern United States, these aircraft have a tactical range of 3,220 km (2,000 miles) compared with the 5,630 km (3,500 miles) of the B-52s. Six AGM-69A air-to-surface missiles can be carried, but a more likely load on long-range strategic operations would be two in the internal weapons bay.

The Command also operates the USAF's flight refuelling force of over 600 Boeing KC-135 Strato-tankers. Most bomb wings are composed of one bomber squadron and one tanker squadron. There are, however, five air refuelling groups and wings wholly composed of KC-135 units. A proportion of the total KC-135 force is operated by part-time airmen, the Air National Guard having 13 squad-rons and the Air Force Reserve three, each with a unit establishment of eight Stratotankers. The

Right: the McDonnell Douglas F-15 Eagle is scheduled to be the second most numerous type in TAC, with over 700 ordered.
Below right: an F-111F of the 48th TFW lands at its Lakenheath, England, base.
Below: a German-based F-4D Phantom of the 52nd TFW leaves the runway

PACIFIC AIR FORCES Headquarters: Hickam AFB, Hawaii
5th Air Force HQ: Yokota, Japan
13th Air Force HQ: Clark Air Base, Philippines

3rd TFW	F-4D, F-5E	Clark AB, Philippines	13AF
8th TFW	F-4E	Kunsan, Korea	5AF
18th TFW	F-4C, RF-4C, F-15A	Kadena, Okinawa	5AF
51st CW	F-4E & OV-10	Osan, Korea	5AF

Abbreviations: CW, Composite Wing; TFW, Tactical Fighter Wing.

primary mission of the tanker fleet is the support of SAC's own bomber and reconnaissance aircraft but in-flight refuelling facilities are also available to other USAF Commands and the US Navy and Marines. In furtherance of this service large numbers of KC-135s are rotated from their bases in the United States to locations in Europe and the Pacific. Some 30 tankers are usually on hand in the United Kingdom, Germany and Spain, while Pacific KC-135s are rotated to Okinawa and Guam.

Strategic reconnaissance is yet another mission embraced by SAC and there are two wings carrying out this highly specialised work. The 55th Strategic Reconnaissance Wing operates RC-135 and EC-135 adaptations of the venerable KC-135 tanker. The EC-135 serves as a flying command post and is equipped with sophisticated control and communication equipment for assisting the SAC strike force. RC-135s are packed with highly-secret electronics for surveillance purposes, usually the monitoring of other nations' radar systems. The 9th Strategic Reconnaissance Wing operates the

Left: the 51st Composite Wing, based at Osan in Korea, undertakes both the tactical fighter role and forward air control duties. For the former commitment it is equipped with F-4E Phantoms. Below: Rockwell OV-10 Broncos form the forward air control element of the 51st Composite Wing. Right: air refuelling units are an important element in Strategic Air Command's bomber force and each bomb wing includes its own tankers. A B-52G of the 319th BW is pictured fuelling from a KC-135

HIGH SPEED BOOM

STRATEGIC AIR COMMAND Headquarters: Offutt AFB, Neb			
8th Air Force HQ: Barkesdale AFB, Louisiana			
15th Air Force HQ: March AFB, California			
2nd BW	B-52G & KC-135A	Barksdale AFB, La	8AF
5th BW	B-52H & KC-135A	Minot AFB, ND	15AF
7th BW	B-52D & KC-135A	Carswell AFB, Tex	8AF
9th SRW	SR-71 A & C, U-2D & R	Beale AFB, Cal	15AF
19th BW	B-52G & KC-135A	Robins AFB, Ga	8AF
22nd BW	B-52D & KC-135A	March AFB, Cal	15AF
28th BW	B-52G & KC-135A	Ellsworth AFB, SD	15AF
42nd BW	B-52G & KC-135A	Loring AFB, Me	8AF
44th SMW	Minuteman II	Ellsworth AFB, SD	15AF
55th SRW	EC-135C RC-135V, S & U	Offutt AFB, Neb	15AF
68th BW	B-52G & KC-135A	Seymour-Johnson AFB, NC	8AF
90th SMW	Minuteman III	Warren AFB, Wyo	15AF
91st SMW	Minuteman III	Minot AFB, ND	15AF
92nd BW	B-52G & KC-135A	Fairchild AFB, Wash	15AF
93rd BW	B-52G & KC-135A	Castle AFB, Cal	15AF
96th BW	B-52D & KC-135A	Dyess AFB, Tex	15AF
97th BW	B-52G & KC-135A	Blytheville AFB, Ark	8AF
100th ARW	KC-135A	Beale AFB, Cal	15AF
301st ARW	KC-135A	Rickenbacker AFB, Ohio	8AF
305th ARW	KC-135A	Grissom AFB, Ind	8AF
307th ARG	KC-135A	Travis AFB, Cal	15AF
308th SMW	Titan II	Little Rock AFB, Ark	8AF

USAF's most intriguing aircraft, the Lockheed U-2 and SR-71. Both types carry highly sophisticated and varied electronic sensing equipment and have been used by the USAF to monitor 'sensitive areas' by operating at heights in excess of 24,400 m (80,000 ft) along territorial borders. They are frequently deployed overseas, although much secrecy surrounds their coming and going and they are usually removed to a hangar immediately after landing at an overseas base.

Apart from the manned bomber force, SAC's part in the nuclear deterrent triad is the intercontinental ballistic missile force. The size of this force has been maintained at 1,054 missiles for several years, each missile being located in a separate underground silo, the silos widely dispersed in the

Below: pressed into service with a rebuilt internal bomb-bay, the Boeing B-52D flew many missions over Vietnam from locations in Guam and Thailand. Right: the pictured Guam-based Stratofortress exhibits the badge of the 43rd Strategic Wing. Below right: the Lockheed SR-71 is flown exclusively by the 9th SRW from Beale AFB. Bottom right: developed from the Boeing 707 airliner, the KC-135 aerial tanker has itself proved the basis for reconnaissance and other variants

319th BW	B-52H & KC-135A	Grand Forks AFB, ND	15AF
320th BW	B-52G & KC-135A	Mather AFB, Cal	15AF
321st SMW	Minuteman III	Grand Forks AFB, ND	15AF
340th ARG	KC-135A	Altus AFB, Okla	8AF
341st SMW	Minuteman III	Malstrom AFB, Mont	15AF
351st SMW	Minuteman II	Whiteman AFB, Mo	8AF
379th BW	B-52H & KC-135A	Wurtsmith AFB, Mich	8AF
380th BW	FB-111A & KC-135A	Plattsburgh AFB, NY	8AF
381st SMW	Titan II	McConnell AFB, Kan	8AF
384th ARW	KC-135A	McConnell AFB, Kan	8AF
390th SMW	Titan II	Davis-Monthan AFB, Ariz	15AF
410th BW	B-52H & KC-135A	Sawyer AFB, Mich	8AF
416th BW	B-52G & KC-135A	Griffiss AFB, NY	8AF
509th BW	FB-111A & KC-135A	Pease AFB, NH	8AF

Holding wings for rotational units

6th SW	RC-135 & KC-135	Eielson AFB, Alaska
11th SW	KC-135	Fairford, England
43rd SW	B-52 & KC-135	Andersen AFB, Guam
306th SW	KC-135	Ramstein, Germany & Mildenhall, England
376th SW	KC-135	Kadena, Okinawa

Abbreviations: ARG, Air Refueling Group; ARW, Air Refueling Wing; BW, Bomb Wing; SMW, Strategic Missile Wing; SRW, Strategic Reconnaissance Wing; SW, Strategic Wing.

AEROSPACE DEFENSE COMMAND Headquarters: Peterson AFB, Colorado

5th FIS	F-106A & B	Minot AFB, ND
48th FIS	F-106A & B	Langley AFB, Va
49th FIS	F-106A & B	Griffiss AFB, NY
57th FIS	F-4E	Keflavik, Iceland
84th FIS	F-106A & B	Castle AFB, Cal
87th FIS	F-106A & B	Sawyer AFB, Mich
318th FIS	F-106A & B	McChord AFB, Wash

Abbreviation: FIS, Fighter Interceptor Squadron.

central United States. The major element is 1,000 LGM-30 Minuteman three-stage solid propellant rocket-powered missiles. Two versions are in service, the Minuteman II with single thermonuclear warhead and the newer Minuteman III which has three separate nuclear warheads which can be targeted independently of each other. The other ICBM is the LGM-25 Titan II, a two-stage liquid-propellant type which carries a considerably higher yield nuclear warhead than the Minuteman. The Titan has a 9,650 km (6,000-mile) range and the Minuteman some 11,260 km (7,000 miles), allowing them to reach most potential targets. Fears as to vulnerability of silo-based missiles in the light of known advances in Soviet strike capabilities have led to a number of developments in launch techniques aimed at minimising this risk.

The Aerospace Defense Command organisation evolved from Air Defense Command and is charged with defending the United States against enemy air attack. At peak strength, in the late 1950s, this command had over 40 regular fighter squadrons backed by a large number of Air National Guard units, as well as controlling a vast network of radar warning installations. With the gradual ascendancy of the intercontinental and submarine-launched missiles, the manned bomber threat has receded; in con-

sequence over the past decade and a half the regular USAF flying element of this Command has been reduced to seven interceptor squadrons. In 1980 Aerospace Defense Command is scheduled to be inactivated, with its flying units being transferred to TAC and missile warning and other surveillance activities passed to SAC.

Six of the fighter squadrons currently fly the Convair F-106 Delta Dart, as do five ANG squadrons under the Command's jurisdiction. These supersonic all-weather fighters have been in service for more than 20 years although often modified and given updated radars. A single interceptor squadron is maintained in Iceland and currently flies F-4E Phantoms. Additionally the Command has the services of three ANG McDonnell F-101 Voodoo and two F-4 Phantom squadrons in the continental United States. The only other major flying squadron in the Command operates Martin EB-57 Canberras, the last remaining USAF unit with this type. These operate in a simulation role, playing the part of enemy bombers and using their electronic jamming equipment to test defensive forces.

As the title of the Command suggests, the missile warning role takes in space. A large part of the surveillance resources are concerned with monitoring all satellites using optical and electronic

Opposite top: a Convair F-106A of the Aerospace Defence Command.
Opposite centre: electronic reconnaissance and instrument training is the province of the ANG's 158th DSEG, flying the Martin EB-57.
Opposite: the 147th FIG of Ellington AFB, Texas, is one of three Air National Guard units to be equipped with the McDonnell F-101B Voodoo.
Top: F-106s of the 94th FIS.
Above: Air National Guard groups retain examples of the Lockheed T-33A trainer for refresher training and other duties. A 147 FIG aircraft is pictured.
Left: known as the 'Deuce', the Convair F-102 was phased out of ANG service in the 1970s as part of the United States' reduction in fighter defence

MILITARY AIRLIFT COMMAND Headquarters: Scott AFB, Illinois, USA
21st Air Force HQ: McGuire AFB, NJ
22nd Air Force HQ: Travis AFB, Calif

60th MAW	C-5A & C-141A	Travis AFB, Cal	22AF
62nd MAW	C-130H & C-141A	McChord AFB, Wash	22AF
63rd MAW	C-141A	Norton AFB, Cal	22AF
89th MAG	VC-6A, VC-9A, VC-12A	Andrews AFB, Md	21AF
	VC-135B, VC-137B, VC-140B		
	CH-3E, UH-1N		
314th TAW	C-130E & C-141A	Little Rock AFB, Ark	22AF
317th TAW	C-130E	Pope AFB, NC	21AF
374th TAW	C-130E	Clark AB, Philippines	22AF
375th AAW	C-9A	Scott AFB, Ill	—
436th MAW	C-5	Dover AFB, Del	21AF
437th MAW	C-141A	Charleston AFB, SC	21AF
438th MAW	C-141A	McGuire AFB, NJ	21AF
443rd MAW	C-141 & C-5	Altus AFB, Okla	22AF
463rd TAW	C-130E	Dyess AFB, Tex	22AF
616th MAG	C-130D, HC-130H & N,	Elmendorf AFB, Alaska	22AF
	CH-3E, HH-3E		

Holding wings for rotational units

435th TAW	C-12 & C-130	Rhein-Main AB, Germany	21AF

Abbreviations: AAW, Aeromedical Airlift Wing; MAG, Military Airlift Group;
MAW, Military Airlift Wing; TAW, Tactival Airlift Wing.

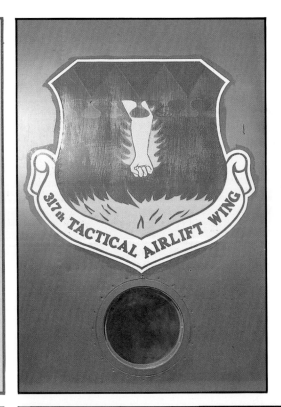

sensors located in many parts of the world. The three Ballistic Missile Early Warning System radars, one of which is located in the United Kingdom, provide 25 minutes' warning of ICBM attack on the United States. New radars, noticeably the 'Pave Pause' type, are coming into operation on the east and west coasts of the United States to give improved warning against submarine-launched missiles. A complete re-organisation of the various aircraft warning radar ground systems is planned to coincide with the dissolution of the Command.

Military Airlift Command provides a regular air freight and passenger service to US military bases in the United States and abroad. In times of crisis, both military and civil, it provides substantial facilities to move large quantities of personnel, materiel or equipment to the scene. There is also a tactical element to provide theatre troop and supply transportation, an air ambulance service, VIP airlift and air rescue and recovery and weather reconnaissance services.

MAC operates a score of different aircraft types with the C-130 Hercules and Lockheed C-141 Starlifter predominating. Most of the Hercules were inherited from Tactical Air Command when its airlift mission was transferred to MAC. Some of

AIR TRAINING COMMAND Headquarters: Randolph AFB, Texas, USA		
12th FTW	T-37B & T-38A	Randolph AFB, Tex
14th FTW	T-37B & T-38A	Columbus AFB, Miss
47th FTW	T-37B & T-38A	Laughin AFB, Tex
64th FTW	T-37B & T-38A	Reese AFB, Tex
71st FTW	T-37B & T-38A	Vance AFB, Okla
80th FTW	T-37B & T-38A	Sheppard AFB, Tex
82nd FTW	T-37B & T-38A	Williams AFB, Ariz
323rd FTW	T-41A	Mather AFDB, Cal
Abbreviation: FTW, Flying Training Wing.		

these aircraft have been used to supplement the larger and longer range transports by performing short-haul cargo flights. A total of 270 C-141s and 76 Lockheed C-5s provide long-range airlift.

MAC is scheduled to activate two squadrons of McDonnell Douglas KC-10As over the next two years to supplement the heavy transport force. These aircraft, based on the DC-10 commercial transport, will serve as in-flight refuelling tankers to increase the range of C-5s and C-141s. They will also be able to carry substantial cargo loads and will be used to support the overseas movement of TAC fighter units by providing air refuelling while transporting the fighter units' ground crews and equipment.

Galaxys and Starlifters fly the regular overseas hauls, chiefly to the main European bases at Mildenhall, UK and Rhein-Main, Germany; Hickam, Hawaii, Kadena, Okinawa and Clark, Philippines in the Pacific are also served. The Command's high rate of operations is made possible by the use of Air Force Reserve Associate units. There is an Associate squadron manned by reservists for

every regular MAC strategic transport squadron. The reservists act as a relief for the regular MAC crews and participate in all types of operation and to all MAC's overseas locations. The reserve units have no aircraft assigned to them, using those of the regular MAC units with which they are associated. In effect, this means that the majority of the large transports are operated by two squadron teams. In addition to the flight crews the reserve Associate units also make up various ground-support teams.

A large proportion of the C-130 force received from TAC has been passed to the reserve. Whereas MAC has only 14 regular squadrons equipped with the type, the Air Force Reserve has 11 and the Air National Guard 18. The ground-support airlift mission also includes four AFR squadrons flying Fairchild C-123 Providers and two squadrons of DHC C-7A Caribou. There is also another Caribou squadron serving with the ANG. Like the Hercules and ageing Providers, the Caribou squadrons are primarily used for troop movement. It is anticipated that in the future even more of the USAF's tactical airlift role will be taken over by reserve units.

Opposite top: the symbolic emblem of the C-130-equipped 317th TAW.
Above far left: the Hercules has been the workhorse of Military Airlift Command since it took over the tactical airlift responsibilities of TAC.
Above left: the massive C-5A Galaxy spearheads Military Airlift Command's long-haul transport force.
Opposite: precursor of the Galaxy was Lockheed's C-141 Starlifter, which commenced operations in April 1965.
Top: Cessna T-37B aircraft of the 82nd Flying Training Wing line up at Williams AFB with the mountains of Arizona in the background. A total of seven training wings currently operate the type.
Above: trainee USAF pilots encounter the Northrop T-38 Talon after primary training on the Cessna T-41 and basic instruction on the T-37. The pictured aircraft is temporarily attached to SAC

Right: sixty FB-111As, the strategic bomber variant of the General Dynamics F-111, equip two wings in SAC. An aircraft of the 380th Bomb Wing is shown. Armed with up to six AGM-69A SRAM stand-off missiles and with a speed of Mach 0·9 at low level, the FB-111A is better-suited to attack heavily-defended targets than the longer-range B-52. Below right: the Air Force Reserve provides a welcome source of reinforcement for the regular USAF, together with the state-organised Air National Guard. A Republic F-105 Thunderchief of the Reserve's 465th Tactical Fighter Squadron, 301st TFW is illustrated

Both the AFR and the ANG also contribute to MAC's rescue and recovery service. This has a varied complement of Bell HH-1 and Sikorsky HH-3 helicopters and HC-130 Hercules specially adapted for the role. Of seven regular MAC Rescue and Recovery squadrons four are based at overseas locations, with detachments at a number of airfields. Most operate a few HC-130H, N or P models with specialised equipment for rescuing downed aircrew and retrieving satellite ejections. These Hercules also act as tankers for rescue helicopters and are equipped with the trailing drogue type refuelling system. Standard helicopter for the regular rescue squadrons is the Sikorsky HH-53, which can carry between 40 and 50 people and has a range of 800 km (500 miles) – although in-flight refuelling equipment is provided.

MAC operates the Air Weather Service supplying world-wide weather information to US military forces. Two special weather squadrons operate three different models of the WC-130 Hercules chiefly for monitoring tropical storms. Their services are regularly called upon to ensure ideal weather conditions before the launching of satellite-carrying missiles from sites in the south-eastern United States. A single aeromedical wing operates from Scott Air Force Base in Illinois, principally with Douglas C-9 Nightingales backed by a few C-130s and C-141s specially fitted out for air ambulance work.

As the remote American state facing the Soviet Union across the Bering Strait, Alaska has a sizeable military presence. These are primarily defence forces but a certain amount of surveillance of Soviet military and naval activity is carried out. The USAF's Alaskan Air Command is a major contributor to the combined US/Canadian North American Air Defense Command which provides warning and interceptor forces to meet any aggressor's air strikes through the Arctic.

The permanently-based flying units in Alaska only number six, although TAC and SAC squadrons regularly use the two main airfields near Anchorage and Fairbanks for training purposes. Elmendorf Air Force Base, at Anchorage, is the Command headquarters and the permanent base of two tactical fighter squadrons equipped with F-4E Phantoms, a single Tactical Airlift Squadron with C-130Es and a Rescue and Recovery Squadron using the fixed-wing Hercules and Sikorsky H-3 Jolly Green Giant helicopters. Elmendorf is also headquarters for the group controlling the many radar warning units operating along the western coast of the State. At Eielson AFB, near Fairbanks, a single Tactical Air

ALASKAN AIR COMMAND Headquarters: Elmendorf AFB, Anchorage		
21st CW	F-4E & T-33	Elmendorf AFB, Alaska
616th MAG	C-130E, HC-130H & HH-3E	Elmendorf AFB, Alaska
25th TASS	O-2A	Eielson AFB, Alaska

Abbreviations: CW, Composite Wing; MAG, Military Airlift Group; TASS, Tactical Air Support Squadron.

AIR FORCE RESERVE Headquarters: Robins AFB, Georgia			
94th TAW	C-7A & C-9A	Dobbins AFB, Ga	MAC
301st TFW	F-105B, D & F	Carswell AFB, Tex	TAC
302nd TAW	C-123K	Rickenbacker AFB, Ohio	MAC
315th MAW(A)	C-141A	Charlestown AFB, SC	MAC
349th MAW(A)	C-5A & C-141A	Travis AFB, Cal	MAC
403rd RWRW	HH-1H, HH-3E, HC-130H & N, WC-130H	Selfridge Fd. Mich	MAC
433rd TAW	C-130B	Kelly AFB, Tex	MAC
434th TFW	A-37B	Grissom AFB, Ind	TAC
439th TAW	C-123K, C-130A&B	Westover AFB, Mass	MAC
440th TAW	C-130A	Billy Mitchell Fd, Wis	MAC
442nd TAW	C-130A & E	Richards-Gebaur AFB, Mo	MAC
445th MAW(A)	C-141A	Norton AFB, Cal	MAC
446th MAW(A)	C-141A	McChord AFB, Wash	MAC
452nd ARW	KC-135A	March AFB, Cal	SAC
459th TAW	C-130A & E	Andrews AFB, Md	MAC
512th MAW(A)	C-5A	Dover AFB, Del	MAC
514th MAW(A)	C-141A	McGuire AFB, NJ	MAC
915th TFG	F-4C	Homestead AFB, Fla	TAC
919th SOG	AC-130A, CH-3E	Eglin AFB, Fla	TAC

Abbreviations: ARW, Air Refueling Wing; MAW(A), Military Airlift Wing (Associate); RWRW, Rescue & Weather Reconnaissance Wing; SOG, Special Operations Group; TAW, Tactical Airlift Wing; TFG, Tactical Fighter Group; TFW, Tactical Fighter Wing.

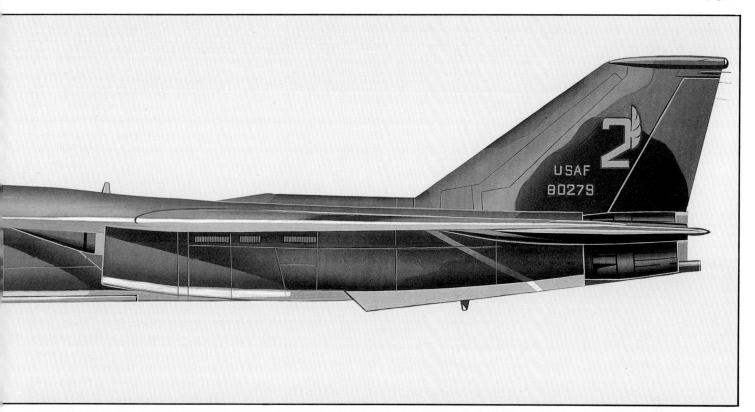

AIR NATIONAL GUARD Headquarters: Andrews AFB, Maryland			
101st ARW	KC-135A	Bangor, Me	SAC
102nd FIW	F-106A & B	Otis AFB, Mass	ADCOM
103rd TFG	F-100D	Winsor Locks, Conn	TAC
104th TFG	F-100D	Westfield, Mass	TAC
105th TASW	O-2A	White Plains, NY	TAC
106th ARR	HC-130H&HH-3E	Suffolk Airport, NY	MAC
107th FIG	F-101B	Niagara Falls, NY	ADCOM
108th TFW	F-105B	McGuire AFB, NJ	TAC
109th TAG	C-130A	Schenectady, NY	MAC
110th TASG	O-2A	Battle Creek, Mich	TAC
111th TASG	O-2A	Willow Grove, Pa	TAC
112th TFG	A-7D	Pittsburgh, Pa	TAC
113th TFW	F-105D	Andrews AFB, Md	TAC
114th TFG	A-7D	Sioux Falls, SD	TAC
115th TASW	O-2A	Traux Field, Wis	TAC
116th TFW	F-105G	Dobbins AFB, Ga	TAC
117th TRW	RF-4C	Birmingham, Ala	TAC
118th TAW	C-130A	Nashville, Tenn	MAC
119th FIG	F-4C	Fargo, ND	ADCOM
120th FIG	F-106A	Great Falls, Mont	ADCOM
121st TFW	A-7D	Rickenbacker AFB, Ohio	TAC
122nd TFW	F-4C	Fort Wayne, Ind	TAC
123rd TRW	RF-4C	Louisville, Ky	TAC
124th TRG	RF-4C	Boise, Idaho	TAC
125th FIG	F-106A & B	Jacksonville, Fla	ADCOM
126th ARW	KC-135A	Chicago, Ill	SAC
127th TFW	A-7D	Selfridge AFB, Mich	TAC
128th ARG	KC-135A	Milwaukee, Wis	SAC
129th ARRG	HC-130H&HH-3E	Haywood, Cal	MAC
130th TAG	C-130E	Charleston, W Va	MAC
131st TFW	F-4C	St. Louis, Mo	TAC
132nd TFW	A-7D	Des Moines, Iowa	TAC
133rd TAW	C-130A	Minneapolis, Minn	MAC
134th ARG	KC-135A	Knoxville, Tenn	SAC
135th TAG	C-7	Baltimore, Md	MAC
136th TAW	C-130E	Dallas, Tex	MAC
137th TAW	C-130A	Oklahoma City, Okla	MAC
138th TFG	A-7D	Tulsa, Okla	TAC
139th TAG	C-130E	St Joseph, Mo	TAC
140th TFW	A-7D	Buckley, Colo	TAC
141st ARW	KC-135A	Fairchild AFB, Wash	SAC
142nd FIG	F-101B	Portland, Ore	ADCOM
143rd TAG	C-130E	Providence, RI	MAC
144th FIW	F-106A	Fresno, Cal	ADCOM
145th TAG	C-130B	Charlotte, NC	MAC
146th TAW	C-130B	Van Nuys, Cal	MAC
147th FIG	F-101B	Ellington AFB, Tex	ADCOM
148th TRG	RF-4C	Duluth, Minn	TAC
149th TFG	F-4C	Kelly AFB, Tex	TAC
150th TFG	A-7D	Kirtland AFB, NM	TAC
151st ARG	KC-135A	Salt Lake City, Utah	SAC
152nd TRG	RF-4C	Reno, Nev	TAC
153rd TAG	C-130B	Cheyenne, Wyo	MAC
154th CG	F-4C	Hickam AFB, Hawaii	PACAF
155th TRG	RF-4C	Meridian, Miss	TAC
156th TFG	A-7D	San Juan, Puerto Rico	TAC
157th ARG	KC-135A	Pease AFB, NH	SAC
158th DSEG	EB-57E	Burlington, Vt	ADCOM
159th TFG	F-4C	New Orleans, La	TAC
160th ARG	KC-135A	Rickenbacker AFB, Ohio	SAC
161st ARG	KC-135A	Phoenix, Ariz	SAC
162nd TFG	A-7D	Tucson, Ariz	TAC
163rd TASG	O-2A	Ontario, Cal	TAC
164th TAG	C-130A	Memphis, Tenn	MAC
165th TAG	C-130E	Savannah, Ga	MAC
166th TAG	C-130A	Wilmington, Del	MAC

Support Squadron flies the Cessna O-2A, a militarised version of the Super Skymaster. Also at this base a number of SAC rotational KC-135s and RC-135s are always on hand.

An extensive organisation with responsibility for recruiting, education, technical and flying training, Air Training Command has more than 85,000 personnel to carry out its programmes at 17 major bases in the United States. Eight of these bases are concerned with flying training and use four main types of aircraft. Prospective pilots' capability is gauged with the T-41A, the USAF's version of the popular Cessna 172 light aircraft, at a civilian contract flight school. Successful candidates then move on to one of seven training bases for 90 hours on the Cessna T-37, the principal primary jet trainer. Each flight training wing has two squadrons, one with the T-37 and the other with the Northrop T-38 Talon. Trainees move from the T-37 to the T-38 for 120 hours in this supersonic aircraft.

Of the approximately 1,500 aircraft operated by ATC, 1,400 are T-37s and T-38s. About 3,000 pilots are turned out by ATC every year, including many for foreign air forces; all pilots for the Luftwaffe, for example, are trained by ATC. Two specialist training wings are those at Mather AFB, California and Randolph AFB, Texas. The former handles navigator training and uses the Boeing T-43A as a flying classroom. The other specialised training wing is for pilot instructors and uses a variety of aircraft.

The USAF has long been heavily dependent upon the services of part-time airmen as is indicated by the fact that of the total of 400 flying squadrons, 144 are manned by reservists. The active personnel strength is 190,000 whereas the regular USAF total is just under 600,000. The Air Reserve is divided into two distinct forces, the Air Force Reserve and the Air National Guard. The AFR is a directly linked USAF reserve and is fortunate in having a large number of experienced people who have had service in the regular USAF. Every state in the union raises its own militia to serve that state in time of emergency or disaster. These state forces may also see federal service during a national emergency. The flying element is organised as the Air National Guard and depends on the USAF for training and supply. Moreover, units–depending upon their equipment–are committed to the various United States Air Force Commands and, like the Air Force Reserve, constitute a highly trained and effective flying reserve.

The ANG is twice as large an organisation as the AFR and has some 92,000 personnel and 91 flying squadrons. Currently these are committed to the four major flying commands in the United States, although with the dissolution of Aerospace Defense Command some two thirds of the ANG force will come under TAC. Both AFR and ANG squadrons take part in USAF manoeuvres and selected units are deployed to Pacific and European locations. In the past both air reserves have depended to a large degree on the regular Air Force's 'cast-offs'; this policy was changed during the mid-1970s and new A-7s went directly to ANG squadrons. The AFR is also to receive new A-10 and F-16 aircraft in a re-equipment programme, while the ANG will receive A-10s and the two-seat A-7K.

Above: an A-7D serving with the 366th Tactical Fighter Wing, which has since relinquished the type in favour of the F-111 – thus releasing A-7s for service with the Air National Guard. This reflects the current trend of supplying the Reserve and ANG with modern warplanes, rather than hand-me-downs from the regular air force.

Left: this A-7D carries the mountain lion emblem of the 120th TFS, 140th TFG, Colorado Air National Guard.

Below left: the 111th Tactical Air Support Group of the Pennsylvania Air National Guard operates Cessna O-2s from Willow Grove. It is one of six ANG units which are equipped with the Super Skymaster in the forward air control role and are committed to Tactical Air Command.

Below: the Cessna A-37B Dragonfly, which is an adaptation of the T-37 trainer for light attack duties, serves with the 175th TFG of the Maryland ANG. Two ANG groups and one Air Force Reserve wing fly the Dragonfly, but this type is not operated by the regular air force

167th TAG	C-130A	Martinsburg, W Va	MAC
169th TFG	A-7D	McEntire Field, SC	TAC
170th ARG	KC-135A	McGuire AFB, NJ	SAC
171st ARW	KC-135A	Pittsburgh, Pa	SAC
172nd TAG	C-130E	Jackson, Miss	MAC
174th TFG	A-37B	Syracuse, NY	TAC
175th TFG	A-37B	Baltimore, Md	TAC
176th TAG	C-130E	Anchorage, Alaska	MAC
177th FIG	F-101B	Atlantic City, NJ	ADCOM
178th TFG	A-7D	Springfield, Ohio	TAC
179th TAG	C-130E	Mansfield, Ohio	MAC
180th TFG	F-100D	Toledo, Ohio	TAC
181st TFG	F-100D	Terre Haute, Ind	TAC
182nd TASG	O-2A	Peoria, Ill	TAC
183rd TFG	F-4C	Springfield, Ill	TAC
184th TFTG	F-105F	McConnell AFB, Kan	TAC
185th FTG	A-7D	Sioux City, Iowa	TAC
186th TRG	RF-4C	Meridian, Miss	TAC
187th TRG	RF-4C	Montgomery, Ala	TAC
188th TFG	F-100D	Fort Smith, Ark	TAC
189th ARG	KC-135A	Little Rock, Ark	SAC
190th ARG	KC-135A	Forbes Field, Kan	SAC
191st FIG	F-4D	Selfridge, Mich	ADCOM
192nd TFG	F-105D	Sandston, Va	TAC
193rd TEWG	EC-130E	Harrisburg, Pa	TAC

Abbreviations: ARG, Air Refueling Group; ARRG, Aerospace Rescue & Recovery Group; ARW, Air Refueling Wing; CG, Composite Group; DSEG, Defense Systems Evaluation Group; FIG, Fighter Interceptor Group; FIW, Fighter Interceptor Wing; TAG, Tactical Airlift Group; TASG, Tactical Air Support Group; TASW, Tactical Air Support Wing; TAW, Tactical Airlift Wing; TEWG, Tactical Electronic Warfare Group; TFG, Tactical Fighter Group; TFTG, Tactical Fighter Training Group; TFW, Tactical Fighter Wing; TRG, Tactical Reconnaissance Group; TRW, Tactical Reconnaissance Wing.

Above: this F-105D Thunderchief, parked on the ramp at Andrews AFB, Maryland, is assigned to the 113th TFW, District of Colombia ANG.
Left: de Havilland Canada Caribous were originally purchased by the US Army to ferry troops and supplies into forward battle areas. In 1967 the USAF assumed responsibility for this mission and acquired the Army's Caribous, designating them C-7s.
Below: Air Force Reserve and Air National Guard units undertake overseas deployments as part of their training commitment. F-105 Thunderchiefs of the Reserve 301st TFW visited Sculthorpe, England, in 1978

CHAPTER TWO

United States Navy

Grumman F-14A Tomcats of VF-32 and VF-14 parked on the crowded flight deck of the aircraft carrier USS John F. Kennedy during a Mediterranean deployment in December 1975. The Tomcat fleet defence fighter is capable of detecting targets at a range of 185 km (115 miles) and can engage up to six enemy aircraft simultaneously

If history has a way of repeating itself, then American naval aviation planners of the late 1970s are justifiably nervous. Just as their predecessors watched with concern as Japan built a superior naval aviation force in the 1930s, so contemporary US Navy air admirals are carefully watching the Soviet Union embark on an aggressive programme to attain air supremacy over the fast-growing Red Navy surface and submarine forces.

A decade ago the US Navy boasted 22 aircraft carriers to reinforce its role of maintaining maritime air superiority to ensure that its power could be projected ashore. At that time the Soviet Union had no aircraft carriers. Today, the US Navy is down to 13 carriers (including one ship used only for naval aviation shipboard training), while the Soviet Union has two carriers in a new force that Western intelligence experts say will eventually comprise 11 or more ships.

Morskaya Aviatsiya, as the Soviet naval air arm is properly known, moved gradually into ship-based aviation. Helicopter deployments aboard normal cruisers were followed in 1967 by operations from the specially-modified cruisers *Moskva* and *Leningrad*, whose entire after sections were flight

decks. In July 1976 the Soviet Union took their first aircraft carrier—technically identified as a through-deck cruiser—into international waters. That ship, the *Kiev*, has since been augmented by another aircraft-carrying vessel, the *Minsk*. Several other ships of the same class are presently under construction.

Viewing this challenge to American naval aviation, Admiral Thomas B. Hayward, the Chief of Naval Operations, notes that the US 'requirement for maritime superiority recognises the strategic realities of our geographic position as an island nation—a nation connected to overseas allies by two broad oceans, confronting a great land power which has chosen, for reasons of its own, to challenge our traditional supremacy on the seas. It is not surprising that the Soviets see benefits in doing so, for they recognise, as we must, that control of the seas is absolutely essential for the survival of the United States as a viable economic entity, as it is to any island nation which wishes to preserve its independence and freedom of action.

'In the past 15 years,' Hayward points out, 'the Soviet Navy has steadily grown from a coastal defence force into a "blue water navy" powerful

*Top: an F-14A of VF-142 makes its
landing approach with wings swept fully
forward and tailhook extended. A
carrier air wing usually includes two
Tomcat-equipped squadrons, although a
number of fleet fighter squadrons still
fly the McDonnell Douglas F-4.
Above: this F-14A carries VF-14's top
hat insignia on its tailfins. It is
anticipated that the US Navy will buy
a total of 521 Tomcats, with procurement
of the type ending in 1985*

enough to challenge the US Navy in most major
ocean areas of the world. It has continually evolved
toward broader missions, a growing ability to
exercise effective command and control on a world-
wide basis, and a capacity to shift forces from one
ocean to another rapidly and to operate them effectively. These impressive developments, when con-
trasted with the considerable reductions in the
size of the US Navy, represent a disturbing trend.'

In addition to its growing aircraft carrier force,
Morskaya Aviatsiya has long maintained a land-
based long-range maritime bomber establishment.

Aircraft and technology improvements make that
establishment a greater threat to US naval aviation.
As Admiral Hayward notes, 'as older aircraft are
retired, they are being replaced by new, longer
range, supersonic [Tupolev Tu-26] Backfire bombers.
If the addition of Backfire bombers to Soviet Naval
Aviation continues, there is every reason to believe
that a very significant proportion of the force will be
composed of such bombers by the mid-1980s and,
thus, the overall capabilities of that force will be
markedly greater.'

Functioning essentially as an island nation, the

United States assigns its Navy to defend and deploy primarily from its two ocean coasts. The same command distinction applies to aviation activities. Hence, east coast aviation activities ultimately report to Vice Admiral George E. R. Kinnear II, currently Commander of the Atlantic Naval Air Force (ComNavAirLant). The aircraft carriers under Kinnear's jurisdiction are the *Lexington*, *Forrestal*, *Saratoga*, *Independence*, *America*, *Kennedy*, *Nimitz* and *Eisenhower*. On the US west coast, Vice Admiral Robert P. Coogan is ComNavAirPac and his command includes the carriers *Midway*, *Coral Sea*, *Ranger*, *Kitty Hawk*, *Constellation* and *Enterprise*, as well as a variety of other Pacific-area naval aviation facilities.

In the days of an abundant carrier force, the US Navy assigned a number of its older *Essex*-class to anti-submarine warfare duties and redesignated the ships CVS to denote that role. However, the CVS vessels were phased out in the early 1970s and today ASW aircraft are part of the tactical Carrier Air Wings (CVW) assigned to all large-deck aircraft carriers.

The normal composition of a contemporary US Navy CVW is: two attack (VA) squadrons of 12 LTV/Vought A-7E Corsair IIs; one VA squadron equipped with 10 Grumman A-6E Intruders and four KA-6D aerial refuelling versions; two fighter (VF) squadrons generally equipped with Grumman F-14A Tomcats – although a good number of VF squadrons continue to operate the McDonnell Douglas F-4J Phantom II or the modernised F-4S variant – one tactical electronic warfare (VAQ) squadron of four Grumman EA-6B Prowlers; one airborne early warning (VAW) squadron of four Grumman E-2C Hawkeyes; a three-aircraft detachment of LTV/Vought RF-8G Crusaders

FLEET FIGHTER SQUADRONS		
No	**Aircraft Type**	**Shore Base**
VF-1	**F-14A**	**Miramar, Ca**
VF-2	**F-14A**	**Miramar, Ca**
VF-11	**F-4J**	**Oceana, Va**
VF-14	**F-14A**	**Oceana, Va**
VF-21	**F-4J**	**Miramar, Ca**
VF-24	**F-14A**	**Miramar, Ca**
VF-31	**F-4J**	**Oceana, Va**
VF-32	**F-14A**	**Oceana, Va**
VF-33	**F-4J**	**Oceana, Va**
VF-41	**F-14A**	**Oceana, Va**
VF-43	**F-5E, A-4E, TA-4J, T-38A**	**Oceana, Va**
VF-51	**F-14A**	**Miramar, Ca**
VF-74	**F-4J**	**Oceana, Va**
VF-84	**F-14A**	**Oceana, Va**
VF-101	**F-14A**	**Oceana, Va**
VF-102	**F-4J**	**Oceana, Va**
VF-103	**F-4J**	**Oceana, Va**
VF-111	**F-14A**	**Miramar, Ca**
VF-114	**F-14A**	**Miramar, Ca**
VF-121	**F-4J**	**Miramar, Ca**
VF-124	**F-14A**	**Miramar, Ca**
VF-126	**TA-4J, T-2C**	**Miramar, Ca**
VF-142	**F-14A**	**Oceana, Va**
VF-143	**F-14A**	**Oceana, Va**
VF-151	**F-4J**	**Atsugi, Japan**
VF-154	**F-4J**	**Miramar, Ca**
VF-161	**F-4J**	**Atsugi, Japan**
VF-171	**F-4J**	**Oceana, Va**
VF-171KW	**F-4N, A-4E, TA-4J**	**Key West, Fla**
	(Gunnery/ACM training)	
VF-211	**F-14A**	**Miramar, Ca**
VF-213	**F-14A**	**Miramar, Ca**

from a photographic reconnaissance (VFP) squadron; one air anti-submarine warfare (VS) squadron of 10 Lockheed S-3A Vikings and one helicopter ASW (HS) squadron equipped with Sikorsky SH-3 Sea Kings.

When not on deployment, all of the carrier squadrons are based at a Naval Air Station (NAS) which either has berthing facilities for aircraft carriers or is close enough to a port facility so that squadrons can easily be loaded aboard the ships. For example, the Norfolk, Virginia area on the east coast is home to NAS Norfolk, where the major F-4 and F-14 squadrons are billeted. Nearby is NAS Oceana at Virginia Beach, Virginia, where the east coast A-6 squadrons are stationed. There are pier facilities at Norfolk for at least two large carriers, which are most likely to be in port when two other carriers are deployed. A Norfolk-based carrier typically deploys with the Sixth Fleet in the Mediterranean, where the US Navy has maintained a strong carrier presence since the end of World War II in 1945.

The NASs, Naval Air Facilities (NAF) and Naval Stations (NS) with aviation facilities in the

Above: an example of the US Navy's 'toned down' markings on a Tomcat of VF-143 makes a striking contrast with a sister unit's colourfully-painted aircraft illustrated opposite. The new finish is intended to make the fighter more difficult to see during air combat. Although in general the Tomcat has performed well since its introduction in late 1972, problems of engine reliability necessitated an improvement programme for the TF-30 turbofans

FLEET ATTACK SQUADRONS		
VA-12	A-7E	Cecil Field, Fla
VA-15	A-7E	Cecil Field, Fla
VA-22	A-7E	Lemoore, Ca
VA-25	A-7E	Lemoore, Ca
VA-27	A-7E	Lemoore, Ca
VA-34	A-6E, KA-6D	Oceana, Va
VA-35	A-6E, KA-6D	Oceana, Va
VA-37	A-7E	Cecil Field, Fla
VA-42	A-6E, TC-4C	Oceana, Va
VA-45	TA-4J	Cecil Field, Fla
VA-46	A-7E	Cecil Field, Fla
VA-52	A-6E, KA-6D	Whidbey Island, Wash
VA-56	A-7E	Atsugi, Japan
VA-65	A-6E, KA-6D	Oceana, Va
VA-66	A-7E	Cecil Field, Fla
VA-72	A-7E	Cecil Field, Fla
VA-75	A-6E, KA-6D	Oceana, Va
VA-81	A-7E	Cecil Field, Fla
VA-82	A-7E	Cecil Field, Fla
VA-83	A-7E	Cecil Field, Fla
VA-85	A-6E, KA-6D	Oceana, Va
VA-86	A-7E	Cecil Field, Fla
VA-87	A-7E	Cecil Field, Fla
VA-93	A-7E	Atsugi, Japan
VA-94	A-7E	Lemoore, Ca
VA-95	A-6E, KA-6D	Whidbey Island, Wash
VA-97	A-7E	Lemoore, Ca
VA-105	A-7E	Cecil Field, Fla
VA-113	A-7E	Lemoore, Ca
VA-115	A-6E, KA-6D	Atsugi, Japan
VA-122	A-7E, TA-7C, T-28B, T-39D	Lemoore, Ca
VA-127	TA-4F, TA-4J	Lemoore, Ca
VA-128	A-6E, TC-4C	Whidbey Island, Wash
VA-145	A-6E, KA-6D	Whidbey Island, Wash
VA-146	A-7E	Lemoore, Ca
VA-147	A-7E	Lemoore, Ca
VA-165	A-6E, KA-6D	Whidbey Island, Wash
VA-174	A-7E, TA-7C, T-39D	Cecil Field, Fla
VA-176	A-6E, KA-6D	Oceana, Va
VA-192	A-7E	Lemoore, Ca
VA-195	A-7E	Lemoore, Ca
VA-196	A-6E, KA-6D	Whidbey Island, Wash

Above right: among the many colourful insignia carried by Navy attack squadrons is the indian chief's head of the Cecil Field, Florida-based VA-87. Opposite top: an A-7E of VA-66 is pictured aboard USS Independence. Final single-seat version of the Corsair to enter Navy service, the A-7E adopted the Allison TF-41 turbofan first utilised by the USAF when they purchased the type. Improved avionics were also fitted. Opposite above: seen at its Lemoore, California shore base, this A-7E of Attack Squadron VA-22 is also attached to CVW-15 aboard Kitty Hawk. The Corsair remains the US Navy's primary light attack aircraft, although no further procurement has been announced. Opposite: a line-up of A-7B aircraft on the flight deck of the USS John F. Kennedy in the Mediterranean, pictured in 1975. The A-7B has since been retired from first line service and equips Reserve units

United States and abroad also support a variety of land-based squadrons, as well as units which serve carriers without being part of the formal Carrier Air Wing. These units are patrol (VP) squadrons of Lockheed P-3C Orions performing long-range maritime patrols; fleet logistics support (VR) squadrons which use various transport aircraft including Grumman C-2A Greyhounds for carrier-onboard-delivery (COD) flights, and aircraft ferry (VRF) squadrons which deliver and pick up aircraft. Helicopter sea control (HSL) squadrons of the US Navy's Light Airborne Multi-Purpose Systems (LAMPS) programme are equipped with Kaman SH-2F Seasprites and generally deploy aboard destroyers to perform ASW and anti-shipping missions. Composite (VC) squadrons use various aircraft for aerial target-towing and miscellaneous support missions, while fleet air reconnaissance (VQ) squadrons perform photographic, electronic reconnaissance and weather reconnaissance missions using specially-equipped F-4s or Douglas EA-3 Skywarriors.

Naval aviators are found in three branches of the American military service. The golden wings of US naval aviation are worn by members of the US Coast Guard and the US Marine Corps. The Coast Guard is a branch of the US Department of Transportation, but its sea-going search and rescue mission is so close to the Navy's maritime mission that its pilots are trained by the Navy. The Marine Corps is part of the Navy Department and in many cases it deploys squadrons aboard aircraft carriers, hence the close USMC affiliation with naval aviation.

Naval aviators, irrespective of branch of service, receive instruction in the six wings of the Naval Air Training Command, whose facilities are in Mississippi, Texas and Florida – all southern locations which allow year-round flying. Training Wings 1, 2 and 3 provide advanced and intermediate strike training; Training Wing (TraWing) 4 offers primary, intermediate and advanced maritime training; TraWing 5 provides primary and intermediate flight training, maritime and helicopter training, and fundamental and advanced helicopter training, while TraWing 6 provides intermediate

and advanced strike training, as well as basic and advanced flight training for non-pilot Naval Flight Officers (NFO).

Today, all naval aviators are officers, although enlisted pilots were trained from 1917 to 1947. Pilot and NFO training is now available to commissioned officers and selected officer candidates. The training cycle begins at NAS Pensacola, Florida, where officer candidates undergo a 12-week course which includes basic military training and officer students take a four-week course to introduce them to aviation.

Upon completion of that portion of the training cycle, prospective pilots go on to a 16-week primary flight training course. A prospective NFO is assigned to training squadron VT-10 for a 26-week period to prepare him for controlling and coordinating the complex electronic weapons systems aboard naval combat aircraft and the communications and navigation functions in a variety of naval aircraft.

After primary flight training, pilots are assigned to one of the Training Wings for intermediate and

advanced training in either strike or maritime aircraft. All naval aviators become qualified in fixed-wing flight operations, even if they eventually elect to undergo helicopter training. Intermediate and advanced training cycles run from 21 to 36 weeks in duration. Upon graduation, the officer candidate is commissioned as an Ensign in the US Naval Reserve and awarded his wings, while officer students are awarded their wings. The new pilots are then sent to a fleet readiness squadron for final training in the aircraft type which they will fly on active duty.

Meanwhile, the prospective NFO has been progressing through training in the various sophisticated navigational aids now available. He has also been making his first flights and operating training instruments similar to those found in Fleet aircraft. Advanced jet navigation is taught at VT-86, which is also the training ground for the Radar Intercept Officers (RIO) assigned to two-place jet fighters, either the McDonnell Douglas F-4J Phantom II or the Grumman F-14A Tomcat. NFO students also receive their gold wings upon graduation from VT-10, from whence they, too, are posted to a fleet readiness squadron.

The fleet readiness squadrons, formerly called Readiness Air Groups, introduce the new pilot or NFO to the conditions he will encounter in Fleet air operations. On the east coast, for example, a prospective LTV/Vought A-7E Corsair II pilot would be assigned to VA-45, the A-7 RAG based at NAS Cecil Field, Florida. On the west coast, a prospective Grumman F-14A Tomcat pilot or RIO would be assigned to VF-124, the F-14 RAG based at NAS Miramar, California. Both of these RAG also fly the two-seat McDonnell Douglas TA-4J Skyhawk to enable pilots and/or RIO to work with experienced pilots before working out in the actual aircraft type to which they have been assigned.

Naval aviation aircrews are also eligible for training in Air Combat Manoeuvring (ACM), a concept which proved especially important during the Vietnam Conflict. Early in that war, American

Left: a Grumman EA-6B Prowler (third from the front) is parked among A-6E attack aircraft of VA-34 aboard USS John F. Kennedy. Production of the all-weather A-6E for the US Navy ended in 1979, but existing aircraft are being upgraded by the fitting of the target recognition and attack multisensor (TRAM), which enables the aircraft to identify and track targets in poor visibility, independently of its powerful radar. Such modifications are part of the Navy's conversion in lieu of procurement programme, which enables substantial savings to be made.
Right: one of Attack Squadron 65's A-6E Intruders illustrated in the markings carried when the unit served aboard USS Independence. A number of Intruders are fitted out as KA-6D tankers for in-flight refuelling and serve alongside the attack version.
Below: the Atlantic Fleet's Intruder training squadron is VA-42, based at Naval Air Station Oceana, Virginia

F-4s were being brought down at a disturbing rate by Soviet-built aircraft considered to be 'inferior' to the F-4. It was soon determined, however, that the older and slower MiG-17s and MiG-19s were using their slower speed and tighter turning radius to outmanoeuvre the faster F-4s. Hence, ACM centres such as VF-43 at NAS Oceana train F-4 and F-14 pilots and RIO how to cope with various Soviet-built aircraft by using American aircraft with similar performance. VF-43 uses the A-4 to simulate the MiG-17 and MiG-19, and the Northrop T-38A Talon and F-5E Tiger II to take the part of the high-performance MiG fighters.

Fleet-experienced pilots receive a variety of assignments to prepare them for naval careers of ever-increasing responsibility. One of the more unusual squadron assignments is with an air development (VX) squadron, of which the US Navy has five, two in the Atlantic Fleet and three in the Pacific. VX-1 at the Naval Air Test Centre (NATC) at NAS Patuxent River, Maryland does a considerable amount of airframe and weapons testing, as does VX-4, which is based at the Pacific Missile Test Centre at Point Mugu, California. One unit, VXE-6, provides logistic and other support functions for the American scientific effort in Antarctica.

For pilots who do not elect to make the Navy a full-time career, culminating in retirement at half-pay following a minimum of 20 years' active service, the Naval Air Reserve Force (NARF) is a way to contribute to national defence while maintaining flight proficiency. Gone are the days when the 'weekend warriors', as the Reservists were known, flew only the hand-me-downs which the Fleet had long since discarded. Today's NARF has essentially the same equipment – albeit earlier modifications – of Fleet aircraft, whether they are McDonnell Douglas F-4 fighters, Vought A-7 attack aircraft or Lockheed P-3 patrol aircraft. While their one weekend a month duty is generally performed at the Reservists' own station, their two-week annual cruises are coordinated with the active

duty Navy to make the exercises as meaningful and as realistic as possible.

The value of the Reserve element is highlighted by Admiral Hayward: 'Presently Reservists train in and contribute to virtually all of the Navy's mission and mission support areas. The Reserve provides all of the Navy's capability in certain functional areas and a major portion in others. This includes 14 per cent of Navy tactical air combat assets, 35 per cent of maritime air patrol . . . and 100 per cent of US-based logistic air capabilities.'

Maintaining an effective Naval Air Reserve Force is part of one of the US Navy's fundamental principles which Hayward believes are essential to a more complete understanding of naval supremacy. He has stated that he believes the US Navy may have to 'fight a major war with essentially what we have at its outset, augmented by the Naval Reserve, which will enhance our capabilities in certain specialised warfare areas and provide some unit and personnel augmentation for active forces. Given the long lead time for production of today's complex ships and aircraft, neither side will have a substantial opportunity to reconstitute major naval units, even if the war is relatively protracted.

'Every major engagement must, therefore, be regarded as potentially decisive in terms of its impact on the naval balance, and every US naval unit must have the maximum offensive capability we can build into it consistent with its mission. It also means that our force structure in peacetime, including the Naval Reserve, must be sufficient in size and capability to prevail in war. There will be little opportunity to expand it significantly once war has begun.'

Under these conditions, it is obvious that the US Naval Air Reserve Force must be maintained in the highest state of readiness. Hence, NARF two-week active duty deployments are being carried out aboard regular aircraft carriers. Earlier, NARF squadrons performed their annual at-sea duties aboard the training carrier USS *Lexington* (CVT-16).

TACTICAL ELECTRONIC WARFARE SQUADRONS		
VAQ-33	EKA-3B, EA-4F, EF-4J	Norfolk, Virginia
	TA-3B, EC-121K	
VAQ-129	EA-6B	Whidbey Island, Wash
VAQ-130	EA-6B	Whidbey Island, Wash
VAQ-131	EA-6B	Whidbey Island, Wash
VAQ-132	EA-6B	Whidbey Island, Wash
VAQ-133	EA-6B	Whidbey Island, Wash
VAQ-134	EA-6B	Whidbey Island, Wash
VAQ-135	EA-6B	Whidbey Island, Wash
VAQ-136	EA-6B	Whidbey Island, Wash
VAQ-137	EA-6B	Whidbey Island, Wash
VAQ-138	EA-6B	Whidbey Island, Wash

But with that ship scheduled soon to be retired and, in view of its relatively small size as an *Essex*-class vessel, there are growing problems in accommodating the newer, larger and heavier aircraft being assigned to Reserve units. Aircraft such as the McDonnell Douglas F-4B Phantom II routinely used by NARF fighter squadrons began the transition to fleet-type carriers and have been followed by all of the other types of ship-based aircraft.

Scheduling NARF units for aircraft carrier flight operations is only one of the pressing requirements of maintaining naval aviation proficiency. Another is the need to provide continual carrier operational activities for active duty units. At a time when the Soviet Navy is experiencing the greatest growth in its history—with a strong new emphasis on ship-based air operations—the US Navy finds itself doubly tested by the rigours of fewer carriers and what is surely an even greater need to deploy aircraft at sea.

For several years the Navy has asked the US Congress to appropriate funds for a fourth large-deck nuclear-powered aircraft carrier to enhance its widespread operations. The Navy argues that the large-deck carrier would give American military power a highly significant and very mobile 'presence' which has been demonstrated time and again since World War II. Adding nuclear power to such a vessel would keep it secure from the spectre of uncertain availability which has marked the use of

petroleum-based fuels in recent times. But, while there is considerable political support for the Navy's position, the opposition led by President Jimmy Carter has so far stifled plans to provide the Navy with a total of five nuclear-powered aircraft carriers (four new and larger ships, plus USS *Enterprise*).

In an attempt to compromise with the Carter Administration, the Navy has requested start-up funding for a smaller carrier, designated CVV. Discussing this request in the fiscal year 1980 shipbuilding programme, Admiral Hayward says that CVV would be 'a conventionally-powered, multi-purpose aircraft carrier designed to the minimum size necessary to operate modern conventional fleet aircraft. It incorporates a number of cost-saving features, such as its smaller size and the reduction in the number of catapults, arresting gear and main engines. The CVV will provide an advanced passive protection system that will enhance the already excellent survivability features that are basic to all carriers.'

The proposed CVV would have neither the great displacement of tonnage nor the large dimensions of the *Nimitz*-class large deck nuclear-powered carriers. It would displace slightly under 60,000 tons and be about the overall size of the modernised *Midway*-class ships. However, the CVV would have greater interior dimensions than the two *Midway*-class ships, as the relatively low ceiling of the older carriers' hangar deck areas is a

A close relation of the A-6 Intruder attack bomber, the four-seat Grumman EA-6B Prowler carries no armament but may be equipped with a variety of sophisticated electronic countermeasures equipment. Electronic jamming pods, one of which is visible on the outboard wing hardpoint of the aircraft pictured left, each contain two transmitters to interfere with enemy radar. Electric power is generated by a ram-air turbine, with a small 'windmill' propeller on the nose. An 'Improved Capability' (ICAP) programme will ensure the Prowler's effectiveness into the 1990s

AIRBORNE EARLY WARNING SQUADRONS		
RVAW-110	E-2B, TE-2A	Miramar, Ca
VAW-112	E-2B	Miramar, Ca
VAW-113	E-2B	Miramar, Ca
VAW-114	E-2C	Miramar, Ca
VAW-115	E-2B	Atsugi, Japan
VAW-116	E-2B	Miramar, Ca
VAW-117	E-2B	Miramar, Ca
RVAW-120	E-2C, TE-2C	Norfolk, Va
VAW-121	E-2C	Norfolk, Va
VAW-122	E-2C	Norfolk, Va
VAW-123	E-2C	Norfolk, Va
VAW-124	E-2C	Norfolk, Va
VAW-125	E-2C	Norfolk, Va
VAW-126	E-2C	Norfolk, Va

Above: first flown in October 1960, the Grumman E-2 Hawkeye embodies the concept of a naval tactical data system. This provides the task force commander with information as to the position of ships and aircraft, together with pinpointing the best-positioned interceptor to meet any threat. In common with many US Navy aircraft, the Hawkeye's wings fold for stowage. Right: an improved avionics system characterised the E-2C, which entered Navy service in late 1973. Seven of the Navy's 12 carrier-based AEW squadrons operate the mark, together with one training squadron. An aircraft of VAW-125 attached to CVW-1 aboard USS John F. Kennedy is pictured landing at a shore base

severely limiting factor in the use of newer aircraft, which have a tendency to have high tails or electronic gear, as seen in Grumman's E-2C Hawkeye. The CVV would deploy with 50 to 55 aircraft, as opposed to the 100-aircraft Carrier Air Wings which operate from the *Nimitz*-class ships.

The CVV is clearly an interim measure. In the face of uncommonly high inflation which tends to de-stabilise any fixing of projected costs, such measures are fast becoming a way of life within national defence systems. Another measure consists of refurbishing existing materials to make them last longer. In the case of its aircraft carriers, the US Navy has begun requesting continual annual funding for its Service Life Extension Programme (SLEP).

Admiral Hayward states: 'The fiscal year 1980 budget continues the advance procurement of materials required in order to commence the 28-month SLEP conversion of USS *Saratoga* (CV-60) in fiscal year 1981. The SLEP programme was developed to extend the service life of the eight conventional *Forrestal* and *Kitty Hawk*-class carriers 15 years beyond their expected service life of 30 years. This is necessary in order to retain a force level of 12 deployable large-deck carriers.'

Once in the SLEP programme, *Saratoga* is not expected to return to active duty until 1983. *Forrestal* would be in the SLEP from about 1982 to

1985. At that point, the two oldest carriers – *Midway* and *Coral Sea* – would be ready for decommissioning, as neither ship has been mentioned as a candidate for SLEP. Indeed, the *Forrestal* and later classes were selected for refurbishing because they are large enough to support adequate numbers of all present anticipated aircraft.

Rising costs are not only affecting shipbuilding. They have also cut into aircraft procurement, which must now be augmented to a considerable degree by upgrading and modifying existing aircraft and aircraft systems. The US Navy's Aircraft Modification Programme consists of two major components: Conversion in Lieu of Procurement (CILOP) and 'other modifications'. CILOP covers any aircraft conversion, service life extension, update, expansion and/or change of mission capability, improvement of combat capability or any combination of these factors. CILOP is performed on existing aircraft for the primary purpose of providing an acceptable alternative to the costly procurement of new aircraft needed to meet or maintain the naval aviation force levels.

The 'other modifications' category applies to changes which enhance the reliability and maintainability of existing aircraft. It also includes safety improvements and changes which modernise systems and components that improve readiness and combat effectiveness.

One of the more significant modification programmes is the CILOP conversion of McDonnell Douglas F-4J Phantom II aircraft to the F-4S by incorporating slats on the leading edge of the wing to give the F-4S greater operational capability in

the air combat role. The Boeing-Vertol CH-46 Sea Knight CILOP medium assault helicopter is being modified for the US Marine Corps to improve operational capability with uprated engines, a crash-worthy fuel system, armoured crew seats and infra-red countermeasures. The USMC, which comes under the Navy Department, receives all aircraft through the Navy. The Sikorsky SH-3A Sea King is being upgraded to SH-3H standard by adding sonobuoy and magnetic anomaly detection (MAD) sensors. Power plant and navigational system improvements will also increase the SH-3's capability in ASW and anti-ship missile defence. The Forward Looking Infra-red (FLIR) radar added to the LTV/Vought A-7E Corsair II proved its value in combat when used by the A-6 during the Vietnam Conflict. A-7Es with FLIR will have a

first-pass visual attack capability at night with a bombing accuracy twice that achieved by radar weapons delivery. The A-7E is also being modified to carry, target and launch the High-speed Anti-Radiation Missile (HARM). Finally, 33 Bell AH-1 Sea Cobra helicopters are being retrofitted with the optically-tracked, wire-guided (TOW) missile system currently being installed on production USMC AH-1Ts.

The impact of conversions and modifications can be seen in remarks made by Vice Admiral Frederick C. Turner at the end of his tenure as Deputy Chief of Naval Operations for Air Warfare. Prior to his retirement in mid-1979, Turner told the US Senate Subcommittee on Defence that the Navy 'procured 13 types in fiscal year 1977 and propose eight types in fiscal year 1980, a significant reduction, but realisation of the economies of this effort is still ahead of us.'

He noted that a 'key element in this effort is the McDonnell Douglas F/A-18 Hornet, which will replace two different aircraft and possibly one or two others. This is our second year of procurement with several years to go before achieving the quantities which will produce significant economies of scale. This aircraft, then, promises to contribute to the solution of our modernisation problem and the maintenance of force levels, while bringing increased capabilities and substantial procurement savings.'

The Navy's procurement plan can barely cover all the diverse needs of US naval aviation, but, as Admiral Hayward points out: 'It has been recognised that the past practice of procuring a relatively wide variety of aircraft in low quantities is not the most economical approach to force modernisation and maintenance. For example, while 13 aircraft types were procured in fiscal year 1977, the number was reduced to 10 in fiscal years 1978 and 1979. Only eight types are requested in fiscal year 1980, with further reduction to seven in subsequent years.

'However, there are two factors which bear on the realisation of the economic benefits of this effort. First, although a relatively significant reduction in types of aircraft being procured can be achieved in the short term, we cannot change our force structure in such a short period of time. As a result,

the Navy is in a period of transition during which procurement is aligned more economically, but the character of the forces remains diverse. This is an evolutionary process and improved return on our investment will be realised only gradually.

'Second, naval aviation is not large enough or sufficiently susceptible to homogeneous organisation to permit the achievement of full economies of scale because our forces are keyed to 12 carriers and their embarked air wings. Improvement is possible and current efforts pursue that goal; for example, the F/A-18 programme promises significant economic benefits when the planned production rate is achieved in a few years.'

The future composition of American naval aviation is a matter of considerable consequence and is being formulated in the Chief of Naval

Above and below: the Kaman SH-2F Seasprite was the ultimate development of a design which originated during 1956. The single-engined UH-2A and B were widely utilised for plane guard duties, while the twin-engined HH-2C and D were gunships. The SH-2D and F fulfil the LAMPS (Light Airborne Multi-Purpose System) requirement and serve aboard anti-submarine destroyers and frigates.
Bottom: the Sikorsky SH-3 was first ordered in 1957 as a submarine hunter/ killer, emerging for service four years later as the Sea King. Various improvements were made during service, including more powerful engines

The Lockheed P-3 Orion started life in the late 1950s as a development of the commercially-unsuccessful Electra airliner. Twenty years later, the type still featured in US Navy procurement plans, the latest model being the P-3C (above). Patrol Squadron VP-50, in whose colours the aircraft is pictured, was one of 15 first-line units flying the 'C' in mid-1979, with a further nine scheduled to relinquish their P-3Bs by 1987. The P-3A (left) and B now mainly equip the Reserve and two training squadrons, VP-30 and VP-31; the former operated the pictured Orion as a VIP transport for the Chief of Naval Operations in Washington, DC

Operations' Sea-Based Master Study Plan. Admiral Hayward says the study plan 'is a broad investigation of alternatives and issues necessary to support decisions concerning the future of sea-based air. These studies include examination of the cost and effectiveness of alternative aircraft systems and platforms to meet the Navy's needs in the 1990s and beyond.' A very firm development in the future of American naval aviation is the use of the cruise missile, an area in which the United States has a firm lead over the Soviet Union. Hence, the US Navy is currently developing the Tomahawk cruise missile to counter the Soviet threat.

Clearly, Admiral Hayward is planning a naval aviation force that will evolve in different ways to meet America's defence needs for the balance of the 20th century and the dawn of the 21st century. That evolution seems headed away from a dependence on large-deck aircraft carriers, as non-conventional take-off and landing (CTOL) aircraft become a more serious consideration for future growth in sea-based air operations. With the arrival of more non-CTOL aircraft, a wider variety of ships – including ships of smaller size – will become part of the overall American naval aviation system. The other major component of future US naval aviation must surely be weapons systems such as the Tomahawk cruise missile, which stresses greater standoff capability with a high degree of strike

PATROL SQUADRONS		
VP-1	P-3B	Barbers Point, Hawaii
VP-4	P-3B	Barbers Point, Hawaii
VP-5	P-3C	Jacksonville, Fla
VP-6	P-3B	Barbers Point, Hawaii
VP-8	P-3B	Brunswick, Maine
VP-9	P-3C	Moffett Field, Ca
VP-10	P-3B	Brunswick, Maine
VP-11	P-3B	Brunswick, Maine
VP-16	P-3C	Jacksonville, Fla
VP-17	P-3B	Barbers Point, Hawaii
VP-19	P-3C	Moffett Field, Ca
VP-22	P-3B	Barbers Point, Hawaii
VP-23	P-3B	Brunswick, Maine
VP-24	P-3C	Jacksonville, Fla
VP-26	P-3B	Brunswick, Maine
VP-30	VP-3A, B and C	Jacksonville, Fla
VP-31	P-3A, B and C	Moffett Field, Ca
VP-40	P-3C	Moffett Field, Ca
VP-44	P-3C	Brunswick, Maine
VP-45	P-3C	Jacksonville, Fla
VP-46	P-3C	Moffett Field, Ca
VP-47	P-3C	Moffett Field, Ca
VP-48	P-3C	Moffett Field, Ca
VP-49	P-3C	Jacksonville, Fla
VP-50	P-3C	Moffett Field, Ca
VP-56	P-3C	Jacksonville, Fla

ANTI-SUBMARINE WARFARE SQUADRONS

VS-21	S-3A	North Island, Ca
VS-22	S-3A	Cecil Field, Fla
VS-24	S-3A	Cecil Field, Fla
VS-28	S-3A	Cecil Field, Fla
VS-29	S-3A	North Island, Ca
VS-30	S-3A	Cecil Field, Fla
VS-31	S-3A	Cecil Field, Fla
VS-32	S-3A	Cecil Field, Fla
VS-33	S-3A	North Island, Ca
VS-37	S-3A	North Island, Ca
VS-38	S-3A	North Island, Ca
VS-41	S-3A	North Island, Ca

HELICOPTER ANTI-SUBMARINE WARFARE SQUADRONS

HS-1	SH-3A, SH-3D, SH-3G, SH-3H	Jacksonville, Fla
HS-2	SH-3D	North Island, Ca
HS-3	SH-3H	Jacksonville, Fla
HS-4	SH-3H	North Island, Ca
HS-5	SH-3D	Jacksonville, Fla
HS-6	SH-3H	North Island, Ca
HS-7	SH-3H	Jacksonville, Fla
HS-8	SH-3H	North Island, Ca
HS-9	SH-3H	Jacksonville, Fla
HS-10	SH-3A, D, G, H	North Island, Ca
HS-11	SH-3D	Jacksonville, Fla
HS-12	SH-3H	North Island, Ca
HS-15	SH-3H	Jacksonville, Fla

LIGHT HELICOPTER ANTI-SUBMARINE WARFARE SQUADRONS

HSL-30	SH-2F	Norfolk, Va
HSL-31	SH-2F	North Island, Ca
HSL-32	SH-2F	Norfolk, Va
HSL-33	SH-2F	North Island, Ca
HSL-34	SH-2F	Norfolk, Va
HSL-35	SH-2F	North Island, Ca
HSL-36	SH-2F	Mayport, Fla
HSL-37	SH-2F	Barbers Point, Hawaii

HELICOPTER MINE COUNTERMEASURES SQUADRONS

HM-12	RH-53D	Norfolk, Va
HM-14	RH-53D	Norfolk, Va
HM-16	RH-53D	Norfolk, Va

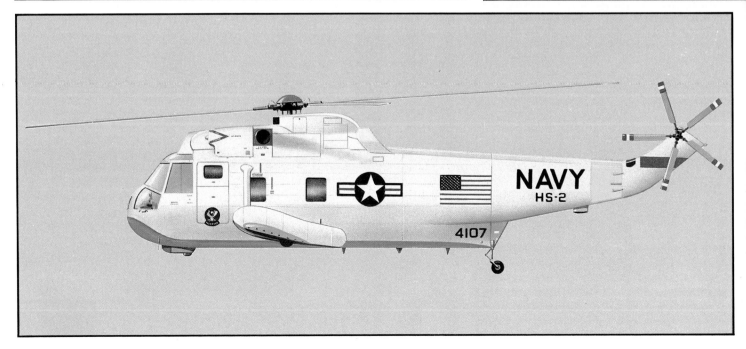

HELICOPTER COMBAT SUPPORT SQUADRONS

HC-1	SH-3G	North Island, Ca
HC-3	CH-46D	North Island, Ca
HC-6	CH-46D, VH-3A	Norfolk, Va
HC-11	CH-46D	North Island, Ca

RECONNAISSANCE ATTACK SQUADRONS

RVAH-3	RA-5C, TA-4J	Key West, Fla
RVAH-7	RA-5C	Key West, Fla
RVAH-12	RA-5C	Key West, Fla

All three due to decommission by autumn 1979.

PHOTOGRAPHIC RECONNAISSANCE SQUADRON

VFP-63	RF-8G, F-8J	Miramar, Ca

FLEET AIR RECONNAISSANCE SQUADRONS

VQ-1	EA-3B, EP-3E, P-3B	Agana, Guam
VQ-1 Det	EA-3B, EP-3E	Atsugi, Japan
VQ-2	EA-3B, EP-3E, P-3A	Rota, Spain
VQ-3	EC-130Q	Agana, Guam
VQ-4	EC-130G, EC-130Q	Patuxent River, Md

RESERVE CARRIER AIR WING TWENTY

VF-201	F-4N	Dallas, Texas
VF-202	F-4N	Dallas, Texas
VA-203	A-7B	Cecil Field, Fla
VA-204	A-7B	Memphis, Tenn
VA-205	A-7B	Atlanta, Ga
VFP-206	RF-8G	Wash, DC
VAQ-208	KA-3B	Alameda, Ca
VAQ-209	EA-6A	Norfolk, Va
VAW-78	E-2B	Norfolk, Va

RESERVE CARRIER AIR WING THIRTY

VF-301	F-4N	Miramar, Ca
VF-302	F-4N	Miramar, Ca
VA-303	A-7B	Alameda, Ca
VA-304	A-7B	Alameda, Ca
VA-305	A-7B	Point Mugu, Ca
VFP-306	RF-8G	Wash, DC
VAQ-308	KA-3B	Alameda, Ca
VAQ-309	EA-6A	Whidbey Island, Wash
VAW-88	E-2B	Miramar, Ca

Top left: the US Navy's standard ship-board ASW aircraft is the Lockheed S-3A Viking. The pictured aircraft, which serves with VS-32, has extended its tail-mounted magnetic anomaly detector boom. Other ASW systems which can be clearly seen are the wing-mounted ESM pods, for detecting enemy radio emissions, and sonobuoy launch tubes on the rear fuselage underside.
Above left: a Sikorsky RH-53D used for mine-countermeasures work by HM-12, the first squadron to be equipped for minesweeping work in 1971.
Far left: the Sikorsky SH-3 Sea King helicopter has been constantly improved and updated with new engines and avionics since its service introduction in 1961.
Left: the principal task of the Boeing-Vertol CH-46D is the 'vertical replenishment' of warships at sea, operating from supply ships

Left: the McDonnell Douglas F-4 Phantom has served the Navy well in both the interceptor and tactical strike roles. The F-4N pictured is flown by VF-302, a reserve unit based at Miramar, California with sister squadron VF-301. Fighter Squadrons 201 and 202 fly the F-4N from Dallas on the east coast as the corresponding units in Reserve Air Wing Twenty.
Below left: the EA-3B Skywarrior is an ECM variant of the carrier-based attack bomber

accuracy. Such changes will be slow, but they are inevitably coming to American naval aviation.

Technically, the US Marine Corps comes under the Navy Department in the American defence line-up. While the Marines do play a key part in fulfilling the Navy's role in projecting military power ashore, they perform their mission in a unique way. In the case of Marine Corps aviation, this involves missions and even aircraft which are clearly distinct from those assigned to US Navy units.

Within the naval establishment, the primary mission of the US Marine Corps is to provide ground assault forces. The relatively small size of the Marine Corps in comparison to other American military branches makes it an élite organisation and it is further distinguished by a spartan devotion to cost-effectiveness. For example, Marine Corps aviation currently provides 12 per cent of American tactical air forces while requiring only 9 per cent of the tactical air budget to operate.

One way the Marine Corps achieves cost-effectiveness is to use aircraft developed primarily for the US Navy. In theory, that should allow the Marines to develop a few of their own aircraft – such as the British Aerospace AV-8A Harrier – but that has not always proved to be the case.

A commentary on US Marine Corps aviation is presented by Lieutenant General Thomas H. Miller, who, until his recent retirement, was Deputy Chief of Staff for Aviation. General Miller notes that in providing air support for Marine ground combat forces 'each aviation weapon system requirement is based on its effectiveness to provide that support. Therefore, in determining its requirements, the Marine Corps strives to reduce the number of different systems needed to accomplish this vital task of saving the lives of our Marines in combat.'

The way to do that, General Miller points out, is to have USMC aviation requirements based chiefly on their ability to be responsive and reliable. 'Responsiveness to the Marine Corps means getting there quickly with enough support to be successful. Reliability means that the Marine on the ground can count on the aviation support being there, regardless of the time of day and weather, and still survive the enemy threat', he states.

While the US Navy does provide carrier-based air support during the early stages of Marine Corps amphibious assault operations, the largest portion of air operations in the amphibious environment is carried out by USMC aviation units. Thus, US Navy aviation can be primarily devoted to maritime air superiority and its allied tactical and strategic efforts. Likewise, US Army aviation can support the non-amphibious ground attack role and the US Air Force can carry out its own defined tactical, strategic and supply missions.

HELICOPTER WING RESERVE		
HAL-4	HH-1K	Norfolk, Va
HAL-5	HH-1K	Point Mugu, Ca
HC-9	HH-3A	North Island, Ca
HS-74	SH-3D	South Weymouth, Mass
HS-75	SH-3D	Lakehurst, NJ
HS-84	SH-3A	North Island, Ca
HS-85	SH-3D	Alameda, Ca
RESERVE PATROL WING ATLANTIC		
VP-62	P-3A	Jacksonville, Fla
VP-64	P-3A	Willow Grove, Penn
VP-66	P-3A	Willow Grove, Penn
VP-68	P-3A	Patuxent River, Md
VP-92	P-3A	South Weymouth, Mass
VP-93	P-3A	Detroit, Mich
VP-94	P-3A	New Orleans, La
RESERVE PATROL WING PACIFIC		
VP-60	P-3A	Glenview, Ill
VP-65	P-3A	Point Mugu, Ca
VP-67	P-3A	Memphis, Ten
VP-60	P-3A	Whidbey Island, Wash
VP-90	P-3A	Glenview, Ill
VP-91	P-3B	Moffett Field, Ca

Right and below right: the Phantom serves with 12 Marine fighter-attack squadrons, epitomising the Marines' policy of modifying and updating current equipment wherever possible. The F-4N is a strengthened and re-equipped F-4B–the mark entered service in 1961–and equips three first-line units including VMFA-531 (right) and the sole attack training squadron, VMFAT-101 (below right). Based at Yuma AFB, Arizona, the latter also flies the F-4J version of the Phantom

At the present time, USMC aviation force levels are based on various past uses of Marine air, with an eye toward current use in a constantly changing geo-political environment. Training exercises and studies have involved such varied scenarios as Northern Europe (NATO), the Persian Gulf and the Korean peninsula. In all cases, each Marine air system and the numbers of those systems have been carefully considered for their effectiveness within the USMC's limited manpower and funding constraints.

In keeping with their spartan philosophy–proudly considered 'lean and mean' by many Marines–the USMC have a long tradition of 'making do' with current aircraft inventory. In General Miller's words: 'It has been the Marine Corps' basic philosophy to utilise current assets as long as they can be updated and modified to meet the threat and perform the assigned task. This has been demonstrated in both fixed-wing and helicopter aircraft systems during the past four years. For instance, the Marine Corps elected to use the McDonnell Douglas F-4 Phantom II and A-4 Skyhawk aircraft rather than the newer types of aircraft used in Navy aviation. This is also true concerning helicopters, as the Marine Corps has elected to retain in service the Boeing-Vertol CH-46 Sea Knight and Bell AH-1 Sea Cobra helicopters, rather than buy newer, more expensive systems being developed primarily for land warfare by the Army.'

The drawback in that philosophy, however, is that it builds in future modernisation problems. If this USMC aviation trend continues, some 76 per cent of Marine Corps aircraft will be over 20 years old by fiscal year 1984. Such a development is 'most disturbing, especially in view of the rapidly rising costs of new and modern systems', General Miller notes. 'If our current plans can be funded, commencing in Fiscal Year 1983,' he continues, 'the Marine Corps should finally start modernisation with the replacement of our fighter-attack F-4s by the McDonnell Douglas F/A-18 Hornet, which is a very important aircraft to us.

'Many analysts question the numbers or size of our fighter force in the Marine Corps. I would like to point out that this force of 144 fighter-attack aircraft not only represents for the Marine Corps a very real air defence capability, which will keep our air and ground forces from being attacked by enemy aircraft. At the same time it also constitutes almost 50 per cent of our attack capability which can be used to interdict ground targets or provide direct close support for the ground forces.'

There is, however, a major concern in the Marine Corps in the area of their attack or offensive weapons systems. Since 1956 the USMC has oper-

MARINE FIGHTER-ATTACK SQUADRONS		
VMFA-115	F-4J	Beaufort, SC
VMFA-122	F-4J	Iwakuni, Japan
VMFA-212	F-4J	Kaneohe Bay, Hawaii
VMFA-232	F-4J	Kaneohe Bay, Hawaii
VMFA-235	F-4J	Iwakuni, Japan
VMFA-251	F-4J	Beaufort, SC
VMFA-312	F-4J	Beaufort, SC
VMFA-314	F-4N	El Toro, Ca
VMFA-323	F-4N	El Toro, Ca
VMFA-333	F-4J	Beaufort, SC
VMFA-451	F-4S	Beaufort, SC
VMFA-531	F-4N	El Toro, Ca
MARINE ATTACK SQUADRONS		
VMA-211	A-4M	El Toro, Ca
VMA-214	A-4M	El Toro, Ca
VMA-223	A-4M	Cherry Point, NC
VMA-231	AV-8A	Cherry Point, NC
VMA-311	A-4M	Iwakuni, Japan
VMA-331	A-4M	Cherry Point, NC
VMA-513	AV-8A	Yuma, Arizona
VMA-542	AV-8A	Cherry Point, NC
MARINE ALL-WEATHER ATTACK SQUADRONS		
VMA(AW) 121	A-6E	El Toro, Ca
VMA(AW) 224	A-6E	Iwakuni, Japan
VMA(AW) 242	A-6E	El Toro, Ca
VMA(AW) 332	A-6E	Cherry Point, NC
VMA(AW) 533	A-6E	Cherry Point, NC

MARINE ATTACK HELICOPTER SQUADRONS		
HMA-169	AH-1T	Camp Pendleton, Ca
HMA-269	AH-1J	New River, NC
HMA-369	AH-1J	Camp Pendleton, Ca

MARINE HEAVY HELICOPTER SQUADRONS		
HMH-361	CH-53D	Tustin, Ca
HMH-362	CH-53D	New River, NC
HMH-363	CH-53D	Tustin, Ca
HMH-461	CH-53D	New River, NC
HMH-462	CH-53D	Futenma, Okinawa
HMH-463	CH-53D	Kaneohe Bay, Hawaii

MARINE LIGHT HELICOPTER SQUADRONS		
HML-167	UH-1N	New River, NC
HML-267	UH-1N	Camp Pendleton, Ca
HML-268	UH-1N	New River, NC
HML-367	UH-1N	Futenma, Okinawa
HML-367 Det	UH-1N	Kaneohe Bay, Hawaii

MARINE MEDIUM HELICOPTER SQUADRONS		
HMM-161	CH-46D	Futenma, Okinawa
HMM-162	CH-46F	New River, NC
HMM-163	CH-46F	Tustin, Ca
HMM-164	CH-46F	Tustin, Ca
HMM-165	CH-46F	Kaneohe Bay, Hawaii
HMM-261	CH-46F	New River, NC
HMM-262	CH-46F	Kaneohe Bay, Hawaii
HMM-263	CH-46E	Quantico, Va
HMM-264	CH-46F	New River, NC
HMM-265	CH-46F	Kaneohe Bay, Hawaii

ated various models of the McDonnell Douglas A-4 Skyhawk series. The A-4 has been a very successful light, highly-responsive attack aircraft. The experience of recent years has begun to show, however, that the A-4 series is reaching its limit for modernisation and must be replaced. Yet, due to limited funds and procurement lead time, the USMC will have to continue to operate Skyhawks well into the 1980s.

The Marines would like to see their A-4s replaced by the British Aerospace AV-8 Harrier. The USMC began operating the Harrier in 1971, but a setback in the programme has been dealt by the delay of research and development funds for Fiscal Year 1979. The Marines were at least temporarily denied further testing of an aircraft which is critically important to their unique mission.

General Miller states: 'Consistent with our experience with the helicopter in Korea and South Vietnam, the "basing flexibility" offered by the helicopter's V/STOL characteristics clearly points the way for improving the responsiveness of our vital heavy airborne fire support for the ground combat Marine. After eight years of operating the AV-8A [the initial variant produced in Britain] in almost every known environment, both in support of ground forces and in many other missions on land and at sea, the Marine Corps considers this type of aircraft system to be most vital and of unqualified success.'

To counter resistance to acquiring combat aircraft from outside the United States and to bring the Harrier's V/STOL experience and tech-

nology to America, arrangements have been made for McDonnell Douglas to produce a licence-built Harrier, the AV-8B. It will also be powered by the same Rolls-Royce turbofan engine, but it will have better lift capability and greater endurance. With the cutback in Harrier funding, however, the USMC's Harrier programme will be delayed at least a year and there is a very real danger that the programme could be forced into eventual termination.

The US Marine Corps currently has three Harrier-equipped attack squadrons: VMA-231 and 542 at the Marine Corps Air Station in Cherry Point, North Carolina, and VMA-513 at the MCAS in Yuma, Arizona. There is also a VMA detachment of AV-8As at the MCAS in Kadena, Okinawa in the Western Pacific. The Marines have been promised additional conventional aircraft most notably the new F/A-18–but none with the Harrier's unique operating capabilities.

Because of the Harrier's basing flexibility, General Miller notes, 'the AV-8 can perform its mission within the shortest possible time and with the least expenditure of fuel. We can put this V/STOL aircraft close to the front lines and shut down the engine. Then, when it is needed, it can be launched immediately, perform the mission and return to its

Opposite: Boeing Vertol CH-46 Sea Knights of HMH-362 line the deck of USS Guadalcanal.
Above left: the A-4M was intended for USMC use.
Left: the North American OV-10A is flown by three front-line and one reserve Marine squadrons.
Below: a Grumman A-6 Intruder attack bomber

MARINE TACTICAL ELECTRONIC WARFARE SQUADRON

VMAQ-2	EA-6A, EA-6B	Cherry Point, NC
VMAQ-2 Det	EA-6B	Iwakuni, Japan
VMAQ-2 Det	EA-6B (under training)	Whidbey Island, Wash

MARINE PHOTOGRAPHIC RECONNAISSANCE SQUADRON

VMFP-3	RF-4B	El Toro, Ca
VMFP-3 Det	RF-4B	Iwakuni, Japan

MARINE OBSERVATION SQUADRONS

VMO-1	OV-10A	New River, NC
VMO-2	OV-10A	Camp Pendleton, Ca
VMO-6	OV-10A	Futenma, Okinawa

MARINE AERIAL REFUELLING/TRANSPORT SQUADRONS

VMGR-152	KC-130F	Futenma, Okinawa
VMGR-252	KC-130F, KC-130R	Cherry Point, NC
VMGR-352	KC-130R	El Toro, Ca

US MARINE CORPS AIR RESERVE

VMFA-112	F-4N	Dallas, Texas
VMFA-321	F-4N	Wash, DC
VMA-124	A-4E	Memphis, Tenn
VMA-131	A-4E	Willow Grove, Penn
VMA-133	A-4F	Alameda, Ca
VMA-134	A-4F	El Toro, Ca
VMA-142	A-4F	Jacksonville, Fa
VMA-322	A-4E	South Weymouth, Mass
VMO-4	OV-10A	Atlanta, Ga
VMGR-234	KC-130F	Glenview, Ill
HMA-773	AH-1J	Atlanta, Ga
HMH-769	CH-53A	Alameda, Ca
HMH-772	CH-53A	Willow Grove, Penn
HMH-777	CH-53D	Dallas, Texas
HML-771	UH-1E	South Weymouth, Mass
HML-776	UH-1E	Glenview, Ill
HMM-764	CH-46D	Tustin, Ca
HMM-767	CH-46D	New Orleans, La
HMM-770	CH-46D	Whidbey Island, Wash
HMM-774	CH-46D	Norfolk, Va

austere base in around 15 minutes. A conventional aircraft, operating from an airfield 50 to 70 miles behind the lines or flying in an orbit overhead, will consume more fuel and not always be as responsive as an aircraft such as the Harrier, which can be launched from a relatively unprepared surface closer to the front lines.'

It remains to be seen whether the AV-8B programme will remain viable, as the Marines hope it will. Meanwhile, US Marine Corps aviation activities with other aircraft continue to be carried out in the three active wings and one Reserve wing which are authorised by law.

The First Marine Aircraft Wing (MAW) is composed of Air Groups MAG-12 and MAG-15 with fixed-wing tactical aircraft, and MAG-36 with helicopters, Lockheed KC-130 Hercules and North American Rockwell OV-10A Bronco aircraft. These units operate from bases in Japan, Okinawa and the Philippines.

The Second MAW is made up of three fixed-wing Air Groups–MAG-14, MAG-31 and MAG-32–as well as two Air Groups of helicopters, MAG-26 and MAG-29, of which the latter is also equipped with OV-10A aircraft. All Second MAW units are based in the southeastern United States. The Third MAW, on the US west coast, comprises two fixed-wing Air Groups–MAG-11 and MAG-13 –and the helicopters of MAG-16 as well as MAG-39 with a mixture of helicopters and OV-10 aircraft, plus the three fixed-wing squadrons in Marine Combat Crew Readiness Training Group MCCRTG-10.

Designated AV-8A in US Marine Corps service, the British Aerospace Harrier has proved a potent and flexible addition to the inventory. Three attack squadrons have been equipped with the type, while a licence-built derivative, the AV-8B, will supplement existing aircraft if funding is obtained. USMC Harriers are operated from both aircraft carriers and helicopter carriers; an AV-8A is pictured during a vertical landing aboard USS Guam

Three fixed-wing and four helicopter squadrons of MAG-24 in Hawaii make up the First Marine Brigade. The Reserve component, known as the Fourth MAW, operates a variety of fixed-wing aircraft and helicopters from locations throughout the continental United States.

At sea, Marine Corps aviation squadrons deploy aboard several types of ships. Standard fixed-wing aircraft–such as the Grumman A-6 and McDonnell F-4–can and do serve aboard US Navy aircraft carriers. Other, more specialised aircraft –helicopters and AV-8As–routinely deploy aboard aircraft carrier-type ships known as LPH (Landing Platform, Helicopter) and LHA (Landing Helicopter, Assault). They can also operate from LSD (Landing Ship, Dock). The helicopter and V/STOL-carrying vessels are particularly important during amphibious landing operations.

The USMC employ the McDonnell Douglas A-4 Skyhawk to attack and destroy surface targets in support of the Landing Force Commander. This jet-powered attack aircraft also escorts helicopters and conducts other air operations. The USMC operates the A-4M and TA-4F models. The A-4M

variant has a more powerful engine and improved weapons system. Five VMA attack squadrons use the A-4M, while one training squadron uses A-4/TA-4 aircraft. Some 30 two-seat TA-4Fs are also used to perform Tactical Air Control Airborne missions.

The McDonnell Douglas F-4 Phantom II is used as an all-weather fighter attack aircraft to intercept and destroy enemy aircraft, as well as surface targets. There are 12 active and two Reserve VMFA fighter attack squadrons equipped with the F-4. The most common model is the F-4J, although under the Conversion In Lieu of Procurement (CILOP) programme, there is an ongoing process to upgrade them to F-4S specifications. The photo-reconnaissance RF-4B variant is flown by VMFP-3 detachments.

The Grumman A-6E Intruder all-weather attack aircraft is now flown by five VMA(AW) all-weather attack squadrons, as well as in one training squadron. The A-6 can be deployed from a US Navy aircraft carrier or from specially-prepared forward airfields and in conjunction with advancing ground forces. Its sister aircraft, the Grumman EA-6 Prowler,

The AV-8A Harrier has provided the Marine Corps with a close air support aircraft which is capable of operating from immediately behind the fighting line. It has also demonstrated to the Navy the potential advantages of V/STOL aircraft at a time when the large aircraft carrier is becoming increasingly costly to build

conducts airborne electronic warfare in support of Fleet Marine Force operations; the EA-6 uses a variety of electronic counter measures (ECM) devices to degrade enemy air defences. The earlier EA-6A is being replaced by the improved EA-6B, which carries a four-man crew.

The four-engined Lockheed KC-130 Hercules turboprop transport is used to provide aerial refuelling service and assault air transport of personnel, equipment and supplies in support of Fleet Marine Forces. Three Marine Aerial Refueller Transport VMGR squadrons and a Reserve squadron operate the Hercules.

Used for aerial reconnaissance, observation and forward air control operations, the twin-turboprop North American OV-10A Bronco serves with two observation VMO squadrons. The night observation surveillance OV-10D variant began appearing in the squadrons in 1979.

The versatile Bell AH-1J Cobra helicopter gunship serves with three light attack HMA units to provide close-in fire support and escort operations. Combat-tested in Vietnam, it carries an array of weapon systems to perform its mission. The larger Bell UH-1N Huey provides utility combat helicopter support in three light helicopter HML squadrons.

Helicopter transport of supplies, equipment and personnel for the landing force is provided by the Boeing-Vertol CH-46D and F Sea-Knight all-weather, twin-engine, tandem-rotor aircraft assigned to 11 medium helicopter HMM squadrons. Each CH-46 can accommodate up to 17 troops.

The Sikorsky CH-53 Sea Stallion is used to

transport up to 6,356kg (14,000lb) of cargo or 35 troops for the landing force during ship-to-shore movement. CH-53D and F helicopters are assigned to six HMM squadrons. Present plans call for the acquisition of CH-53E Super Stallion helicopters which will transport 14,620kg (32,200lb) of cargo or 55 ground troops.

With continual modernisation and adequate funding—especially for the Harrier programme—US Marine Corps aviation can be expected to keep pace with developments which may require its use. During times of turmoil a favourite American expression is 'send in the Marines' and it is anticipated that they will go in and perform their mission under a spartan but effective air cover.

Top: the McDonnell Douglas F-18 Hornet is intended to serve in the fighter and attack roles with the US Navy and the Marine Corps. Current plans envisage the procurement of 1,366 Hornets, including 153 two-seat variants, of which 30 are earmarked for tactical air control work with the Marines. Above: this Bell AH-1J SeaCobra serves with Medium Helicopter Squadron HMM-262 at Kaneohe Bay, Hawaii. The unit's main equipment comprises the CH-46F, but it flies the UH-1N and AH-1J as part of the 1st Marine Brigade's quick-reaction force

CHAPTER THREE

United States Army

Below: the Bell UH-1 'Huey' formed the backbone of US Army aviation during the Vietnam War, where it operated as a troop transport and largely pioneered the helicopter gunship concept. The UH-1 remains the most widely used US Army helicopter, but it will gradually be replaced by the Sikorsky UH-60 from 1980 onwards. Bottom: medical evacuation is one of the secondary duties undertaken by the 'Huey', an UH-1H of the 68th Medical Detachment, based on Hawaii, being pictured. The transport of wounded troops from the battle area to well-equipped base hospitals by helicopter was one of the innovations of the Korean War, which has greatly reduced the number of fatalities among seriously wounded men

With approximately 10,000 fixed-wing aircraft and helicopters on charge in 1979, the United States Army is easily the largest aviation-oriented element of the US Armed Forces. Although this aspect forms only a relatively small part of Army operations as a whole, the emphasis placed on aviation is assuming an ever-greater importance, in particular, the armed helicopter or 'gunship' is seen as one of the most valuable weapons in the conventional warfare inventory. The years which have passed since the end of the Vietnam conflict have seen a remarkable shift in forward planning and the US Army now recognises Europe as being the most likely theatre in which it might be called to fight in future. Consequently a major re-equipment programme has been underway in this area for the past ten years, the pace of this accelerating as activity in South-East Asia declined. The European Theatre now has some of the most modern weapons systems in the US Army's inventory.

The traditional role of Army aviation has, despite numerous advances in technology, remained virtually unchanged for many years, it is still seen as that of providing support to combat forces operating in the field of battle. This encompasses a variety of tasks and to enable these to be fulfilled satisfactorily a variety of equipment is operated. Medium-lift helicopters such as the Boeing-Vertol CH-47 Chinook permit artillery to be moved rapidly from combat zone to combat zone, a facility of immense value when operating in areas of poor terrain. Smaller scout helicopters can provide valuable intelligence information to, for example, field commanders, thus enabling them to deploy forces in the most effective manner. Medical evacuation, liaison, reconnaissance and surveillance are just a few of the other areas in which the helicopter and, to a lesser degree, fixed-wing aircraft have proved to be of value.

The US Army is broadly organised into four separate Army commands. These comprise two in the USA – the 1st Army at Fort Meade, Maryland and the 6th Army at Presidio, San Francisco, California – and two overseas – the 7th Army at Heidelberg, West Germany and the 8th Army at Youngsan, South Korea. Each controls a number of divisions; there are 16 in all, although these are not equally distributed. For example, the 7th Army in Europe has two infantry and two armoured divisions while the 8th Army in Korea has only one infantry division. Although the division represents the basic combined arms formation, it is the battalion which is the primary manoeuvre unit, this fighting as part of a brigade. Each division has three brigade headquarters, with each controlling from three to five battalions.

In most cases a division includes a combat aviation battalion as part of the manoeuvre elements. In Europe, for instance, each of the heavy divisions has such a unit, these generally operating 42 AH-1S Hueycobra, 51 OH-58 Kiowa and 23 UH-1H Iroquois helicopters on duties including reconnaissance, anti-tank warfare and minelaying. The two armoured cavalry regiments stationed in West Germany also have an attack helicopter company assigned, this being similarly equipped. Infantry Divisions invariably include several troop lift helicopter companies offering enhanced mobility and range, these generally operating the Bell UH-1H Iroquois; approximately 20 to 25 such helicopters are normally assigned to a specific company.

In addition, it is common for specialist units such as artillery and engineer Regiments to operate a small number of helicopters on miscellaneous

duties including liaison and observation while there is also a large number of non-aligned aviation companies engaged on tasks including transport, surveillance and medical evacuation. Types in use include the CH-47 Chinook, OV-1 Mohawk and UH-1 Iroquois. These generally report to a separate Army aviation headquarters within the overall Army chain of command. Thus, the HQ US Army Europe at Heidelberg has a separate aviation headquarters to control the activities of such units.

At present the most numerous helicopter type is the Bell UH-1 Iroquois. Several variants are operated, the bulk being of the UH-1H variety. This is still the principal troop-carrying helicopter, being used extensively by combat support aviation companies, air cavalry units and casualty evacuation companies. One of the most successful helicopters yet conceived, the Huey was widely used in Vietnam and benefited greatly from that conflict in that it was progressively updated to meet the ever-increasing demands being placed on it. Several thousand remain in service but these are to give way to Sikorsky's UH-60A Black Hawk, which will enter full operational service with the 101st Airborne Division at Fort Campbell, Kentucky in 1980. US Army Hueys should, however, remain a common sight for many years to come with surplus machines being used to continue the modernisation of US-based Reserve forces.

The first helicopter to be designed solely for offensive combat operations, Bell's Hueycobra entered service in Vietnam during 1967 and over 1,000 AH-1Gs were supplied to the US Army. Recent years have seen the introduction of a major modernisation programme, which will, when it terminates in 1984, have resulted in the updating of some 986 Hueycobras to full AH-1S standard with provision for eight Tow missiles, two 19-tube rocket pods and a chin turret housing either a 20 mm M197 Gatling gun or a Hughes 30 mm chain gun. The programme also involves re-engining the basic AH-1G with the more powerful Lycoming T53-L703 power plant, strengthening outer pylons, providing heat-dissipating exhaust efflux nozzles, revising cockpit windows and installing Tow optical sighting systems. New tactics are being developed to take full advantage of the modernised Cobra, these mainly involving so-called 'nap-of-the-earth' techniques, in which the Cobras and their accompanying scouts maintain very low altitude when in the vicinity of the battlefield. This entails the use of every available scrap of cover and would be instrumental in reducing combat losses when operating in a high-ECM (electronic countermeasures) environment, such as that encountered in the presence of enemy surface-to-air missiles.

The standard medium-lift helicopter is Boeing-Vertol's CH-47 Chinook, several hundred of which remain in service. The most numerous and most advance model is the CH-47C but examples of the earlier CH-47A and CH-47B are still in the active inventory. Boeing-Vertol are currently engaged in the early stages of a major modernisation project which, if forthcoming tests prove successful, should lead to most surviving Chinooks being updated in the early eighties, thus extending their service lives until well into the next decade. Among the changes planned for the CH-47D, as it will be known, are

provision of an automatic flight control system, revision of existing hydraulics systems, fitting of fibreglass rotor blades and an improved drive system and, finally, fitment of multiple cargo hooks intended to provide both increased flight stability of external loads and the capability of multiple point delivery of such loads. One example of each of the three major Chinook variants is to be converted to YCH-47D prototype configuration for flight trials and Boeing-Vertol anticipate further contracts for modification of existing Chinooks when initial testing is completed in 1980.

Observation and scouting duties are performed by two principal types of helicopter, namely the Bell OH-58 Kiowa and the Hughes OH-6A Cayuse, although none of the latter are stationed in Europe. Some 2,200 Kiowas were acquired in the late 1960s and early 1970s and the majority remain in use, although a substantial number now serve with Reserve units. Although normally unarmed, they can be fitted with two XM27 7·62 mm minigun pods

Top: this Boeing-Vertol CH-47C Chinook is assigned to the German-based 180th Aviation Company. The Chinook can carry 44 troops in the assault transport role and can also lift bulky underslung loads.
Above centre: the US Army in Germany flies the Bell OH-58 Kiowa on observation and scouting duties, this example serving with the 2nd Support Brigade based at Hanau.
Above: the Hughes OH-6 Cayuse is another Army scouting helicopter.
Right and inset right: the Bell AH-1G HueyCobra is armed with a 7.62 mm Minigun and a 40 mm grenade launcher in the nose turret. The stub wings have four hardpoints for rocket pods, anti-tank missiles and other ordnance. The gunner and pilot are seated in tandem in the narrow fuselage

on side sponsons. This variant, known as the OH-58C, also features a more powerful 420 hp Allison 250-C20B engine, infra-red suppressors and flat, glint-reducing windows. The smaller Hughes OH-6A on the other hand is permanently armed and approximately 1,000 remain in use.

Other types in the active inventory include the Grumman Mohawk multisensor tactical observation and reconnaissance aircraft, the CH-54A Tarhe heavy-lift helicopter, the twin-engined Beech U-21, several versions of which exist for such duties as liaison and electronic surveillance, and the Beech C-12A Huron which is used for communications. Basic training is mainly accomplished on the fixed-wing Cessna T-41 and rotary-wing Hughes TH-55, aircrews then progressing to aircraft and helicopters similar to those that they will fly when they join an operational unit.

Looking to the future, the US Army will soon receive the first full production Sikorsky UH-60A Black Hawk. At least 1,100 are to be acquired by the mid-1980s and these will replace the UH-1H as the principal squad-carrying helicopter, bringing significantly improved capability while simultaneously offering enhanced survivability in combat zones. In addition to the basic utility role it seems highly likely that specialist versions will also appear; one derivative now under consideration is the EH-60 signal intelligence and electronic warfare platform.

Offensive operations should also receive a boost with the arrival of the Hughes AH-64 AAH (Advanced Attack Helicopter) in the early eighties, assuming production go-ahead is given at the end of the Army's evaluation in December 1980. Potentially the most lethal anti-tank helicopter yet, it has been designed to operate with a new specially developed 'fire and forget' missile known as Hellfire and will be able to carry 16 of these plus rocket pods and the almost-obligatory Hughes 30 mm chain gun. Two YAH-64 prototypes have been flying since 1975 and they will be joined by a further three during 1979. An impressive array of target detection, designation and aiming systems are planned for the AH-64; if production is authorised, some 536

examples are expected to be delivered between the end of 1982 and March 1989.

With both the UH-60A and AH-64 making good progress and the modernised CH-47D Chinook coming along, the US Army's immediate needs for the mid- to long-term future seem to be adequately catered for. Although numerically inferior to the Soviet forces, it undoubtedly has a qualitative advantage when it comes to equipment currently operated and planned for the future. One can only hope that, in the event of war, this superiority will prove adequate to meet the challenge.

Below left: a Sikorsky CH-54 Tarhe 'flying crane' demonstrates its capabilities with an underslung armoured fighting vehicle. The Tarhe is also used to retrieve crashed aircraft and to assist army engineers.
Below: this CH-54A operates from Germany with the 295th Aviation Coy.
Bottom: the Grumman OV-1 Mohawk operates in the battlefield observation role. This OV-1C of the 122nd Aviation Coy carries a side-looking air radar pod

CHAPTER FOUR

United Kingdom

Royal Air Force/Fleet Air Arm/Army Air Corps

At the threshold of the 1980s the prospects for the air arms of the British armed Services appear more settled than for some time past. The last two decades have been marked by accelerating changes in military aviation which have greatly affected the operations and planning of the Royal Air Force, the Fleet Air Arm and the Army Air Corps. In addition to the demands of technical progress, and the consequent need to re-equip with a new generation of aircraft and systems, there has been the counter-weight of demands for increased thrift in defence expenditure. The strain induced by these opposing trends has disrupted forward planning, and has been especially galling for the Services when they observe the rapid expansion in the quantity and

A Vulcan B Mark 2 bomber of No 44 Squadron, based at Waddington in Lincolnshire, shows the aircraft's distinctive delta wing planform to advantage. Since the Royal Navy took over the strategic nuclear deterrent role in 1969, the RAF's Vulcans have operated at low-level with conventional and tactical nuclear weapons

quality of the Soviet and Warsaw Pact forces which constitute the threat to the security of the West.

A further factor has shaped the recent history of the Forces, namely the fundamental re-alignment of the UK's defence commitments from their previously worldwide responsibilities to their now almost-exclusive orientation towards Europe. Although Britain, as a founder member of the NATO Alliance 30 years ago, has all along participated to a major degree in the defence efforts of the Organisation, the abandonment of the Services' far-ranging intervention capability and the need to adjust to the requirements of European defence was not an easy process.

Nor was the situation in the European operating environment entirely straightforward. In place of NATO's previous strategy of nuclear retaliation to any attack–the so-called 'trip-wire' policy–the Alliance decided in 1967 that a strategy of 'flexible response' would be implemented against any attack on Western Europe. This meant that the NATO reaction would attempt to match the aggressor

NO 1 GROUP, STRIKE COMMAND

No 9 Sqn		Vulcan	Waddington
No 12 Sqn		Buccaneer	Honington
No 13 Sqn		Canberra	Wyton
No 27 Sqn		Vulcan	Scampton
No 35 Sqn		Vulcan	Scampton
No 39 Sqn		Canberra	Wyton
No 44 Sqn		Vulcan	Waddington
No 50 Sqn		Vulcan	Waddington
No 51 Sqn		Canberra	Wyton
		Nimrod	
No 55 Sqn		Victor	Marham
No 57 Sqn		Victor	Marham
No 101 Sqn		Vulcan	Waddington
No 208 Sqn		Buccaneer	Honington
No 216 Sqn		Buccaneer	Honington
No 360 Sqn		Canberra	Wyton
No 617 Sqn		Vulcan	Scampton
No 230 OCU		Vulcan	Scampton
No 231 OCU		Canberra	Marham
No 232 OCU		Victor	Marham
No 237 OCU		Buccaneer	Honington
		Hunter	

Abbreviation: OCU, Operational Conversion Unit

Above right: Vulcans of No 101 Squadron fly over the Ballistic Missile Early Warning Station at Fylingdales during a training sortie.
Below: operational training on the Buccaneer is undertaken by No 237 OCU at Honington, Suffolk

with conventional forces, if appropriate, rather than resort to nuclear weapons at the outset. This change required the Alliance to increase its investment in non-nuclear forces and to up-grade their readiness states as a guard against surprise attack.

The UK is one of two European countries which provides forces for all three major NATO Commands–Allied Commands Europe, Atlantic, and Channel, respectively known by the abbreviations ACE, ACLANT, and ACCHAN. In the Central Region of Europe, the UK supplies a major part of NATO's 2nd Allied Tactical Air Force and the UK-based combat and reserve forces of the RAF would likewise be available in their differing roles to ACE as well as to ACLANT and ACCAN. In the Eastern Atlantic and Channel areas the Royal Navy contributes its front-line warships and the aircraft aboard them to the Alliance. The British Army of the Rhine and Army reserve forces in the UK are important elements in ACE's Northern Army Group, and their attached aviation component includes Army Air Corps helicopters assigned in support roles.

While the identity of Britain's individual armed Services, and the traditions of their air forces, remain as distinct as ever, their operations are less independent than before. Whether at sea, or on

land, or in the air, the military aviation effort is an interlocking and multi-faceted one, demanding extensive co-ordination of the specialist contributions of all three Services.

The combat and second-line forces of the RAF are deployed today in two home-based commands and one overseas command, as a result of the wide-ranging organisational changes which have taken place in the past ten years. All the operational forces based in the UK and the remaining overseas units outside the RAF in Germany, are controlled by Strike Command (STC), the largest of the three. The other UK-based command is Support Command, which is responsible for flying training and the provision of the miseellany of support services to the front-line forces which its name implies. These include engineering, communications, supply, administration and all aspects of ground operations training. The third command, and the only other operational formation apart from STC, is RAF Germany (RAFG).

Between them, STC and RAFG operate about 1,050 aircraft of which almost 700 are in the front line, the remainder being used for operational training and support functions–although available as wartime reinforcements. All the front-line aircraft, together with certain of the training aircraft,

Below: the Panavia Tornado will replace the RAF's Vulcans and Buccaneers in the low-level strike/attack role.
Below centre: a Victor K Mark 2 from the Marham Wing refuells two BAe Lightning interceptors of No 5 Squadron flying from Binbrook, Lincs.
Bottom: No 39 Squadron undertakes high-altitude reconnaissance duties from Wyton, Huntingdonshire, flying the Canberra PR Mark 9

are assigned to NATO or available in support of NATO operations. In wartime these aircraft and their squadrons would be controlled by the Alliance as part of the British contribution to NATO; and though in peacetime they remain under national control, they exercise and train continuously as part of the NATO air forces.

The aircraft of the UK-based Strike Command are assigned to four operational Groups, each with differing responsibilities which partly reflect the former functional commands of the RAF – Bomber, Fighter, Transport, Coastal etc. These Groups also contain the Operational Training Units whose task is to undertake the final phase of aircrew training on the aircraft types operated by the Group's front line squadrons and their counterparts in RAF Germany. The operating roles of these STC Groups are: No 1 Group, strike/attack (overland and maritime), air-to-air refuelling and reconnaissance; No 11 Group, air defence; No 18 Group, maritime patrol/anti-submarine warfare (ASW), and search & rescue (SAR); No 38 Group, attack/reconnaissance and transport (fixed-wing and helicopters).

Royal Air Force Germany occupies four stations in the Federal Republic and forms part of the NATO 2nd Allied Tactical Air Force along with the Belgian and Dutch Air Forces and elements of the German Air Force in the northern half of the country. RAFG is Britain's main air contribution to NATO air defences on the Continent, and its operational squadrons are tasked in air defence, strike/attack, reconnaissance and battlefield transport roles to support NATO land and air operations.

The air defence forces of the RAF consist of fighter aircraft and surface-to-air missiles (SAMs) which are based both in the UK with Strike Command and in RAF Germany. The home-based interceptor units comprise Phantom and Lightning squadrons of No 11 Group, and there are two Phantom squadrons in RAFG. Supplementing the fighters are two squadrons of Bloodhound Mark 2 medium-range missiles, one deployed in the UK and the other shielding the so-called 'clutch' of RAF stations – Laarbruch, Bruggen and Wildenrath – just west of the Rhine. RAF Regiment squadrons equipped with low-level Rapier SAMs provide point defence facilities for RAF airfields. In peacetime the air defence squadrons are responsible for maintaining the integrity of national air space over the UK and over the northern half of West Germany. Together with the USAF and French Armée de l'Air, the RAF in Germany is also obliged to maintain access to Berlin in the three air corridors.

The importance of the UK's geographical position to NATO, as a base for air operations over the Eastern Atlantic and Central Europe; and as a staging point for trans-Atlantic reinforcements in any European conflict was underlined by the creation of a new major NATO command in 1975 – UK Air Forces (UKAIR), whose C-in-C is also AOC-in-C RAF Strike Command. UKAIR is one of the only five principal commands in Allied Command Europe (ACE) and its responsibilities include the air defence of the UK and the surrounding sea areas forming the UKADR (UK air defence region). Air defence information is built up by the operation of ground-based radars and computerised data links with other components of the NATO Air Defence Ground Environment (NADGE) system in Europe.

An important adjunct to these facilities, which also provides a vital seawards extension of the low-level radar cover, are the Shackleton AEW Mark 2 airborne early-warning aircraft of No 8 Squadron at Lossiemouth. The venerable Shackletons, due to be replaced in the early-1980s by the Nimrod AEW aircraft, fulfil a variety of tasks in addition to warning of low-level intrusions into the UKADR. They control interceptors and ·can

act in support of maritime strike/attack aircraft and naval forces.

The mainstay of the RAF fighter squadrons since the mid-1970s has been the McDonnell Douglas Phantom; the Service operates both the F-4K and F-4M variants–respectively designated the FG Mark 1 and FGR Mark 2, the former being the aircraft originally procured for the Royal Navy and the latter the version which the RAF initially used in the strike/attack and reconnaissance roles. With the entry of the Jaguar into squadron service in 1974, the Phantom FGR Mark 2s were progressively switched to air defence duties, thereby permitting the retirement of most of the BAC Lightnings which had occupied the air defence front-line during the preceding 15 years. There are currently five squadrons of Phantoms in No 11 Group (Nos 23, 29, 43, 56 and 111) together with a residual two-squadron force of Lightnings at Binbrook and two further Phantom squadrons (Nos 19 and 92) at RAF Wildenrath in Germany.

The availability of the more heavily armed longer-range Phantom brought about a transformation in the hitting power of UK air defences; the aircraft can carry eight air-to-air missiles (AAMs)–four radar-homing Sparrow and four infra-red seeking Sidewinder AAMs, as against the Lightning's two Red Top or Firestreak missiles, plus the six-barrelled 20mm Vulcan cannon in place of the Lightning F Mark 6's twin 30mm Aden cannon. More notably the Phantom has a much enhanced air interception (AI) radar and fire control system, offering a look down/shoot down capability which enables it to detect and attack low-level targets

without the radar ground return problems suffered by earlier AI sets. These characteristics plus the aircraft's longer radius of action and the benefit of its two-crew operation have furnished the RAF with an interceptor which is more appropriate to the task of defending the large area of the UKADR. Despite its advancing years, the Lightning is still effective, excelling in the climb rate and manoeuvrability required for rapid reaction against high-level and supersonic targets. Its range limitations on internal fuel can be extended by joint operations

Below: four Sparrow radar-guided missiles are carried by this Phantom FGR Mark 2 of No 111 Squadron. Infra-red Sidewinder missiles and a gun pod can also be carried on air defence sorties by Phantom interceptors.
Bottom: one of No 23 Squadron's Phantoms awaits its crew at its Wattisham base. The interceptor's radar is operated by the navigator from the rear cockpit

with the Victor K Mark 2 air-to-air refuelling tankers of No 1 Group's No 55 and 57 Squadrons which are likewise available to the Phantoms.

The RAF's UK air defence interceptor stations are located in three tiers along the east coast of Scotland and England – the northern sector being covered by Leuchars, Fife, the central area by Binbrook and Coningsby, Lincs, and the south by Wattisham, Suffolk. At these bases Battle Flights of fully-armed aircraft are maintained at continuous readiness to intercept any unidentified or potentially hostile intrusions into the UKADR. The squadrons at Leuchars draw the largest share of this 'trade' and the Interceptor Alert Force there is regularly scrambled to monitor the activities of Soviet aircraft approaching UK air space.

In wartime the UK's functions as a rear base for Allied Command Europe and a forward operating location for Allied Command Atlantic would be very likely to attract the attention of Warsaw Pact bomber and fighter bomber sorties on deep interdiction and counter-air tasks. This contingency has become more certain with the qualitative performance gains – in range, warload, speed, avionics/weapons systems and ECM capability – assessed for Soviet aircraft in recent years. Furthermore, the numbers of Soviet aircraft being produced are such as to indicate a greater weight of attack on the UK.

In a war situation the essential task of the UKAIR air defence formations would be to engage these threats which, given their stand-off weapons, would often need to be met at several hundred miles range from their targets. Hence the significance of AEW aircraft and air-to-air refuelling tankers working in conjunction with the interceptors respectively to give advance warning of low-level intruders hidden from ground radars and to extend the range and on-station endurance of combat air patrols mounted by friendly fighters. Apart from the home defence responsibilities, one of the No 11 Group Phantom squadrons, No 43, is primarily tasked in the maritime air defence role to give cover to the Fleet at sea – a commitment which can involve long sorties out to distant patrol lines.

Enemy aircraft taking the shortest route between Eastern Europe and the UK would also have to track through the NATO air defences in the Central Region, where RAFG Phantom squadrons are just one element of Allied Air Forces Central Europe's interceptor force. These are geared to deal not only with the counter-air and interdiction threats, but also the main weight of the opposing tactical fighter-bomber strength and its air superiority cover. At Wildenrath a Battle Flight mounted by Nos 19 and 92 Squadrons is held at a constant 5 minute's readiness and in wartime these units would be assigned to SACEUR for the defence of a sector of Central Region air space in collaboration with F-104G Starfighters of the Belgian Air Force.

Training exercises have also evaluated the efficacy of RAF Phantoms providing combat air patrol/escort facilities for strike/attack aircraft, which are limited in their air-to-air defensive armament. The Jaguar, for example, is fitted with twin 30mm Aden cannon – but these are not entirely appropriate in the air-to-air mode – while the Buccaneer is bereft of gun or missile self protection, relying instead on its very considerable ability to evade or out-run the defences at low-level and high speed.

The implications of the Warsaw Pact's counter-air capability and the lessons derived from the 1967 Arab-Israeli War have not been lost on NATO, which has implemented many measures to improve the survivability of its airfields. Hardened shelter accommodation for aircraft (and other vital base installations, such as combat operations centres and stores) has been constructed for the entire RAF Germany combat strength, and is being extended to Strike Command bases in 1979. Additional airfield survival steps have included the 'toning down' of runways, taxiways and other surfaces to make them less visible from the air; the painting of vehicles, buildings and equipment in an inconspicuous drab green colour; and the provision of multiple taxiway routes between aircraft shelters and runways. Taxiways parallel to the runways can be employed as alternative take-off/landing surfaces and deployments have been made to stretches of highway which could be used as standby runways. The aircraft too have been toned down with smaller or less visible squadron markings, national insignia with only red and blue in the roundels, and the extension of aircraft upper surface camouflage patterns to fuselage and wing under surfaces as well.

If the best form of defence is attack, then this aspect of the RAF's capability has received much attention during the 1970s. The introduction of new aircraft and weapon systems has bolstered the front line elements, whose mission is more properly labelled 'strike/attack'. In NATO parlance the role description 'strike' applies to air operations involving the delivery of nuclear weapons, while the 'attack' category of missions relates to a variety of tasks in which the ordinance would be 'conventional'. The combined reference to strike/attack in portraying the role of many STC and RAFG squadrons reflects the fact that their assignments encompass both nuclear and conventional weapon options and in these cases the units are described as 'dual capable'. Some squadrons are charged with a primary maritime role, others are assigned overland tasks, but in either case the strike/attack classification remains appropriate. The terminology is complicated by the fact that some RAF squadrons are charged solely to develop an attack capability, and in such cases their role is separately described as 'ground support' or 'offensive support'.

The RAF's overland strike/attack and ground support capability is generally divided into three mission categories – counter-air, interdiction, and offensive support. Counter-air operations are those directed against the enemy air force by attacking it in the air or on the ground. The weight of the RAF's counter-air capability is aimed at the attack of ground installations and facilities, rather than the exercise of air superiority by fighters operating over hostile territory. The main targets of counter-air strike/attack missions would be enemy airfields and principally the runways, taxiways and other vulnerable operating surfaces on them.

Interdiction missions would be targeted against transportation systems, stores and logistic areas at some distance behind the battlefield with the aim of

NO 11 GROUP, STRIKE COMMAND

No 5 Sqn	Lightning	Binbrook
No 7 Sqn	Canberra	St Mawgan
No 8 Sqn	Shackleton	Lossiemouth
No 11 Sqn	Lightning	Binbrook
No 23 Sqn	Phantom	Wattisham
No 29 Sqn	Phantom	Coningsby
No 43 Sqn	Phantom	Leuchars
No 56 Sqn	Phantom	Wattisham
No 85 Sqn	Bloodhound	Deployed UK
No 100 Sqn	Canberra	Marham
No 111 Sqn	Phantom	Leuchars
No 228 OCU	Phantom	Coningsby
LTF	Lightning	Binbrook
No 1 TWU	Hawk, Hunter, Jet Provost	Brawdy
No 2 TWU	Hunter	Lossiemouth

Abbreviations: OCU, Operational Conversion Unit; LTF, Lightning Training Flight; TWU, Tactical Weapons Unit.

Above: the F Mark 6 was the ultimate production version of the Lightning, one of No 11 Squadron's aircraft being depicted. This Mark introduced an enlarged ventral fuel tank to alleviate the Lightning's poor range characteristics. However, the interceptor's endurance can be usefully extended when it operates with Victor tankers

disrupting lines of communication and re-inforcement. This would seal off the enemy's front line ground forces and prevent the reinforcement and resupply of their forward troops. A distinction is usually drawn between 'deep interdiction' tasks such as these and the 'battlefield interdiction' mission, in which sorties would be mounted directly against forces in the enemy's rear echelon or units moving up to the front. Battlefield interdiction is in effect a half-way area of operations between attacks on rearward targets and close air support. This third category of attack roles is aimed at enemy armour, infantry and transport on the battlefield.

Stationed in the UK and West Germany the RAF presently has 18 squadrons of Vulcan, Buccaneer, Jaguar and Harrier aircraft to fulfil this spectrum of overland strike/attack missions as their primary role. In addition there are further units assigned to the equivalent maritime tasks and other squadrons which can undertake attack missions as a secondary role. In wartime, these front line forces would be supplemented by aircraft drawn from the appropriate Operational Conversion Units in STC.

The heavyweight component of Strike Command's strike/attack inventory is formed by the six squadrons of Vulcan B2 bombers operating as part of No 1 Group and based at Waddington and Scampton in Lincolnshire. The last in a long line of famous four-engined 'heavies' to have served with the RAF, the Vulcan was until 10 years ago one of the V-bomber types which provided the UK's strategic nuclear deterrent – a mission switched to the Royal Navy's Polaris missile armed submarines in mid-1969. Subsequently the Vulcan has been deployed in the tactical strike role and in this capacity the B Mark 2-equipped squadrons are assigned to SACEUR. The aircraft are capable of operating also on conventional attack missions,

for which they could carry up to 9,525 kg (21,000 lb) of high-explosive bombs. Although destined to be replaced by the Panavia Tornado GR Mark 1 from 1981 onwards, the 20-year old Vulcan has been a potent element of RAF offensive airpower, its range, all-weather low-level penetration performance and ECM systems ensuring that it would give a good account of itself on counter-air and deep interdiction tasks.

Purpose built for the high-speed, low-level strike/attack role is the Buccaneer, which originally started life as a carrier-borne aircraft with the Royal Navy and was later procured for the RAF after the successive abandonment of plans to produce the TSR 2 and buy the General Dynamics F-111K from the United States. Although the Buccaneer's second career ashore thus got off to a late start, its subsequent service with the RAF has been a considerable success story. A rugged airframe – optimised for fast and low attack profiles – plus

long-range and sizeable weapons carrying capacity have made the Buccaneer a formidable strike/attack system. The aircraft's effectiveness has been enhanced by the adoption of the Martel air-to-surface missile and latterly it has been the trials platform for the RAF's future introduction of precision guided munitions in the form of the American Paveway bomb guidance kit and the associated Pavespike laser designator.

The first squadron to form on the Buccaneer was No 12 at Honington, Suffolk in 1969, followed by Nos 15 and 16 Squadrons in RAFG and, later, another STC unit in the UK – No 208 Squadron. This year has also seen the appearance of a further front-line Buccaneer squadron, No 216. This was made possible by the availability of ex-Royal Navy aircraft after the withdrawal of HMS *Ark Royal* late in 1978 and the disbandment of the FAA's last Buccaneer unit, No 809 Squadron. Nos 12 and 216 Squadrons are assigned to the maritime strike/

Below left: crewmen wearing nuclear, biological and chemical protective clothing work on one of No 29 Squadron's Phantom FGR Mark 2 aircraft. The RAF periodically exercises the techniques necessary to operate in a contaminated environment.
Below: the British Aerospace Hawk T Mark 1 serves with No 1 Tactical Weapons Unit at Brawdy, Dyfed. This unit instructs pilots who have completed advanced flying training (also on the Hawk) in low flying and weapon delivery before they progress to an operational conversion unit.
Bottom: No 29 Squadron began to fly the Phantom FGR Mark 2 in the air defence role in December 1974. The Squadron is based at Coningsby, Lincs, alongside the Phantom OCU

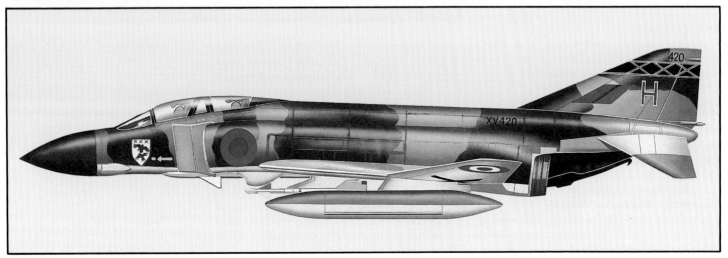

attack role in support of SACLANT (the Buccaneer OCU is also earmarked to undertake a similar maritime mission), while the other STC squadron, No 208 and the two RAFG units are assigned to SACEUR in the overland strike/attack role. The Buccaneer is able to undertake a secondary reconnaissance role and can be fitted with a bomb-bay mounted camera/sensor pack.

Very much aware of the improvements which have taken place in Warsaw Pact gun and missile ground defences in recent years, the RAF's approach to strike/attack missions stresses the benefits of operating fast and low, with avionics and weapons systems which offer a high probability of first pass attack accuracy. In this respect the Service's tactics are noticeably distinct from that of other NATO air arms which place greater reliance on active defence suppression, such as supporting strikes on AAA/SAM sites and their radars, and the provision of a dense, protective ECM 'umbrella' for the bombing force. While the RAF strike/attack capability is not without defence suppression and ECM facilities, there is not the same accent on this task as there is, for example, in the US air arms with their special purpose aircraft dedicated to these missions. Nonetheless the Buccaneer, to take one case, can carry the radar-homing version of the Martel ASM and the Westinghouse AN/ALQ-101 jamming pod; and the NATO air forces functioning as part of an allied attack force would presumably count on the assistance of specialist support elements based in Europe. USAFE 'Wild Weasel' defence suppression F-4D Phantoms were noteworthy participants at the 1978 RAF Tactical Fighter Meet.

Given high-speed/low-level tactics, the consistent upgrading of ground defences in the Warsaw Pact area has tended to emphasise the importance of a surprise attack. In the ECM field this priority has led to the widespread adoption of passive warning receivers (PWR) which can alert aircrew to ground and air radar threats in time to take avoiding action and without revealing the presence of the PWR-carrying attack aircraft. Another requirement for mission effectiveness at low level, especially in bad weather or at night, is the achievement of first pass accuracy, so navigation and weapons aiming systems have been in the forefront of recent equipment developments.

Exemplifying the increased utilisation of avionics systems and the enhanced capability they offer is the Sepecat Jaguar which now equips three squadrons in STC's No 38 Group and five squadrons in RAFG. All these units are tasked for a range of strike/attack missions, but in two cases (No 2 Squadron at Laarbruch, Germany, and No 41 Squadron at Coltishall, Norfolk,) these are secondary to a tactical reconnaissance role. There is a difference too between the assignments of the other four Jaguar squadrons based in Germany and the two in the UK. The former are available to SACEUR in the dual capable strike/attack role, while the UK-stationed squadrons train for a primary offensive support role.

The front-line versions (as opposed to the two-seat trainer versions) of the Anglo-French-built Jaguar were conceived essentially as a ground attack aircraft operating in the low-level environment and with an extended radius of action. The single-seaters serving with the RAF are fitted with nose-mounted laser rangefinder and marked target seeker (LRMTS) in addition to a sophisticated, computerised navigation and weapons aiming system which is connected to a moving map display and a head-up display (HUD). These devices confer on the Jaguar a significant first-pass attack accuracy, and its endurance at low-level is sustained by two fuel-conserving turbofan engines. Apart from the nuclear strike option, weapons carried by the Jaguars include 1,000lb HE bombs and BL755 cluster bomb units, the latter being a dispenser of sub-munitions which are scattered for area effect, notably against tanks and other armoured vehicle formations.

NO 18 GROUP, STRIKE COMMAND

No 22 Sqn A Flt	Whirlwind	Chivenor
No 22 Sqn B Flt	Wessex	Leuchars
No 22 Sqn C Flt	Wessex	Valley
No 22 Sqn D Flt	Whirlwind	Brawdy
No 22 Sqn E Flt	Wessex	Manston
No 42 Sqn	Nimrod	St Mawgan
No 120 Sqn	Nimrod	Kinloss
No 201 Sqn	Nimrod	Kinloss
No 202 Sqn A Flt	Sea King	Boulmer
No 202 Sqn B Flt	Whirlwind	Leconfield
No 202 Sqn C Flt	Whirlwind	Coltishall
No 202 Sqn D Flt	Sea King	Lossiemouth
No 206 Sqn	Nimrod	Kinloss
No 236 Sqn	Nimrod	St Mawgan
SKTU	Sea King	Culdrose

Abbreviations: OCU, Operational Conversion Unit; SKTU, Sea King Training Unit.

NO 38 GROUP, STRIKE COMMAND

No 1 Sqn	Harrier	Wittering
No 6 Sqn	Jaguar	Coltishall
No 10 Sqn	VC10	Brize Norton
No 24 Sqn	Hercules	Lyneham
No 30 Sqn	Hercules	Lyneham
No 32 Sqn	Andover, HS125, Gazelle, Whirlwind	Northolt
No 33 Sqn	Puma	Odiham
No 41 Sqn	Jaguar	Coltishall
No 47 Sqn	Hercules	Lyneham
No 54 Sqn	Jaguar	Coltishall
No 70 Sqn	Hercules	Lyneham
No 72 Sqn	Wessex	Odiham
No 115 Sqn	Andover	Brize Norton
No 207 Sqn	Devon	Northolt
Det 1		Wyton
Det 2		Turnhouse
No 230 Sqn	Puma	Odiham
No 226 OCU	Jaguar	Lossiemouth
No 233 OCU	Harrier	Wittering
No 240 OCU	Puma, Wessex	Odiham
No 241 OCU	VC10	Brize Norton
No 242 OCU	Hercules	Lyneham
Queen's Flight	Andover, Wessex	Benson
AHQ Cyprus		
No 84 Sqn	Whirlwind	Akrotiri
Hong Kong		
No 28 Sqn	Wessex	Sek Kong

Abbreviations: OCU, Operational Conversion Unit; Det, Detachment.

Top: the RAF's standard long-range maritime patrol aircraft is the BAe Nimrod MR Mark 1. The Nimrod fleet is to be progressively modified to Mark 2 standard, with improved ASW systems, from 1980 onwards.
Above centre: the Sea King HAR Mark 3 search and rescue helicopter entered service with No 202 Squadron in 1978. Its operating radius is more than three times that of the Whirlwind.
Above: four squadrons of Lockheed Hercules C Mark 1 tactical transports operate from Lyneham, Wiltshire, some 48 of the original 66 remaining

The RAF's offensive support capability features the unique Harrier – two large squadrons, Nos 3 and 4, are based at RAF Gutersloh in Germany and there is one No 38 Group squadron, No 1, plus an OCU in the UK. Ten years after its entry into RAF service, the Harrier remains the only fixed-wing aircraft operational in the Western World which is capable of vertical take-offs and landings. As such it has the vital ability to function without depending on the use of airfield runways and this makes the aircraft a superb solution to the problem of coping with hostile counter-air operations. There are, of course, a host of other advantages attaching to the Harrier in the front line combat inventory, but most important is this ability to fly the aircraft from dispersed

field sites away from the main bases which would undoubtedly be the subject of enemy attack in wartime.

The main roles of the RAF Harrier squadrons are offensive support and battlefield interdiction – tasks which would be conducted in close liaison with the Army. Indicative of the mission link with ground forces, and the requirement to have the Harriers operational from locations close to the FEBA (forward edge of the battle area), was the repositioning of the RAFG Harrier squadrons east of the Rhine in 1977 when they moved from Wildenrath to Gutersloh. From their new base, 240 km (150 miles) further east, Nos 3 and 4 Squadrons could deploy more rapidly to pre-planned field sites in support of the forward elements of NATO's Northern Army Group and, moreover, without the sortie warload restrictions which would have attended initial operations from points further west.

As with all the UK-based No 38 Group's tactical units (including the Jaguars of Nos 6, 41 and 54 Squadrons and the battlefield support helicopters of Nos 33, 72 and 230 Squadrons), the Harriers of No 1 Squadron are organised for maximum mobility to fulfil wartime reinforcement roles. In fact No 1 Squadron's stated task is that of joining the ACE Mobile Force (AMF) to provide close air support wherever the AMF may be directed to operate in NATO flank areas. The No 38 Group Jaguar squadrons and the aircraft in the Jaguar and Harrier OCUs, on the other hand are variously earmarked for wartime assignment to the UK Mobile Force and as reinforcements which can be acquired by RAF Germany:

Continuous improvements in the power output of the Harrier's Pegasus vectored thrust engine have successively increased the 'payload' potential of the aircraft, and its weapons carrying capacity has been augmented also by the short take-off and vertical landing (STOVL) method of operation rather than the vertical take-off and landing (VTOL) procedure with which the machine first became renowned. Operating from country roads or short natural strips in the STO mode the extra wing-generated lift from rolling take-offs allows a useful increment in weapons loads as well as saving fuel. While the Harrier's warload is less than that which can be uplifted by other fixed-wing types, this is offset by the high sortie rate which would be expected of the force. From their concealed locations close to the battlefield, the Harriers offer a quick response to requests for ground support, and their short haul to the target areas makes for a productive pattern of operations at sortie rates which can be better than one per hour per aircraft.

It would be difficult to imagine a greater contrast in operating environments than that between battlefield support in Europe and the ocean areas of the North Atlantic, but the RAF's strike/attack capability extends to cover both ends of this wide mission spectrum. Although the RAF has all along been heavily committed to the conduct of maritime air operations, this aspect of its activity has increased in significance during the past decade. In part this has been a reaction to the build-up of the Soviet Navy which poses a growing threat to the security of Europe's seaborne lines of trade and communication.

But it has also reflected a requirement to mitigate the rundown in carrier-based fixed-wing naval aviation by substituting shore-based air support. Operating in conjunction with air-to-air refuelling and airborne early warning aircraft, land-based aircraft can offer long range attack and air defence facilities to the Fleet at sea. So today the RAF's maritime-assigned airpower encompasses elements of the strike/attack forces of No 1 Group and the air defences of No 11 Group, as well as the resources of No 18 Group, whose Nimrod aircraft are the inheritors of the traditional maritime patrol and anti-submarine mission formerly undertaken by RAF Coastal Command.

The No 1 Group maritime strike/attack force earmarked for anti-surface vessel warfare (ASVW) is provided by the Buccaneers of Nos 12 and 216 Squadrons. These units are presently based at Honington, Suffolk but are shortly due to move north to Lossiemouth, Scotland, making way for the Tornado weapons training unit and relocating to a station which places them closer to their likely area of operations in the North and Eastern Atlantic and covering the entry routes thereto from Soviet Northern Fleet bases. With its naval ancestry, the Buccaneer is certainly very much 'at home' in the maritime role and would be a potent wartime contribution to SACLANT. The aircraft's attack ordinance includes bombs and the TV-guided version of the Martel air-to-surface missile. It is with these two squadrons that the RAF is introducing the Paveway/Pavespike laser guided weapons system this year, as well as making additional improvements to the navigation systems of their aircraft. The maritime Buccaneers will also be among the first aircraft to be armed with a new generation stand-off anti-ship missile, the BAe P3T which has the capability of approaching its target at very low level.

Working very much in conjunction with the Buccaneers in their maritime role (and with naval sea and air forces generally) would be another No 1 Group squadron, No 27 at Scampton. This squadron has the special task of maritime radar reconnaissance and is equipped with the SR Mark 2 version of the Vulcan. The need for accurate and timely pre-strike information on the location and strength of enemy surface units is obvious, and the mission of the Vulcan SR Mark 2s is to generate this intelligence from surveillance and shadowing patrols.

The support picture for the air defence and strike/attack front line is complemented by two more No 1 Group squadrons, Nos 55 and 57, which provide air-to-air refuelling facilities with Victor K Mark 2 tankers for long-range interception and maritime strike/attack sorties by Phantoms, Lightnings and Buccaneers of Strike Command. Similar air-to-air tanking practice is a recurrent feature of other STC squadron training and the Victors are utilised particularly by Jaguar and Harrier aircraft during overseas deployments.

The most substantial RAF contribution to Britain's maritime forces is provided by No 18 Group/STC which commands both the Service's long-range anti-submarine and patrol squadrons and its search and rescue (SAR) helicopters. Of all the roles in which air support is available to naval forces, the

anti-submarine warfare (ASW) and maritime patrol task is of singular importance. The disruptive potential of the Soviet submarine fleet is a threat which embraces both the strategic nuclear capability of ballistic missile-armed boats and the anti-shipping submarines which could attack traffic in Atlantic sea lanes.

While land-based maritime patrol aircraft are just one element of the varied forces – including destroyers, frigates, ship-based ASW helicopters and hunter-killer submarines – which are deployed to counter the submarine menace, they discharge an especially significant role in view of their ability to range far and fast over huge areas of ocean. In the Nimrod the RAF has arguably the finest maritime patrol ASW aircraft in the world. It is capable of rapid transit to its operating locations, long endurance on-station (usually cruising on only two of its four engines) and it is fitted with an

Below: the VC10 aircraft of No 10 Squadron at Brize Norton operate in the strategic transport role.
Bottom: UK-based Harriers of Strike Command's No 38 Group practice operating from dispersed sites away from their base at Wittering.
Right: the RAF's Puma tactical assault and transport helicopters are based at Odiham, Hants. An aircraft of No 230 Squadron is pictured with an underslung load.
Below right: No 2 Squadron flies the Jaguar GR Mark 1 on tactical reconnaissance duties in Germany,
Bottom right: a bomb-laden Jaguar of No 54 Squadron from Coltishall shows the camouflaged undersurfaces adopted for low-level operations

array of submarine detecting sensors, a fully computerised navigation and attack system, and a capacious bomb bay for ASW torpedoes and bombs (plug wing pylons for the carriage of additional air-to-surface missiles). Due to enter service with the RAF in 1980 is the Mark 2 version of the Nimrod incorporating the more advanced and computer controlled Searchwater radar, a new digital acoustic processor designed to handle data from innovatory types of air-dropped sonobuoys, and revised navigation and tactical computer systems.

No 18 Group operates four Nimrod squadrons from Kinloss in Scotland and St Mawgan, Cornwall, plus an OCU at the latter station. The Group's AOC is also the NATO maritime air commander in the Eastern Atlantic and Channel commands and in wartime the Nimrods would provide anti-submarine support for SACLANT and maritime reconnaissance sorties for his Striking Fleet Atlantic.

In peacetime the Nimrods' role is to maintain surveillance of potentially hostile surface units and submarines at sea in the ocean areas of interest to the UK and NATO, monitoring their movements and updating information on their capabilities. The aircraft have other tasks as well, including the so-called 'offshore tapestry' mission of fishery protection and airborne surveillance of Britain's oil and gas fields in the North Sea.

Search and rescue operations figure in the Nimrod's peacetime range of duties, its on-board navigation, detection and communication systems making it ideal for the job of incident location and rescue co-ordination. No 18 Group also controls the RAF's rotary-wing SAR elements, two squadrons of helicopters being deployed in nine detached Flights at coastal locations around the UK. No 202 Squadron is in the process of re-equipping with the Mark 3 version of the Sea King, which has effected a notable improvement in the operating radius and all-weather capabilities of the SAR force. No 22 Squadron flies the Wessex and Whirlwind. No 202 Squadron's SAR 'parish' extends down the UK east coast from Scotland to East Anglia, with two of No 22's Wessex Flights occupying additional locations at Leuchars and Manston; C Flight, No 22 Squadron looks after the central area of the Irish Sea and North Wales from its base at Valley, while the Squadron's two Whirlwind Flights are located either side of the Bristol Channel at Chivenor, Devon and Brawdy, Dyfed.

RAF SUPPORT COMMAND		
No 1 FTS	Jet Provost	Linton-on-Ouse
No 2 FTS	Whirlwind, Wessex	Shawbury
Det	Whirlwind	Valley
No 3 FTS	Jet Provost, Bulldog	Leeming
No 4 FTS	Hawk, Hunter	Valley
No 6 FTS	Dominie, Jet Provost	Finningley
No 7 FTS	Jet Provost	Church Fenton
METS	Jetstream	Finningley
RAF College	Jet Provost, Dominie	Cranwell
CFS	Jet Provost, Bulldog	Leeming
	Hawk	Valley
	Gazelle	Shawbury
'Red Arrows'	Gnat	Kemble
CATCS	Jet Provost	Shawbury
UAS	Bulldog	various

Abbreviations: FTS, Flying Training School; Det, Detachment; METS, Multi-engined Training Squadron; CFS, Central Flying School; CATCS, Central Air Traffic Control School; UAS, University Air Squadron.

Whether in the maritime or overland environment the effective use of airpower depends on the accurate and prompt flow of information regarding enemy targets, their location and identity. Without such intelligence data, air defence and strike/attack resources can become scarce and costly assets. RAF reconnaissance aircraft are accordingly distributed throughout front line forces in support of the particular tasks which would be undertaken by related combat units in each Group or Command. The specialist reconnaissance units include the two Canberra photo-recce squadrons and the Vulcan SR2 squadron in No 1 Group, and the Jaguar squadrons in No 38 Group and RAF Germany. No 1 Group also deploys the Canberra and Nimrod equipped No 51 Squadron, which undertakes electronic countermeasure reconnaissance duties. In addition Buccaneer and Harrier aircraft can be fitted with sensor pods to conduct reconnaissance missions as a secondary role linked to their primary strike/attack and tactical air support tasks.

This flexible employment of reconnaissance aircraft is also a feature of the missions conducted by Nos 2 and 41 Squadrons with the Jaguar, whose role encompasses secondary duties which make use of attack/offensive support capability of the aircraft. In this respect their overall function is akin to the traditional fighter-recce role, being able to mount specialist reconnaissance or armed reconnaissance sorties or reverting to the mainstream of attack operations. The key to the Jaguar's versatility is a detachable centre-line reconnaissance pod, which houses cameras and infra-red linescan–the latter system representing a vital aid to night and/or bad weather missions.

Above: the Jetstream T Mark 1 trains pilots in the techniques of multi-engined flying at Finningley, Yorks.
Right: the RAF's basic trainer is the Jet Provost, a T Mark 3A of No 1 Flying Training School being shown

RAF GERMANY (RAFG)		
No 2 Sqn	Jaguar	Laarbruch
No 3 Sqn	Harrier	Gutersloh
No 4 Sqn	Harrier	Gutersloh
No 14 Sqn	Jaguar	Bruggen
No 15 Sqn	Buccaneer	Laarbruch
No 16 Sqn	Buccaneer	Laarbruch
No 17 Sqn	Jaguar	Bruggen
No 18 Sqn	Wessex	Gutersloh
No 19 Sqn	Phantom	Wildenrath
No 20 Sqn	Jaguar	Bruggen
No 25 Sqn	Bloodhound	Deployed RAFG
No 31 Sqn	Jaguar	Bruggen
No 60 Sqn	Pembroke	Wildenrath
No 92 Sqn	Phantom	Wildenrath

The RAF's air transport force (ATF) of fixed wing aircraft and battlefield support helicopters is largely the responsibility of No 38 Group, with whose squadrons resides the essential task of airlifting Army and RAF reinforcements into whatever part of the NATO operating theatre they are required. With the re-orientation of Britain's military commitments more exclusively to NATO in the 1970s and the withdrawal of UK forces from garrisons around the world, the RAF's long range transport fleet has been greatly reduced in numbers and the types of aircraft it operates. The mid-1970s retirement of the Comets, Belfasts, Britannias and Andovers has

A Jaguar GR Mark 1 of No 20 Squadron is parked in front of its hardened aircraft shelter on Bruggen airfield in Germany. In common with the other RAF bases in Germany, in recent years Bruggen has implemented many measures designed to blunt an enemy air attack. These include the emplacement of missile and anti-aircraft gun defences, the protection of aircraft, stores and key personnel in hardened shelters and the toning down of runways, airfield installations and vehicles

left the Group with only VC-10s and Hercules in its fixed-wing front-line support inventory, the Hercules squadrons also having been cut back to four in number as part of the economy measures.

The main role of the fixed-wing ATF in wartime would be the rapid reinforcement of BAOR and RAFG from the UK, with ancilliary tasks comprising the air dispatch of the UK contribution to the ACE Mobile Force and the deployment to forward and dispersed operating bases of STC's home-based squadrons. All these missions form part of the ATF's intra-theatre airlift capability within Europe, but No 38 Group is also obliged to retain the ability to operate worldwide either for the limited war deployment of air-mobile army and RAF forces–as was the case in Belize, for example– or to conduct emergency operations such as the supply of disaster relief and the evacuation of UK and other nationals from overseas trouble spots. The UK's armed services undertake extensive training programmes outside Europe and in peacetime the ATF supports the cargo and personnel airlift requirements of such detachments. Two of the Hercules squadrons based at Lyneham, Nos 47 and 70, train for an additional task–the 'transport support' of ground forces, which involves the air dropping of parachute troops or stores and the ability to fly into short semi-prepared strips.

Tactical transport and logistic support for the Army in the field is the responsibility of the RAF's helicopter squadrons, most of which are part of

No 38 Group. The centre of these operational rotary wing activities is RAF Odiham, the base for two Puma squadrons, a Wessex HC Mark 2 squadron and the OCU for both types. One Wessex HC Mark 2 squadron is stationed in RAF Germany, and there is another in Hong Kong. These helicopters provide battlefield airlift for troops and supplies, and they can carry underslung equipment loads ranging from artillery pieces to fuel and vehicles. The Wessex squadron in Germany is exclusively earmarked for a wartime role in support of No 1 (British) Corps and it would be joined there by its opposite number from Odiham. The Pumas would be subject to more mobile Alliance tasking, some being for use by the ACE Mobile Force and others ready to join the UK Mobile Force in its reinforcement of the Central Region air forces.

Backing up the RAF's combat forces in Strike Command and RAF Germany is the appropriately named Support Command, the new organisation created by the merger of the former Training and Support Commands in mid-1977. The work of the Command covers an enormous range of activities embracing the training of ground and aircrew; the provision of maintenance support to the front line (which in itself encompasses a diversity of aircraft/systems engineering and major servicing, communications and signals maintenance, and supply and stores management); administration; and medical services.

On the aircrew training side the task of producing fully qualified personnel is split between Support Command itself and RAF Strike Command, the dividing line being between the student pilot or navigator's completion of advanced flying training and the start of operational flying training, which is conducted by the STC Operational Conversion Units. Pilot flying training is carried out in two stages, basic and advanced. The introductory courses are conducted at the RAF College Cranwell and at Nos 1 and 7 Flying Training Schools on Jet Provost trainers. At the successful completion of this phase pilots are streamed for a trio of advanced courses. Potential fast-jet pilots – those assessed as suitable for the air defence and strike/attack roles – proceed to No 4 FTS at RAF Valley, where the training is now conducted on the Hawk, the last of the Gnat trainers having recently been withdrawn. The multi-engine stream – those pilots selected for maritime, transport and medium bomber aircraft – are posted to the Multi-Engine Training Squadron at RAF Finningley and rotary-wing pilots go to No 2 FTS at RAF Shawbury for helicopter training on the Whirlwind – to be replaced in late 1979 by the Gazelle.

At the conclusion of these courses, the pilots pass out of the Support Command system and join Strike Command for the final training which precedes postings to operational squadrons. In the case of the fast jet stream, this is a two-stage process as the pilots go firstly to the Tactical Weapons Units, currently re-equipping with the Hawk in place of the Hunter, and then to an Operational Conversion Unit. Other aircrew – navigators, air electronics operators and air engineers – receive their basic and advanced training at No 6 FTS on the Jet Provost and Dominie, and then join the pilots in type conversion at the Operational Conversion Units.

The guardian of the RAF's flying training standards is the Central Flying School (CFS) and, through its role as the training organisation for qualified flying and helicopter instructors, it maintains a close supervision of pilot training performance and techniques as well as the quality of the end-product. Numerous supplementary flying courses are conducted by Support Command, these ranging from pre-entry training available at University Air Squadrons for undergraduates planning an RAF career, to refresher flying courses at No 3 FTS (and the STC Tactical Weapons Units) for aircrew returning to flying duties after ground postings.

The most notable addition to the RAF's front line in the 1980s will be the Panavia Tornado multi-role combat aircraft, which has been developed jointly by the UK, West Germany and Italy. The RAF will acquire 220 of the basic interdictor/strike (IDS) version of the aircraft – the Tornado GR Mark 1 – plus a further 165 examples of an air defence variant (ADV) to be designated Tornado F Mark 2. Incorporating variable geometry wings, fuel-efficient turbofan engines, an elaborate avionics fit (including terrain following radar, a combined nav/attack radar and both passive and active electronic warfare equipment), the Tornado GR Mark 1 is optimised for high-speed, low-level target penetration by day or night and in all weather conditions. It also has an impressive short take-off and landing performance.

Starting in 1981 the Tornado GR Mark 1 will progressively replace the Vulcan and the Buccaneer in the overland and maritime strike/attack roles. In view of the preponderant number of Warsaw Pact aircraft facing NATO, the counter air mission has assumed increased importance, and it is planned that the IDS Tornadoes will carry a new airfield attack weapon, the JP233, which incorporates a runway cratering capability and other cluster-bomb-type sub-munitions which would have an area denial effect in delaying enemy attempts to repair their operating surfaces. As well as ECM, the Tornadoes' penetration may be assisted by new, active defence suppression weapons currently under study; and in the maritime role the aircraft is expected to carry the P3T anti-ship missile – the successor to Martel. The Tornado GR 1 will also replace the Canberra in its reconnaissance role and, equipped with new sensor systems, offer a considerable step forward in capability – particularly in night and/or adverse weather conditions. Plans have been announced to station a Tornado unit in RAFG in the primary reconnaissance role as an addition to the in-theatre capability of the Command's Jaguar GR1s.

The first prototype of the Tornado F2 interceptor, designed to meet the RAF's special requirement for the large area air defence of the UKADR, is due to make its maiden flight later this year and will enter squadron service in the mid-1980s. It differs principally from the IDS version in having a longer fuselage, incorporating extra fuel capacity and a revised nose configuration housing an advanced AI radar with the facility to track multiple targets at long-range. The ADV Tornado will be armed with a British re-design of the American Sparrow AAM, the Skyflash, featuring a new homing system offering

Above right: a Harrier of No 3 Squadron undergoes servicing in a temporary hangar at a dispersed site. Right: air defence in RAF Germany is the responsibility of two Phantom FGR Mark 2 squadrons based at Wildenrath, a No 92 Squadron aircraft being shown. Below right: the BAe Buccaneer S Mark 2 squadrons in Germany train for interdiction missions deep into enemy territory, operating at low level and flying in all weathers. Bottom right: No 16 Squadron is No 15's sister Buccaneer unit at Laarbruch

much better 'snap-down' engagement of low-level targets, the ability to discriminate between multiple targets and resistance to enemy electronic counter-measures. The Skyflash missile is currently entering RAF service on the Phantom and it has been decided to procure an improved version of the Sidewinder AAM – the AIM-9L, for these aircraft.

The operational deployment and capability of the Tornado F Mark 2 in the long-range, area air defence role will be assisted by the acquisition of the AEW Mark 3 version of the Nimrod and a more recent decision to supplement the Victor K Mark 2 fleet of air-to-air refuelling tankers by the purchase of ten ex-airline VC10s modified to this role. The AEW Nimrods will provide long-range, early-warning of hostile aircraft, missiles and surface fleet movements in sea areas around the UK allowing interception of these threats at distances outside the operating range of their offensive weaponry.

The RAF's tactical mobility and battlefield airlift mission will benefit in the short-term from the procurement of the Boeing-Vertol Chinook medium-lift helicopter and plans to 'stretch' the cargo holds of 30 of the No 38 Group ATF's Hercules by the insertion of a 4·5m (15ft) fuselage plug, thus boosting their freight and personnel carrying capacity. The 33 Chinook helicopters on order as replacements for the Wessex will greatly increase the support potential for Harrier operations in the field and the Army's air mobility. In the latter role the Chinook's 44 troop or 12 ton load-carrying performance represents an enormous advance compared with the 12 troops/1½ tons capability of the Wessex.

For the longer-term future, the UK is discussing with other NATO countries the possibility of a collaborative programme leading to a new tactical combat aircraft which for the RAF, would represent an eventual replacement of the Jaguar and Harrier. In the interim, of course, there is still an extended lifetime ahead of both types, which continue to be up-dated by power plant and weapons systems refits. The Harrier, in particular, is presently the subject of studies into the feasibility of a larger wing re-design which would confer notable improvements in the aircraft's warload, range and manoeuvrability.

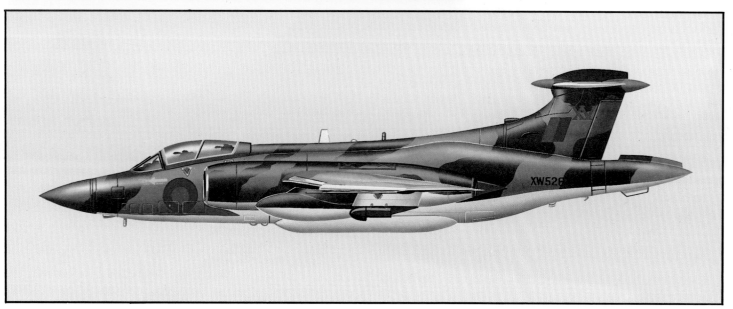

FLEET AIR ARM		
No 702 Sqn	Lynx	Type 42 GMD
		Type 21 GPF
		'Leander' GPF
No 703 Sqn	Wasp	Portland
No 705 Sqn	Gazelle	Culdrose
No 706 Sqn	Sea King	Culdrose
No 707 Sqn	Wessex	Yeovilton/RFA
No 737 Sqn	Wessex	'County' GMD
No 750 Sqn	Jetstream	Culdrose
No 771 Sqn	Wessex	Culdrose
No 772 Sqn	Wessex	Portland
No 781 Sqn	Sea Devon	Lee-on-Solent
	Sea Heron	
	Wessex	

Two squadrons of Westland Wessex HU Mark 5 assault transport helicopters of the Fleet Air Arm operate in support of the Royal Marine Commandos. A Commando Brigade and supporting ships are intended to reinforce NATO's northern flank in time of crisis and they periodically train under Arctic warfare conditions. The Wessex helicopters will fly from the ASW carriers HMS Bulwark and HMS Hermes and they are to be reinforced by a version of the Sea King helicopter which has been specially developed for heavy lift work

For the last 40 years the Fleet Air Arm has been an integral part of the Royal Navy. Its shipborne airpower, formerly concentrated aboard aircraft carriers, has become increasingly diversified in recent years with the development of the naval helicopter into a multi-role weapons system, capable of operation from almost every type of ship in the Fleet. There have been few facets of naval warfare in this period which have been uninfluenced by airpower, and the importance of naval aviation has never been more apparent than it is today. The Royal Navy is reliant on the Fleet Air Arm (FAA) for a fast-moving and wide-ranging extension of its ship-based anti-submarine warfare (ASW)

and anti-surface vessel warfare (ASVW) systems, and for the air mobility support of seaborne assault landing forces. These tasks and others are performed by helicopter squadrons of the FAA, some in conjunction with shore-based maritime air units.

At present, the FAA's operational front line entirely consists of rotary-wing aircraft. Following the late-1978 withdrawal from service of the last fleet carrier, HMS *Ark Royal*, and the disbandment of the FAA's final fixed-wing squadrons of air defence McDonnell Douglas Phantoms, strike/attack British Aerospace Buccaneers, and airborne early-warning BAe Gannets in the *Ark*'s air group, the Royal Navy is temporarily without its own ship-based air support of the type which these units offered. Although land-based units of the RAF provide cover in these roles and in the vital arena of ASW, they are not necessarily a complete substitute for aircraft stationed on the spot with the Fleet. The number of aircraft needed to mount combat air patrols over the Fleet, for example, inevitably increases at greater ranges from land and as the area of the Northern Atlantic to be defended is very considerable, the capability of shore-based maritime air defence is likely to diminish with distance unless supported by a sizeable force of air-to-air refuelling tankers.

Naval aircraft, on the other hand, can maintain an assured continuous presence with the Fleet and it is partly a reflection of the advantages which stem from this fact that has led to the decision to procure the navalised version of the Harrier for the Royal Navy. This aircraft will mark the return of fixed-wing airpower to the FAA and it will serve aboard

the new *Invincible* class of through-deck cruisers and a modified HMS *Hermes* in the fighter, reconnaissance and strike roles. However, the numbers of Sea Harrier aircraft planned for deployment aboard each of the new ships is limited, and the Fleet will continue to be dependent on additional air support plus its existing and future anti-air warfare (AAW) missile systems.

The Naval Air Squadrons of the FAA are numbered in the 700 or 800 series, the distinction generally being that the former are second-line or training units while the latter are front-line squadrons. In some instances, however, the squadrons in the 700-series have operational commitments. With practically every ship in the fleet of 2,000 tons displacement and upwards capable of acting as a helicopter operating platform, the FAA's pattern of operations is widespread and co-ordination through shore-based 'parenting' squadrons is necessary for the detached flights (often of a single aircraft).

The shore-based HQ or operating locations for the 18 squadrons comprising the main force of the FAA first- and second-line units are mainly on three Naval Air Stations at Culdrose, Yeovilton and Portland, with the remaining units at Lee-on-Solent and Prestwick. In total numbers, the FAA presently operates some 270 aircraft of which approximately 170 serve with the front-line helicopter squadrons and about 100 with the training and support units.

The prospect of a continuing build-up in the size of the Soviet submarine fleet – by the mid-1980s it is expected to number over 300 units, most of

No 814 Sqn	Sea King	HMS Hermes
No 819 Sqn	Sea King	Prestwick
No 820 Sqn	Sea King	HMS Blake
No 824 Sqn	Sea King	Royal Fleet Auxiliaries
No 826 Sqn	Sea King	HMS Bulwark
No 829 Sqn	Wasp	'Leander', 'Tribal', 'Rothesay' and Type 21 GPF
No 845 Sqn	Wessex	Yeovilton
No 846 Sqn	Wessex	HMS Bulwark
Other units		
FRADU	Hunter, Canberra	Yeovilton
BRNC	Chipmunk	Plymouth-Roborough
SAR Flight	Wessex	Lee-on-Solent

Abbreviations: GMD, Guided Missile Destroyers; GPF, General Purpose Frigates; RFA, Royal Fleet Auxiliaries; FRADU, Fleet Requirements and Direction Unit; BRNC, Britannia Royal Naval College.

them nuclear-powered – has magnified the undersea warfare threat to NATO and made anti-submarine warfare (ASW) one of the Royal Navy's top priorities. As well as the increase in numbers, the Soviet Navy is expected to be able to noticeably upgrade the performance of its submarines in the near future; they will be better propelled, capable of faster speeds and of diving to greater depths, and equipped with greater numbers of longer-range weapons – ballistic missiles, anti-ship cruise missiles and anti-submarine weapons. This threat is countered by a wide spectrum of maritime ASW forces – ships, hunter/killer submarines, land-based maritime patrol/ASW aircraft and helicopters – all of

No 706 Squadron's Westland Sea King helicopters are lined up at Royal Naval Air Station Culdrose, Cornwall. This unit is responsible for training crews on the Sea King before they are posted to an operational squadron. The Sea King's primary role is anti-submarine warfare, for which it carries a search radar, mounted atop the fuselage aft of the rotor head, and sonar equipment. Offensive armament comprises homing torpedoes or depth charges. Search and rescue is a secondary role of the helicopter in FAA service

which are capable of operating in concert.

The Royal Navy's most significant contribution to NATO's anti-submarine forces in the future will be the ASW Task Groups centred on the new *Invincible* class cruisers. These ships will carry a squadron of heavy ASW helicopters, and their escorting surface ships, as at present, will carry their own light ASW helicopters.

The FAA currently deploys four of its five helicopter types aboard Royal Navy and Royal Fleet Auxiliary (RFA) surface units in the ASW role. Operating from the largest vessels – the helicopter carriers and cruisers, *County* class guided missile destroyers and, most recently, the RFAs – are the Sea King HAS Mark 2 and Wessex HAS Mark 3, both types fitted with dipping sonar for the on-board detection of enemy submarines and ASW weapons for their attack. The capability of the Sea King in this role is the subject of continued improvement and the type is being modified by the installation of a passive sonobuoy processor. The five operational squadrons include one permanently shore-based unit at Prestwick in Scotland which is classed as a maritime anti-submarine warfare support squadron. Its duties include the defence of RN nuclear submarines based in the Clyde area and it also maintains an SAR Flight on standby for the support of civilian rescue services.

The helicopter Flights aboard the Royal Navy's destroyers and frigates are furnished by the Wasps of No 829 Squadron and the Lynx of No 702 Squadron. The earlier generation Wasp still serves in large numbers–31 ship's flights at a recent count– aboard general purpose frigates of the *Leander*, *Rothesay*, *Tribal* and Type 21 *Amazon* classes, and Type 42 destroyers of the *Sheffield* class. Its replacement on most of these vessels, the Lynx, is entering service in increasing numbers and has represented a major advance in ASW capability. This helicopter is particularly designed to cope with the exacting nature of operations in rough sea states off the small ships' none-too-large decks. In the ASW role it carries Mark 44 or Mark 46 homing torpedoes and Mark 11 depth charges, and consideration is being given to retrofitting the aircraft with some form of submarine detection device.

The task of engaging enemy surface ships is one which is best conducted at the greatest possible range from threatened naval forces or merchant shipping in order to keep the hostile units beyond the operating range of their missile or gun armament. The early identification by maritime patrol aircraft (or other means) of the enemy vessels would ideally be followed by aircraft or submarine attacks, then at closer range by shipborne aircraft attacks and only thereafter by surface action.

The Fleet Air Arm's anti-surface vessel warfare (ASVW) force is currently rotary-wing-borne, but the advent of the Sea Harrier in the early 1980s will restore the fixed-wing anti-ship capability formerly exercised as part of the Fleet carrier air groups' responsibilities. The Wasp helicopter armed with the AS12 air-to-surface missile (ASM) fulfills a limited ASVW role, but the more hard-hitting rotary-wing operator in this form of warfare is the Westland Lynx. Equipped with its own search radar, tactical navigation system and armed with Sea Skua ASMs, the Lynx provides the Fleet with an over-the-horizon all-weather extension of destroyer and frigate ASVW weapons systems.

As part of the UK's reinforcement commitments to NATO's northern flank areas, which include Norway and the Baltic approaches, a Royal Marine Commando Brigade is fully-trained for Arctic warfare in a combined amphibious force. This includes five battalion-sized units–four RM Commandos and one Royal Netherlands Marine Corps Amphibious Combat Group–plus support elements, and is known as the UK/NL Amphibious Force. The formation would be transported by a mixed fleet of naval and commercial ships including the Royal Navy's assault ships and the carriers HMS *Bulwark* and HMS *Hermes* which, although assigned a primary ASW carrier role, also have the

Left: firefighters watch a Sea King of No 820 Squadron preparing to take off from HMS Blake.
Below: a Sea King of No 826 Squadron leads a fly-past of the type at the Jubilee Fleet Review in 1977, with two Royal Australian Navy Sea Kings bringing up the rear. Five front-line units of the FAA fly the Sea King

A Westland Wasp HAS Mark I shipboard helicopter flies over a Scimitar-class fast patrol boat during exercises. Armed with the AS 12 missile, this helicopter can provide its parent ship with a measure of defence against fast missile and torpedo boats. The later Lynx, equipped with radar and Sea Skua missiles, will be even more effective in this role than its predecessor.
Opposite: the Westland Lynx HAS Mk 1 will replace the Wasp aboard all but the smallest destroyers and frigates.
Opposite below: the Sea Harrier FRS Mark 1 will operate from Invincible-class cruisers and ASW carriers

ranges than the Sea Dart.

The Fleet Air Arm maintains rescue alert commitments along the south and southwestern coasts of England from Sussex to the Bristol Channel and in western Scotland and the outer islands. The search and rescue task is handled by the permanent SAR Flight at Lee-on-Solent and by Nos 772 and 771 Squadrons at Portland and Culdrose respectively, as well as by SAR duty crews with the longer range Sea Kings of Nos 706 and 819 Squadrons at Culdrose and Prestwick.

The training of pilot and observer aircrew for the Fleet Air Arm involves the use of fixed-wing aircraft in its initial stages before a transition to helicopters during the intermediate or, in the case of the observers, advanced phases of the process. Pilot training starts with a basic course on the Bulldog T Mark 1 at the RN Elementary Flying Training Squadron at RAF Leeming, followed by rotary wing conversion at the RN Flying Training School at Culdrose, where flying training is conducted on the Gazelles of No 705 Squadron and ground courses are operated by the RN Helicopter School. On successful completion of this stage, advanced flying training and operational conversion is undertaken on the front line helicopter types serving with second line training squadrons. Observer aircrew training courses start on the Jetstream T Mark 2 with No 705 Squadron at Culdrose and continue with an advanced phase on the Wessex HAS Mark 3 with No 737 Squadron and three months of operational flying training before postings to front-line units. All the flying stages of the process are paralleled and interspersed with ground courses.

On the support side, the FAA operates its own 'commuter airline' of de Havilland Sea Devon and Sea Heron aircraft and Westland Wessex helicopters with No 781 Squadron based at Lee-on-Solent. The important task of providing target facilities for Fleet AAW training is handled by Fleet Requirements and Direction Unit at Yeovilton flying single-seat and two seat Hawker Hunters and various marks of the English Electric Canberra. The operation of these aircraft is contracted out by the Royal Navy to Airwork Services Ltd.

The handover of the first Sea Harrier to the Royal Navy in June 1979 marked the introduction of V/STOL capability to naval aviation and the foundation of a new era in fixed-wing aircraft operations. The FAA will receive 34 Sea Harriers plus one trainer version to equip an HQ/training squadron and three front-line squadrons, Nos 800, 801 and 802. A derivative of the land-based Harrier GR Mark 3, but with a much altered front fuselage/cockpit configuration and greatly modified avionics, the Sea Harrier FRS Mark 1 will fulfill three basic missions for the FAA. As an air defence system armed with Sidewinder missiles and 30mm cannon the aircraft will provide AAW air defence for the Fleet, notably against long-range reconnaissance aircraft and aircraft providing command and guidance facilities for stand-off missile attacks. It will carry out visual and electronic reconnaissance at low or high level and will offer a potent ASVW capability armed with bombs, rockets or the new P3T anti-ship missile being developed as a follow-on to the Martel. The P3T is a 'fire and forget' weapon which is fitted with an active radar homing head.

secondary task of amphibious support.

The Fleet Air Arm has earmarked Nos 845 and 846 Squadrons to support these forces, equipped with the Wessex HU Mark 5 helicopter. Although this aircraft has served satisfactorily as a tactical troop transport, it is unable to carry the heavier equipment coming into service with the Royal Marines–the 105mm light gun, heavy over-snow vehicles and other larger transport. To increase its heavy lift capability, the FAA will receive a new variant of the Sea King helicopter the HU Mark 4, with an underslung load-lifting capacity of 3,600kg (8,000lb)–more than double that of the Wessex– and seats for 27 troops.

The Royal Navy's anti-air warfare (AAW) capability–the role description is that used by NATO to describe naval air defence operations–is not one in which the Fleet Air Arm presently participates, although the capability to do so will be regained once the front-line Sea Harrier squadrons are operational. In the meantime the airborne defence of the Fleet at sea relies on land-based interceptors and AEW aircraft, while the shipborne surface-to-air defences incorporate a range of missile and gun systems backed up by radar jamming countermeasures. The primary area air defence vessel in future RN Task Groups will be the Type 42 *Sheffield*-class destroyers armed with the Sea Dart medium/long-range missile, which also has a secondary ASVW capability. Recently accepted into RN service with the commissioning of the first Type 22 ASW frigate HMS *Broadsword* is the new point defence SAM, the Sea Wolf. This is a rapid-reaction weapon designed to engage fast low-flying anti-ship missiles and aircraft at shorter

The flight of the missile, is just above sea level in the final phase of the attack.

The 'platforms' for the Sea Harriers deployment with the Fleet will be the *Invincible* class ASW cruisers and the carrier HMS *Hermes*. All are being fitted with the revolutionary 'ski-jump' take-off aid, which allows increased payload and/or a shorter take-off distance.

For the longer-term future, the Royal Navy is working to define the specification for a new helicopter needed as an ASW Sea King replacement later in the 1980s. A Naval Staff Requirement has been agreed for a Westland proposal designated the WG34, which is likely to be a three-engined machine capable of ship-borne deployment, independent operation at considerable ranges from its parent ship, with sonobuoy (rather than dipping sonar) detection and supporting sensors, and automatic data handling. It is anticipated that the aircraft will be developed in collaboration with other European aviation industries and in such a manner as to suit it for other roles, such as troop transport.

The Royal Marine Commando forces – a Brigade HQ, four battalion-size Commandos (Nos 40, 41, 42 and 45) plus artillery, engineer and logistic support – are backed up by a light helicopter air element comprising five flights grouped in the 3rd Command Brigade Air Squadron. These flights operate the Gazelle AH Mark 1 in the liaison and communications role. The Royal Marines are also to receive the Lynx AH Mark 1 helicopter armed with the TOW anti-tank missile in the early-1980s.

The Army Air Corps' transformation to full Corps status in late-1973 – placing Army aviation on a par with the Royal Artillery, Royal Engineers, etc – emphasised the growing importance attached to this branch of the service and the need to establish a permanent cadre of aircrew personnel, many of whom had previously been seconded from other corps on temporary tours of flying duty. Thenceforth not only aircrew but gound crews were to be recruited directly into the Army Air Corps (AAC) to maintain an increasingly complex inventory of aircraft, and their avionics and weapons systems.

As a result of a re-organisation carried through in 1978, which particularly affected the AAC squadrons based in Germany with the British Army of the Rhine (BAOR), the strength of the Corps is mostly grouped into regiments. Each regiment comprises two helicopter squadrons, a light squadron for communications and liaison work, and a utility and anti-tank guided weapons (ATGW) squadron which performs a number of battlefield and air mobility support roles. The UK-based squadrons of the AAC are distributed among the Field Forces stationed in Yorkshire and southern England – the 5th Field Force elements not based in BAOR, plus the 6th and 7th Field Forces based in the South East and Eastern Districts of UK Land Forces (UKLF) – and in flight-sized detachments elsewhere, notably at Netheravon and in support of the security forces in Northern Ireland. The hub of the Corps' operations is the AAC Centre at Middle Wallop, Hants, which is responsible for Army flying training and military aviation development flying. Outside UKLF and BAOR the AAC has units deployed in Cyprus, Hong Kong and Brunei, Belize and Canada.

ROYAL MARINES

3rd Commando Brigade
Air Squadron

Brunei (B) Flt	Gazelle	**Plymouth-Coypool**
Dieppe (D) Flt	Gazelle	**Plymouth-Coypool**
Salerno (S) Flt	Gazelle	**Plymouth-Coypool**
Kangaw		
(K) Flt	Gazelle	**Plymouth-Coypool**
Montfortebeek		
(M) Flt	Gazelle	**Arbroath**

ARMY AIR CORPS

UK-based units

No 655 Sqn	Scout, Gazelle	**Topcliffe**
No 656 Sqn	Scout, Gazelle	**Aldershot**
No 657 Sqn	Scout, Gazelle	**Oakington**
No 658 Sqn/		
6 Flt	Gazelle	**Netheravon**
No 658 Sqn/		
8 Flt	Scout	**Netheravon**
No 2 Flt AMF	Gazelle, Scout	**Netheravon**
No 3 Flt	Gazelle, Scout	**Omagh**
Beaver Flt	Beaver	**Aldergrove**
AAC Centre	Chipmunk, Bell 47G,	**Middle Wallop**
	Gazelle, Lynx, Scout,	
	Chipmunk, Beaver	

Like the Fleet Air Arm, the AAC's aircraft inventory is made up largely of helicopters; almost 300 are currently in service, more than half of which are Gazelle AH Mark 1s used for the communications/liaison task and for the first stage of advanced flying training. The long-serving Scout AH Mark 1 helicopter operated on utility and anti-tank duties is in the process of being joined by the Army version of the Westland Lynx in this role. The Lynx will have completely replaced the earlier type by the mid-1980s. The rotary-wing force is completed by a small number of Sud Alouette AH Mark 2s serving with the two AAC Flights in Cyprus. A complement of Bell 47s, akin to the AAC's former Sioux helicopters, are stationed at Middle Wallop and operated by a civilian contractor on basic rotary-wing flying training. The AAC also retains a few fixed-wing de Havilland Canada Beavers and Chipmunks, for a range of second-line duties including transport and forward air controller training.

The AAC's main function in support of front-line ground forces is to provide a highly-mobile form of transport for command personnel and a superior observation platform from which to assess the evolution of the land battle. However, there has been a significant addition to these tasks during the last decade in the anti-tank role, a mission pioneered by the US Army in South-east Asia with machine gun, cannon and rocket-equipped helicopter 'gunships' supporting ground troops and covering helicopter-borne assault landings.

The light squadrons of the AAC regiments and the independently-operating flights have in the Gazelle a compact, manoeuvrable, and fast-moving vantage point, the main roles of the aircraft being communications, liaison and observation. The airborne movement of battlefield commanders and other staff personnel between sectors of their command is of unquestioned advantage in enabling them to assess the situation 'on the ground' and to co-ordinate the progress of individual units by personal visits and briefings. The observation task involves the surveillance and shadowing of enemy forces, making use of terrain masking and concealment. The Gazelle also has an underslung load-lifting capability and can be employed to move small numbers of troops between key positions.

The Army has five squadrons of Gazelles in West Germany each with a 'light squadron' establishment of 12 helicopters. These, like their counterpart utility and utility/ATGW Squadrons, are assigned to the support of No 1 (British) Corps HQ and the four Armoured Divisions in BAOR. In the UK the Gazelles make up half the strength of four more squadrons, as well as No 2 Flight AAC which is assigned to the support of NATO's ACE Mobile Force. One additional, partly Gazelle-equipped flight is stationed in Northern Ireland and aircraft on temporary duty from BAOR support firing range training in Canada.

Above right: an SS 11-armed Scout AH Mark 1 is refuelled during an exercise at Netheravon.
Right: a Gazelle pictured at AAC Centre Middle Wallop, where the type is operated by the Advanced Rotary Wing Squadron's Conversion Flight for advanced flying training and type conversion and also by the Demonstration and Trials Squadron

BAOR-based units		
No 651 Sqn	Scout, Lynx	Hildesheim
No 661 Sqn	Gazelle	Hildesheim
No 652 Sqn	Scout, Lynx	Bunde
No 662 Sqn	Gazelle	Munster
No 653 Sqn	Scout, Lynx	Soest
No 663 Sqn	Gazelle	Soest
No 654 Sqn	Scout, Lynx	Detmold
No 664 Sqn	Gazelle	Minden
No 659 Sqn	Scout, Lynx	Detmold
No 669 Sqn	Gazelle	Detmold
No 7 Flt	Gazelle	Berlin-Gatow
No 12 Flt	Gazelle	Wildenrath
Other Units Overseas		
16 Flt	Alouette	Dekhelia
AAC Flt-UNFICYP	Alouette	Nicosia
Hong Kong		
No 660 Sqn	Scout	Sek Kong
Brunei		
No 660 Sqn/Det	Scout	detached from Sek Kong
Belize		
(TDY ex-UK)	Scout	Belize City
Canada		
(TDY ex-BAOR)	Gazelle, Scout	Suffield, Alta
(Deployed on site)	Beaver	

Abbreviations: ATGW, Anti-Tank Guided Weapons; ACE, Allied Command Europe; AMF ACE Mobile Force; FAC, Forward Air Control; BAOR, British Army of the Rhine.

The other AAC regiment rotary-wing squadrons perform a wide variety of missions. These can encompass not only the type of communications and liaison work which is the speciality of the light squadrons, but such additional tasks as artillery air observation posts, forward air controllers for ground support aircraft, airborne communications relay and command posts, casualty evacuation, and troop lift and stores movement. These squadrons are presently receiving the Lynx AH Mark 1 helicopter, which will make up half their 12 aircraft strength during the intermediate stages of the re-equipment programme eventually leading to the complete replacement of the Scout.

For the moment, however, the Scout AH Mark 1 continues as the mainstay of the anti-tank guided weapons capability in the utility/ATGW squadrons of BAOR. Armed with Nord SS 11 wire-guided missiles and using a gyro-stabilised sight system, these helicopters can provide a quick-reaction anti-tank force to contain enemy armoured threats.

The Scouts will continue in this role until withdrawn in favour of the next generation anti-tank missile and helicopter combination – the Lynx armed with the Hughes BGM-71A TOW (Tube-launched, Optically-tracked, Wire-guided) missile. TOW-equipped Lynx deliveries are due to begin in BAOR during 1980 and the ATGW Scout AH Mark 1s are expected to have been phased out by 1984.

Army Air Corps flying training courses are conducted at the AAC Centre Middle Wallop, where the *ab initio* trainees begin with an introductory course of three months on the Basic Fixed-Wing Flight learning to fly the Chipmunk. The second phase of basic training introduces the students to rotary-wing flying using a version of the Sioux helicopter which at one time served in considerable numbers with the AAC until replaced by the Gazelles. The aircraft and instructors for this course are supplied under contract by Bristow Helicopters.

After 60 hours each on the basic fixed-wing and rotary wing courses, the student moves on to a 115 hour/15-week advanced phase on Gazelles which introduces the essentials of flying in varying weather conditions and terrain, and teaches the techniques of air observation, forward air control and armed action. At the successful attainment of wings standard, the pilots proceed to a conversion course on the type of helicopter they will fly in the operational squadrons.

For pilots changing from one type to another in the course of their flying careers, the Advanced Rotary Wing Squadron Conversion Flights also provide re-training facilities, as well as refresher flying courses for aircrew returning from ground tours. An advanced Fixed Wing Flight continues to take care of the pilot training requirements needed to support the Beaver and Chipmunk-equipped units and looks after the training of forward air controllers.

Top left: a Scout AH Mark 1 armed with Nord SS 11 wire-guided anti-tank missiles.
Above left: the Lynx will have completely replaced the Scout as the AAC's utility/ATGW helicopter by the mid-1980s. A German-based Lynx is shown.
Left: a number of DHC Beavers remain in AAC service for communications duties

Below: a Gazelle AH Mark 1 of No 656 Squadron, AAC pictured flying over a junk in Victoria Harbour, Hong Kong in 1977. The Gazelle's main roles are liaison, communications and observation. The current AAC squadron in Hong Kong is No 660, which operates the Scout AH Mark 1 helicopter.
Left: a UK-based AAC Gazelle pilot checks his cockpit instruments before take-off

CHAPTER FIVE

France
Armeé de l'Air/Aéronavale/ALAT

For the past thirty or so years, Western Europe has been slowly moving towards economic, defensive and – ultimately – political unity along a path beset with predominantly self-made difficulties. Each alliance has its share of supposed 'special cases', but no one nation occupies a more ambiguous position than France. Although wholly devoted to economic integration, France insists upon treading a parallel but separate path to her European allies in the sphere of Western defence.

The French departure from NATO in 1966 was no abandonment of national defence commitments, but the product of a strengthened defensive posture linked to the availability of the Mirage IV nuclear strike force, and disillusionment with the progress made by the other NATO countries in countering the threat posed by the Warsaw Pact. For ten years, the French defensive posture was based on the Ailleret 'tous azimuts' concept, which unrealistically envisaged an attack upon the country from any direction; only recently has operational planning tacitly acknowledged that the threat lies only from one quarter.

This aside, France still professes herself a non-aligned nation, a stance which is accepted by the Soviet Union and confirmed by courtesy visits of VVS fighter aircraft to French bases. Liaison with NATO is maintained at the highest levels only, although aircraft of the Armée de l'Air do participate in NATO exercises on a 'guest' basis. NATO forces have also provided the 'enemy' during national air defence exercises. Undeniably, French unwillingness to station air forces in West Germany has

Below: like other escadrons, GC 1/7's aircraft carry earlier French units' badges on their fins. SPA 77's is to starboard (illustrated) and SPA 15's plumed knight's helmet is on the port side.
Bottom: Jaguar Es of GC 1/7 are pictured in flight

FRANCE

L'ARMEE DE L'AIR

Unit	Role	Aircraft type	Base
EC 2	Tactical fighter	Mirage IIIE	Dijon
EC 3	Strike	Mirage IIIE/Jaguar	Nancy
EC 4	Strike	Mirage IIIE	Luxeuil
EC 5	Air superiority	Mirage F1	Orange
EC 7	Strike	Jaguar	St Dizier
ECT 8	Strike OCU	Mystère IVA	Cazaux
EC 10	Air superiority	Mirage IIIC	Creil
EC 11	Strike	Jaguar	Toul
EC 12	Air superiority	Mirage F1	Cambrai
EC 13	Strike	Mirage IIIE/5	Colmar
EC 30	Air superiority	Mirage F1	Reims
ER 33	Tactical recce	Mirage IIIR/RD	Strasbourg
ELA 41	Communications	Broussard/Paris/N262	Metz
ELA 43	Communications	Broussard/Paris/N262	Bordeaux
ELA 44	Communications	Broussard/Paris/N262	Aix
ELAS 1/44	SAR	Puma/Alouette/Noratlas	Solenzara
ETOM 50	Comms/transport	Alouette II/Transall	Réunion
EE 51	ECM	DC-8	Evreux
ETOM 52	Comms/transport	Alouette II/Puma	Tontouta
EE 54	ECM	Noratlas	Metz
ETOM 55	Comms/transport	Alouette II/Noratlas	Dakar
GAM 56	Comms/transport	Noratlas/Broussard	Evreux
EC 57	Calibration	Noratlas	Villacoublay
ETOM 58	Comms/transport	Alouette/Broussard/Puma	Point-à-Pitre
ET 60	Communications	several	Villacoublay/Roissy
ET 61	Transport	Transall	Orleans
ET 63	Transport OCU	Noratlas/N262	Toulouse
ET 64	Transport	Noratlas	Evreux
ET 65	Communications	several	Villacoublay
EH 67	Transport/comms/SAR	Puma/Alouette II & III	several
EC 70	Aircraft ferrying	several	Chateaudun
ETOM 82	Transport	Caravelle	Papeete
ETOM 88	Comms/transport	Broussard/Alouette/Noratlas	Djibouti
EB 91	Nuclear deterrent	Mirage IVA	several
ERV 93	Air refuelling	Boeing C-135F	several
EB 94	Nuclear deterrent	Mirage IVA	several
GI 312	Flying training	Magister/CAP10 & 20	Salon
GE 313	Flying training	Magister/CAP10	Aulnat
GE 314	Flying training	T-33/Alpha Jet	Tours
GE 315	Flying training	Magister	Cognac
GE 316	Air navigation school	Noratlas/Flamant	Toulouse
GE 319	Multi-engine school	Flamant	Avord
CIFAS 328	Strategic OCU	Mirage III/IV/Noratlas/T-33	Bordeaux
CEVSV 338	Instrument flying	T-33	Nancy
CPIR 339	Radar training	Mystere 20SNA	Luxeuil
CIEH 341	Helicopter OCU	Alouette/Puma	Chambery

Abbreviations: EC, Escadre de Chasse; ECT, Escadre de Transformation; ER, Escadre de Reconnaissance; ELA, Escadrille de Liaison Aériennes; ELAS, Escadrille de Liaison Aériennes et de Sauvetage; ETOM, Escadron de Transport d'Outre Mer; ET, Escadre de Transport; EE, Escadrille Electronique; GAM, Groupe Aérienne Mixte; EH, Escadre de Hélicoptères; EB, Escadre de Bombardement; ERV, Escadre de Ravitailement en Vol; GI, Groupement d'Instruction; GE, Groupement Ecole; CIFAS, Centre d'Instruction de Force Aérienne Stratégique; CEVSV, Centre d'Entrainement en Vol Sans Visibilité; CPIR, Centre de Prediction et Instruction Radar; CIEH, Centre d'Instruction des Equipages d'Hélicoptères.

severely weakened NATO, although it has been officially acknowledged that in the event of war, the French front line would be regarded as the East/West German border and not the Rhine.

Frustrating though the policy may be to otherwise willing allies, this individualistic approach must be accepted as no mere whim. Resolved never again to suffer the humiliation of defeat, France began the postwar years with a strong determination to establish a viable and efficient defence force. During the five years of occupation, world aeronautical progress had moved ahead by leaps and bounds, relegating the French aircraft industry to the status of sub-contractor for the German war machine. Left far behind in design technique, France was forced to rely first on Britain and then the United States for her first-line equipment while frantically working to regain lost ground.

The quality and selling power of France's present-day aviation products, which provide all three flying services with the major proportion of their requirements, cannot be denied. Only Sweden can equal the French achievement of a front-line composed entirely of indigenous or joint-production aircraft.

During 1979, France will spend 3.26 per cent of the gross domestic product on the armed forces, an increase in real terms of 14 per cent. Despite the NATO shift towards conventional capability, France will continue to devote 33 per cent of equipment credits to the nuclear deterrent provided by the Armée de l'Air and the submarines of the Marine Nationale. Continuity of programmes is assisted by the implementation of a long-term plan covering a five year period, currently envisaging an average growth rate of 14.8 per cent. While the Armée de l'Air will receive only 22 per cent of these appropriations, this will be sufficient to continue the policy of replacement of its older aircraft types, at the same time maintaining numerical strength and increasing conventional weapons capability.

Present strength of the Armée de l'Air is some 40 combat squadrons and 30 or so transport, helicopter and support units, manned by 104,000 personnel. A total of 600 modern combat aircraft equip the front line, and continuing deliveries of a further 140 will begin replacement of the older Mirages before 1982. Support squadrons contain more than 1,000 aircraft placing the French air force as the fourth largest in the world and to approximately equal status with the Luftwaffe and the Royal Air Force.

The air elements of the Army and Navy are additionally undergoing a comprehensive programme of renewal and improvement. Cognisant of Warsaw Pact armoured superiority, the Aviation Légère de l'Armée de Terre, (ALAT) is rapidly expanding its anti-tank helicopter fleet and has recently completed a large-scale re-grouping of its aerial forces to match the new ground deployment in which the brigade has been eliminated as a battle formation.

Previously static territorial units of the Défense Opérationelle du Territoire will achieve mobility and combine with the Force de Manoeuvre (equivalent to the NATO ACE Mobile Force) for rapid deployment to trouble spots. The most significant improvement will be in anti-tank helicopter capa-

bility, where 70 Alouette IIIs equipped with the Nord/Aérospatiale SS 11 missile will be augmented by an eventual total of 240 Gazelles fitted with up to six Euromissile HOT anti-tank missiles. The ALAT, which became independent in 1954, now comprises 565 helicopters and a further 105 aircraft, although all but a few of the latter are due for imminent withdrawal. From a total Army strength of 324,400 men, some 5,600 are attached to the Aviation Légère de l'Armée de Terre.

Naval aviation is vested in the Aéronautique Navale (Aéronavale), an integral part of the Marine National with the main duties of protection of sea communications, coastal defence and participation in the inter-service Force d'Intervention. A small but efficient force of 270 front-line aircraft and 100 support types comprises carrier-borne strike, air defence and anti submarine elements on two aircraft carriers, *Clemenceau* and *Foch*. Each displaces 27,300 tons, has a speed of 32 knots and embarks 36 aircraft. A further helicopter carrier, the 11,000 ton *Jeanne d'Arc*, may carry between 5 and 8 rotorcraft for ASW or assault duties, but is operated as a cadet training ship during peacetime, although a nuclear-powered helicopter vessel of 18,000 tons has been delayed and will not now enter service until 1988. This will accommodate combinations of up to 25 Lynx, 10 Super Frelon or 15 Puma helicopters and will be armed with navalised Crotales or a new anti-aircraft missile. V/STOL strike aircraft are being considered, yet in view of the long-term nature of the project, some time will elapse before a final decision is made.

The Marine Nationale shares with the Armée de l'Air the responsibility of the French nuclear deterrent, the first generation of which was completed in March 1968 with the delivery of the last of 62 Dassault Mirage IVA high-level bombers for the air force's strategic arm, the Force Aérienne Stratégique. Headquartered in an underground command post at Base Aérienne 921, Taverny, near Paris, the FAS formed on the first day of 1964, and is currently commanded by General Delaval. Colloquially referred to as the 'Force de Frappe', but now more approvingly known as the 'Force de Dissuasion', the FAS has undergone progressive reductions in recent years with the availability of nuclear missile forces.

As originally constituted, nine squadrons, each with four Mirage IVA bombers at separate bases formed three wings with attached squadrons of four Boeing C-135F tankers for airborne replenishment. Disbandment of one wing in June 1976 resulted in the present deployment of six squadrons of four aircraft and a separate tanker wing with the remaining 11 C-135Fs. The basic Armée de l'Air formation is an *escadre*, or wing, to which squadrons

Top left: the Mirage IVA strategic bomber serves with two escadres de bombardement as part of France's Force de Dissuasion. Each armed with a 60 kiloton nuclear bomb, the current force of 47 aircraft will serve on into the mid-1980s.
Above left: the range of the Mirage IV can be extended by refuelling from Boeing C-135F tankers.
Left: the main interceptor in French service is the Mirage F 1C, 225 of which are being delivered. An aircraft of EC 3/12 is illustrated

Top: this Mirage F 1C serves with the 2ᵉ (Normandie) Escadron of the 30ᵉ Escadre de Chasse Tous Temps, based at Reims. It is armed with a Matra R530 missile of 18km (11 mile) range. Above: the Jaguar A strike fighter serves with the 7ᵉ and 11ᵉ Escadres

(escadrons), each distinguished by a name, are attached. Mirages are allocated to 91 Escadre de Bombardement with escadrons EB 1/91 'Gascogne' at Mont-de-Marsan, EB 2/91 'Bretagne' at Cazaux and EB 3/91 'Cevennes' at Orange and 94 Escadre de Bombardement with EB 1/94 'Guyenne' at Istres, EB 2/94 'Marne' at St Dizier and EB 3/94 'Arbois' at Luxeuil. The three tanker squadrons of 93 Escadre de Ravitailement en Vol are ERV 1/93 'Aunis' at Istres, ERV 2/93 'Landes' at Mont-de-Marsan and ERV 3/93 'Sologne' at Avord. Two former bases at Creil and Cambrai retain their support installations for emergency dispersal.

Each FAS base has a Depot-Atelier de Munitions Speciales to store, maintain and assemble the 60 kiloton free-falling French nuclear bomb which is the primary weapon of the Mirage IVA, and performance has been improved by a virtual halving of weapon weight by progressive improvements. The Mirage IVA has capability for up to 45 minutes' flight at Mach 2 at altitude and an unrefuelled range of 2,485 nautical miles, but modifications have been incorporated for low-level penetration as part of the mission profile; from 1975, aircraft have received a coat of camouflage.

Training requirements of the FAS are catered for by operational conversion unit Centre d'Instruction de FAS 328 at Bordeaux, which has four Mirage IVA aircraft modified to accommodate a 1,000kg (2,200lb) reconnaissance pod for long-range surveillance. This unit also operates a dozen or so Lockheed T-33As for deployment among FAS squadrons for instrument rating and communications work, five Mirage IIIB conversion trainers, ten Mirage IIIB-RV flight refuelling trainers and nine Nord 2501 Noratlas SNB crew trainers with Mirage IVA radar in enlarged nose radomes.

Some 47 Mirage IVAs remain in the FAS inventory, scheduled for retention until 1985 when they will be replaced by ground or air-launched (cruise) missiles. The C-135F tankers, nicknamed

the sousmarin, or submarine, because of their grey paint scheme and lack of windows, will be retained thereafter for the tactical elements of the Armée de l'Air. A programme of underwing skin-panel renewal has recently been completed to assuage the effects of fatigue, while it was announced late in 1978 that all would be re-engined with the SNECMA/GE CFM56 turbofan which should see them in continued service up to the latter years of the century.

Airborne deterrence is augmented by 1ᵉʳ Groupement de Missiles Stratégiques situated on the Plateau d'Albion, near Avignon. This became operational in 1971 with two squadrons each of nine silo-based Aérospatiale SSBS (sol-sol balistique stratégique) or two-stage, solid rocket IRBMs. Third generation nuclear delivery is allocated to the Marine Nationale and its growing fleet of missile-launching submarines, soon to reach the planned total of six. The first of these became operational in 1973. A longer-term requirement of the Force Océanique Stratégique is the second-generation MSBS (mer-sol balistique stratégique) M4 at present under development by Aérospatiale, with three-stage configuration bestowing a range of 2,160 nautical miles and six or seven MIRV warheads with an individual yield of 150kt. The M4 is scheduled for deployment in the sixth nuclear submarine from 1985, and four of the earlier vessels will have been modified to carry this larger diameter missile by 1991.

Army nuclear weaponry is of a tactical nature, centred on the Aérospatiale Pluton missile which began replacement of the Honest John in 1971 and now equips five Régiments d'Artillerie. The Pluton has an accuracy of 150–300 metres over ranges of eight to 64 nautical miles, and 30 are allocated to a regiment. Prospective targets for the Pluton are pinpointed and reconnoitred by an army-operated force of 41 Aérospatiale R20 battlefield reconnaissance drones, which will be replaced by the Canadair CL-89 from 1980 onwards.

The air force tactical nuclear force–Armament Nucléaire Tactique (ANT)–is concentrated within 1er Commandement Aérien Tactique, commonly abbreviated to CATac. Previously based in Germany on attachment to NATO's 4th Allied Tactical Air Force, CATac was withdrawn to its present HQ at Metz in 1966, and its squadrons took up residence on many of the airfields vacated by departing US and Canadian units. Numerically, the largest command in the Armée de l'Air, CATac has 21 squadrons comprising over 300 strike and reconnaissance aircraft equipped with AN52 nuclear weapons and Martel and AS30 missiles for ground strike, principally against enemy communications, plus some 13,500 personnel.

As with the strategic bomber units, *escadrons* (squadrons) are grouped into *escadres* (wings) one to each base, each *escadre* also having a Section de Liaison et de Vol Sans Visibilité (SLVSV) equipped with a small number of Holste MH1521 Broussards Potez Magisters and Lockheed T-33As for communications and instrument rating work. The majority of these *escadres de chasse* (fighter wings) are equipped with the Dassault Mirage IIIE, IIIR or 5F.

The Mirage I, ancestor of over a thousand aircraft produced for air forces throughout the world, flew for the first time in June 1955. In 1961, the Armée de l'Air received the first of 95 Mirage IIIC strike fighters initially for the 2e and 13e Escadres de Chasse and in the following year took delivery of the first of an eventual total of 61 tandem-seat Mirage IIIB trainers. The final 20 of these were to IIIBE standard, equivalent to the Mirage IIIE, of which 180 were received between 1964 and 1972 for the 2^e, 3^e, 4^e and 13^e Escadres de Chasse. Mirages replaced the American-supplied F-84F Thunderstreak and F-100D Super Sabre, the earlier IIIC aircraft passing later to the air defence role.

Fifty reconnaissance Mirage IIIR aircraft were delivered to 33 Escadre de Reconnaissance from 1963 onwards, and augmented in 1967–68 by a further 20 IIIRD aircraft with improved sensors. Last to be received were 50 Mirage 5F aircraft, a simplified version of the III series originally destined for Israel under the designation Mirage 5J. All were impounded by the French Government in 1970 before they could be flown to Israel, and from 1972 onwards were issued to two squadrons of the Armée de l'Air.

The remaining eight CATac squadrons fly the Anglo-French SEPECAT Jaguar, a light strike-fighter which is also in large-scale service with the RAF. France has 200 Jaguars on order of which 160 'A' attack version, and the remainder two-seat 'E' Entrainement (trainer) aircraft. Jaguars entered service with 7 Escadre in May 1973, and with priority given to production of the Jaguar E, all 40 had been delivered by March 1976, when only 54 A versions had been built. By the end of 1978, 118 strike aircraft were on Armée de l'Air charge with remaining batches of 16 and 26 due for funding in the following two years, the final examples for storage against attrition.

As presently constituted, CATac squadrons possess between 15 and 18 aircraft each, with two, three or four squadrons to a wing. At Dijon-Longvic, 2 Escadre de Chasse has two escadrons of Mirage IIIEs, EC 1/2 'Cicognes' and EC 3/2 'Alsace' for normal strike duties. A third, over-size squadron, Escadron de Chasse et d'Entrainement 2/2 flies with eight Mirage IIIBs and 15 IIIBEs as the Mirage Operational Conversion Unit (OCU), training pilots from both the Armée de l'Air and overseas forces which have bought the type. 3 Escadre de Chasse is based at Nancy-Ochey with a further two squadrons of Mirage IIIEs, EC 1/3 'Navarre' and EC 2/3 'Champagne', and a third escadron, 3/3, of Jaguars. EC 3/3 'Ardennes' previously operated the Mirage 5F from 1974 to 1977, but these were then forwarded to EC 13 in favour of

Escadre de Chasse 10 flies the Mirage IIIC in the air defence role. One of the 1^e Escadron's aircraft is pictured at its base at Creil in 1974. The Mirages of EC 3/10 have been deployed overseas at Djibouti since late 1978

ability from newly-qualified to highly-experienced.

The second Jaguar wing, 11 Escadre de Chasse at Toul-Rozières, has four escadrons for a variety of tasks. EC 1/11 'Roussillon' is allocated conventional strike roles in support of the French Army. CATac electronic countermeasures are delegated to EC 2/11 'Vosges', while EC 3/11 'Corse' is responsible for the support of units overseas. All aircraft of EC 11 have A-series aircraft fitted with the laser rangefinder, adopted from the 81st production example. The latest addition to the growing ranks of Jaguar squadrons is EC 4/11 'Jura' which was until December 1978 equipped with the F-100D Super Sabre in the former French colony of Afars and Issas. 'Jura' is detached to Bordeaux-Merignac where it replaces the recently retired Vautours of 92 Escadre de Bombardement as the sole permanent air element of 2 CATac.

With headquarters at Nancy, 2 CATac is the Armée de l'Air contribution to the tri-service Force d'Intervention, to which any other unit can be added as a situation demands. The Vautours were used for ECM and further Jaguar deliveries will form a fourth escadron for 7 Escadre de Chasse, detached to Istres-le Tube. In the past two years, EC 11 has based several aircraft in the former West African colonies to assist the governments of Chad and Senegal in their wars against guerilla forces, and at least three aircraft were lost during the course of operations in 1978.

Following the transfer of one squadron from EC 3, Armée de l'Air Mirage 5Fs are now concentrated within 13 Escadre de Chasse at Colmar-Mayenheim. The two squadrons concerned, EC 2/13 'Alpes' and EC 3/13 'Auvergne', are partnered by a third, EC 1/3 'Artois' which retains the Mirage IIIE. At the tactical reconnaissance base of Strasbourg-Entzheim, 33 Escadre de Reconnaissance has two squadrons of Mirage IIIRs, ER 1/33 'Belfort' and ER 2/33 'Savoie' and ER 3/33 'Moselle' with the infra-red sensor-equipped Mirage IIIRD. A reconnaissance version of the Mirage F1 air superiority fighter designated F-1R will replace the aircraft of ER 33 from 1983, and EC 2 will be the first to receive the new Mirage 2000 which will enter service in the same year. CATac is destined to receive up to 200 long-range strike versions of the Mirage 2000 in the late 1980s, armed with the Aérospatiale ASMP (*air-sol moyenne portée*) 100–150 kt air-to-surface missile powered by an integral ramjet for supersonic performance at all altitudes. A batch of 127 Mirage 2000s is on order, of which the initial ten will be delivered before the end of 1982 in preparation for the formation of the first *escadron* the following year. The Mirage IIIE is scheduled for total replacement by 1985, the IIIRD by 1986 and the 5F by 1991.

Naval aviation is similarly undergoing a process of improvement following deliveries of yet another Dassault product, the Super Etendard, which began in June 1978. Derived, as its name suggests, from the Etendard shipborne fighter and reconnaissance aircraft, 71 Super Etendards are due for delivery before the end of 1981 to re-equip five Aéronavale *flottilles* or operational squadrons. The Aéronavale order of battle comprises 16 *flottilles*, supported by 11 second-line units designated *escadrilles de servitude* with an 'S' suffix to

Top: a total of 54 Mirage IIIB two-seat conversion trainers remained in French service in 1979. This Mirage IIIB-RV is used by 328 Centre d'Instruction to train Mirage IV crews in flight refuelling procedures.
Centre: the Armée de l'Air's sole tactical reconnaissance unit is the 33ᵉ Escadre de Reconnaissance, based at Strasbourg. A Mirage IIIR of the 1ᵉ Escadron is illustrated.
Above: the Mirage IIIE tactical fighter currently serves with four escadres de chasse and 136 of the 185 aircraft originally purchased remain in service. EC 1/2 at Dijon perpetuates the Cigognes insignia

the present equipment.

Anti-radar strike with the Martel missile is practised by 4 Escadre de Chasse at Luxeuil-St Saveur and its two escadrons of Mirage IIIEs, EC 1/4 'Dauphine' and EC 2/4 'La Fayette'. After working-up as the first two Jaguar squadrons in the Armée de l'Air, EC 1/7 'Provence' and EC 3/7 'Languedoc' at St Dizier-Robinson adopted the role of nuclear strike early in 1975 with the conversion of their aircraft to carry the AN52 weapon. The training role for Jaguar pilots was continued within 7 Escadre by the formation in May 1974 of EC 2/7, an OCU with nine of both the A and E versions in contrast to the one or two trainers attached to operational squadrons. EC 2/7 'Argonne' converts an average of 50 pilots per year to the Jaguar in four courses catering for pilots of varying

their number. Personnel strength of 13,000 represents more than one-fifth of the entire naval manpower.

Among the first-line units, the carrier-based ASW element comprises 4 F and 6 F, each operating 12 Breguet Alizé armed with Nord/Aérospatiale AS 12 missiles. Between May 1979 and February 1984, 28 will be re-worked with Iguane radar, Doppler navigators and improved armament for continued service up to 1990 when the aircraft carriers *Foch* and *Clemenceau* are scheduled for retirement. The remaining dozen aircraft will be transferred to India, where they will join the survivors of 15 Alizés supplied during the 1960s. The modified aircraft will first go to 6 F at its shore base of Nîmes-Garons to cover the Mediterranean area from March 1980 onwards to be followed by 4 F at Lann-Bihoué for Atlantic operations.

The Alizé has been supplemented aboard the carriers by the Aérospatiale SA-321G Super Frelon heavy ASW helicopter of 32 F, which is based at Lanveoc-Poulmic for Atlantic deployments and defence of the submarine base at l'Ile du Longue. Fourteen aircraft ordered in 1963 as partial replacements for the Sikorsky H-34J have been joined by a further 10, some of which are based in the Pacific missile test area. Replacement of the remaining ASW H-34s within 31F at St Mandrier began in September 1978 with the arrival of the first Westland Lynx HAS Mark 2 (FN) modified to accommodate dunking sonar. The two initial Lynx batches, totalling 26 aircraft, have been delivered, and a requirement exists for up to 55 to replace AS 12-armed Alouette IIIs.

Helicopter platforms are now standard equipment on newly-built vessels of the Marine Nationale, and out of thirty C 70-class destroyers scheduled for service by the end of the eighties, 21 will be allocated to anti-submarine duties. The three F 67-class destroyers each have accommodation for two helicopters, and three more vessels of earlier classes can deploy helicopters when required. One cruiser and nine frigates have been similarly converted to accommodate a helicopter, but lack of hangarage restricts operations. There are, in addition, 19 further support ships and landing craft with helicopter provision. The latter category occasionally receives detachments of Sikorsky H-34s from 33F at St Mandrier, which are armed with SS 12 missiles for commando assault work, but the majority of helicopters afloat with the F 67 and C 70-classes of vessel are Alouette IIIs of 34 F at Lanveoc-Poulmic.

For fixed-wing carrier strike, the Aéronavale has until recently been operating two squadrons, 11 F and 17 F at Landivisiau and Hyères respectively, with the Etendard IVM. The Etendard IVP tactical reconnaissance aircraft is flown by 16 F at Landivisiau, which co-operates with ER 33 of the Armée de l'Air when ashore. Etendards have 'buddy-pack' flight refuelling equipment and a limited nuclear capability and this has been enhanced from 1978 onwards by deliveries of the Super Etendard powered by the SNECMA/8K-50 and fitted with Agave radar and SKN 2602 inertial navigation equipment. By January 1979, 11 F had received 17 aircraft in preparation for embarkation on the carrier *Clemenceau* following a 14-month refit.

Currently undergoing conversion as the second

Super Etendard unit is 14 F, previously one of two carrier interceptor squadrons equipped with the LTV F-8E(FN) Crusader at Landivisiau. Out of 42 Crusaders delivered from 1964 onwards, only 25 remain, and these are now being concentrated in 12 F, also at Landivisiau. All have been converted to F-8J standard with modified mainplanes for extension of fatigue lives into the eighties, while the 20 selected for operation by 12 F are to have new afterburners.

With the later re-equipment of 17 F to the Super Etendard it is likely that the Crusaders will be withdrawn from carrier operations, allowing the former to take-over interceptor as well as strike roles, equipped with the MATRA R.530 and Sidewinder air-to-air missiles. At present armed with the AN52 atomic weapon, the Super Etendard may

Top: the Mirage 5 is a simplified variant of the Mirage III series, optimised for close-support missions. They were originally produced against an order from Israel, but the deal was embargoed and the Mirage 5 now serves with Escadre de Chasse 13, a fighter of EC3/13 'Auvergne' being shown.
Centre: a Super Mystère B2 of EC1/12 'Cambrésis' finished in the striking markings seen at the NATO Tiger Meet in 1977. The Escadron was re-equipped with the Mirage F 1 later that year.
Above: the Fouga Magister is the Armée de l'Air's basic jet trainer and the mount of the Patrouille de France aerobatic team

Top: the N2501 Noratlas remains CoTAM's numerically most important tactical transport and will serve until the mid-1980s. An N2501 of ET 64 is pictured.

Above: ET 61, based at Orleans-Bricy, is one of the escadres of CoTAM's tactical transport component, operating all the C160 Transalls in Armée de l'Air service. A further 25 have been ordered and will probably equip ET 63

later be equipped with the medium range nuclear air-to-surface missile, the ASMP, under development for the Mirage 2000. Conventional warload includes the Martel or AM-39 air-to-surface weapons for anti-shipping strike.

Five shore-based maritime reconnaissance *flottilles* complete the Aéronavale front line. Of these, four units operate the AMD-Breguet Atlantic ASW patrol aircraft, while a further *flottille* continues to use the Lockheed SP-2H Neptune, all combining to provide anti-submarine and surface strike capability over the Mediterranean, Channel and Atlantic under the administrative control of the Commandement de l'Aviation de Patrouille Maritime (abbreviated as PAT-MAR).

Mediterranean squadrons 21 F and 22 F at Nîmes-Garons have seven Atlantics each, the former being the initial recipient of the type in December 1965. The requirements of northern waters are catered for by 23 F and 24 F with a similar establishment of aircraft at Lann-Bihoué. A total of 41 Atlantics was delivered to the Aéronavale up to 1972 as replacements for earlier marks of Neptune, and 42 further examples of an improved model, the Atlantic Nouvelle Génération or ANG, are ear-

marked for delivery in the eighties. ANGs will be armed with AM 39 air-to-surface missiles in place of the current AS 12s and will have a new Iguane radar, improved inertial navigation equipment, ECM, ESM and acoustic detection systems, together with TRT/SAT forward-looking infra-red opto-electronics.

First priority for the ANG will be the replacement of the ageing Neptunes in 25 F at Lann-Bihoué, and ultimately of the first-generation Atlantics which will be approaching the end of their service lives by 1990. Interest is currently being expressed in a simpler maritime surveillence aircraft to replace a small number of Atlantics and Neptunes detached from their home bases or serving with second-line *escadrilles*. The Aéronavale maintains a single Atlantic at Dakar to search-out possible guerilla activity at sea, a role for which the aircraft is grossly over-equipped. France is considering the purchase of some 16 British Aerospace 748s for paratroop dropping and navigation training, four of which would be the Coastguarder version, ideally suited to monitoring less sophisticated forms of shipping.

The air defence of France is administered by the Armée de l'Air's Commandement Air des Forces Défence Aérienne (CAFDA) and it has an integrated system of manned interceptors, surface-to-air missiles and computerised warning and communications network, Système de Transmission et Représentation des Informations Radar (STRIDA), which is interconnected with the other armed services and the NATO NADGE system. CAFDA disposes some 7,000 personnel in eight home-based *escadrons* and one further flight overseas with 150 front-line interceptors plus support types.

Principal CAFDA aircraft is the Mirage F1C, of which 225 are planned for delivery by 1983 to complete conversion of the entire defence force. The remainder will pass to CATac for ER 33 in the reconnaissance role, but CAFDA is ultimately to receive the Mirage 2000 by the mid or late 1980s.

Deliveries of the Mirage F1 temporarily ceased during the Spring of 1978 to allow production to be increased for several overseas customers, at which time CAFDA had received 105 aircraft, including 24 re-worked to F1A standard with fixed flight-refuelling probes. At least 14 of those remaining on order will be the tandem-seat Mirage F1B, which has already been exported to Kuwait and Morocco, due for acceptance in 1980 to form the initial equipment of a proposed operational conversion unit.

The squadrons of CAFDA are grouped in four *escadres*, each with 15 Mirage F1s or a varied number of Mirage IIICs. At Orange-Caritat, 5 Escadre de Chasse converted its two squadrons to the Mirage F1C late in 1974, but these were replaced by flight-refuelling-capable F1As in the spring of 1978; EC 5 is, as yet, the sole wing with this variant. In 1980, component squadrons EC 1/5 'Vendee' and EC 2/5 'Ile de France' will be joined by Escadron de Chasse et de Transformation 3/5 which will become the Mirage F1 OCU, equipped with most, if not all, the 14 Mirage F1B trainers currently on order. Fifty remaining Mirage IIICs, due for withdrawal in 1980, remain at Creil-Senlis on charge to 10 Escadre de Chasse. Two squadrons, EC 1/10 'Valois' and EC 2/10 'Seine' have 20 aircraft each, while a third unit, EC 3/10 'Vexin', was formed in September 1978 and deployed to Djibouti two months later to relieve the Super Sabres of EC 4/11. 'Vexin' comprises only one flight, or *escadrille*, of ten aircraft, as opposed to the majority of *escadrons de chasse* which have two.

Most recent recipient of the Mirage F1 is 12 Escadre de Chasse at Cambrai-Epinoy, with its two components EC 1/12 'Cambresis' and EC 2/12 'Cornouaille', the former unit relinquishing its last Super Mystere B2 in September 1977. At Reims-Champagne, 30 Escadre de Chasse Tous Temps retains the 'all-weather' title as a relic of the days in which it flew the Gloster Meteor NF Mark II. ECTT 2/30 began re-equipment with the Mirage F1

in December 1973, ECTT 3/30 following in the spring of the next year. In common with the *escadres* of CATac, each wing additionally possesses an SLVSV of Potez Magisters, Lockheed T-33As and Holste Broussards. The CAFDA defence organisation is completed by Thomson-CSF MATRA R.440 Crotale low-level SAM batteries. A defensive network covering strategic and nuclear bases will be completed in 1980.

Airborne early warning, as provided by the Boeing E-3A AWACS, has received recent attention, but although France was included in the NATO AWACS discussions, the E-3A has been rejected on the grounds that its intended deployment will not adequately cover all French areas of concern. The recent development of the Thomson-CSF Aladin low-looking radar along the north-east and south-east borders of France provides greatly improved low-level surveillance and tactical control in conjunction with Centaure aquisition radars, installation of which will be completed during 1982–83. France is, however, mindful of the threat of an under-radar attack from the North African coast which the NATO AWACS will not cover, and has formulated an independent programme based on the purchase of approximately five early-warning aircraft for service from 1982. A military derivative of the Airbus Industrie A300 wide-bodied transport is under consideration, but the short time-scale will dictate equipment with off-the-shelf avionics, although the Grumman E-2C Hawkeye would provide an immediately-available substitute. This apart, communications with NATO air defence networks are being strengthened from 1979 onwards by a special system using landlines, satellites and tropospheric scatter equipment.

During 1979, the ALAT will complete a three year programme of unit re-grouping, in which the emphasis will be on anti-armour capability, although some time will elapse before all the planned missile-equipped helicopters are delivered. Army Aviation made its first use of the helicopter on

The Alouette II has been operated by all three French air services since 1956 in various roles. The main user is now ALAT, with which 180 machines form the main equipment of its light helicopter squadrons with the Aérospatiale Gazelle SA 341F. The Alouette II has been relegated within ALAT to liaison duties, having been replaced in the anti-tank role by the Alouette III and Gazelle SA 342M. CoTAM's 67 Escadre de Hélicoptères uses the Alouette II for liaison, security and training, while the Aéronavale uses the type for communications and search and rescue. An Alouette II of the Armée de l'Air is pictured at Nancy-Ochey in 1973

AVIATION LÉGÈRE DE L'ARMÉE DE TERRE 1979

1 RHC	Anti-tank/liaison	Gazelle/Puma	Phalsbourg
2 RHC	Anti-tank/liaison	Gazelle/Puma	Freiburg
3 RHC	Anti-tank/liaison	Gazelle/Puma	Etain
4 RHC	Anti-tank/liaison	Gazelle/Puma	Trier
5 RHC	Anti-tank/liaison	Gazelle/Puma	Compiègne
6 RMC	Anti-tank/liaison	Gazelle/Puma	Pau
1C GHL	Liaison	Gazelle/Alouette II & III	Phalsbourg
2C GHL	Liaison	Gazelle/Alouette II & III	Friedrichs-hafen
1 GHL	Liaison	Gazelle/Alouette II	Les Mureaux
2 GHL	Liaison	Gazelle/Alouette II	Lille-Lesquin
3 GHL	Liaison	Gazelle/Alouette II	Rennes
4 GHL	Liaison	Gazelle/Alouette II	Bordeaux
5 GHL	Liaison	Gazelle/Alouette II	Lyon-Corbas
11 GRL	Liaison	Gazelle/Alouette II	Nancy
DHA	Support	Alouette III/Puma	Djibouti
ECCFFA	Liaison	several	Baden-Baden
GALSTA	Trials	several	Valence
GMEA	Transport OCU	Puma	Aix
ES	Flying training	Gazelle/Alouette II	Dax
EA	Advanced training	several	Le Luc

Abbreviations: RHC, Régiment d'Hélicoptères de Combat; GHL, Groupe d'Hélicoptères Légers; DHA, Détachement d'Hélicoptères Armes; ECCFFA, Escadrille de Commandement Centrale des Forces Français Allemande: GALSTA, Groupement ALAT de Section Technique de l'Armée de Terre; GMEA, Groupe de Manoeuvre de l'Ecole d'Applications; ES, Ecole de Specialisation; EA, Ecole d'Application.

becoming an independent formation in November 1954. The Bell 47 and Sud Djinn were joined in the liaison role by the world-famous Alouette II which served an an anti-tank helicopter when fitted with the SS10 missile. In the higher weight category, the Sikorsky H-19 was replaced by over one hundred Piasecki H-21s, which served as assault transports until replaced by the Aérospatiale SA 330 Puma from 1969 onwards.

Flown for the first time in April 1965, the Puma has achieved creditable sales totals, with some 600 built thus far, and an improved variant undergoing flight trials. The initial ALAT order for 130 of the SA 330B version was subsequently increased to 140, the last ten to SA 330H standard with improved avionics, and the 1979 Defence Budget includes funding for a further small batch. A further 20 have been delivered to the Armée de l'Air as general duties transports and SAR helicopters, while a small number are attached to various trials establishments.

The ALAT retains some 180 of the 230 Alouette IIs which it received from 1956, the final 55 fitted with the higher-powered Astazou engine in place of the original Artouste. The helicopter has been relegated to liaison work only, and was replaced in the anti-tank role by the Alouette III, of which some 70 remain. Fitted with the SS 11 missile, the larger Alouette III has borne the responsibility of airborne anti-armour strike, but is now giving way to the Gazelle in its SA 342M configuration.

First flown in August 1967, the Gazelle entered production in 1971, and by early 1978, the initial order for 166 ALAT examples had been completed. These are intended for liaison and observation duties and designated SA 341F, but such is the urgent requirement for missile helicopters that 110 of these are being retrospectively fitted with Euromissile HOT anti-tank missiles on two outriggers, each holding two rounds, with associated guidance sight mounted above the cockpit roof; these are designated SA 341M.

Due to enter service this year, the SA 342M Gazelle will have HOT equipment fitted as standard, and provision for up to six rounds to be released per sortie. A total of 128 SA 342M models will be delivered up to 1983, each powered by the uprated Astazou XIVH giving a 20 per cent increase in all-up weight. Other revisions will include all-weather avionics, intake filter and a jet-pipe deflector for reducing the infra-red signature to minimise danger from heat-sensing anti-aircraft missiles.

Both models of the Gazelle are, however, seen as an interim measure pending the arrival of a purpose-built helicopter gunship to satisfy the HAC (*helicoptère anti-char*) requirement, and the French and Germans are near to agreement on a

Above left: fifty aircraft remain of a total of 316 Max Holste MH1521 Broussards delivered, partly equipping six communications and transport units. Left: the ALAT is currently undergoing re-organisation and re-equipment. Under the new system seven helicopter escadrilles will be grouped into a regiment, undertaking liaison, anti-armour and transport duties for the army. Two of these will be equipped with the Aérospatiale Puma (illustrated) for battlefield troop transport

Top: the remaining Dassault Etendard IVM carrier strike aircraft are in service with 11F and 17F, but these units are in the process of re-equipment with the Super Etendard.
Above: Aéronavale LTV Crusader carrier fighters are armed with 20 mm cannon and Matra R 530 AAMs. They have been modified to US Navy F-8J standard to extend their lives into the 1980s

AERONAUTIQUE NAVALE 1979

4 F	Anti-sub/SAR	Alizé	Lann-Bihoué
6 F	Anti-sub/OCU	Alizé	Nîmes-Garons
11 F	Strike	Super Etendard	Landivisiau
12 F	Air superiority	Crusader	Landivisiau
14 F	Air superiority	Super Etendard	Landivisiau
16 F	Photo-recce	Etendard IVP	Landivisiau
17 F	Strike	Etendard IVM	Hyères
21 F	Naval patrol	Atlantic	Nîmes-Garons
22 F	Naval patrol	Atlantic	Nîmes-Garons
23 F	Naval patrol	Atlantic	Lann-Bihoué
24 F	Naval patrol	Atlantic	Lann-Bihoué
25 F	Naval patrol	Neptune	Lann-Bihoué
31 F	Anti-submarine	SH-34J/Lynx	St Mandrier
32 F	Anti-submarine	Super Frelon	Lanvéoc-Poulmic
33 F	Assault transport	SH-34G	St Mandrier
34 F	Anti-submarine	Alouette III	Lanvéoc-Poulmic
2 S	Communications	several	Lann-Bihoué
3 S	Communications	several	Hyères
9 S	Naval patrol/SAR	C-47/C-54/Neptune	Tontouta
12 S	Naval patrol/SAR	Neptune	Faaa
20 S	Helicopter trials	several	St Raphael
22 S	Communications/SAR	Alouette II & III	Lanvéoc-Poulmic
23 S	Communications/SAR	Alouette II & III	St Mandrier
27 S	Transport	Super Frelon	Mururoa
55 S	Multi-engine school	Nord 262	Aspretto
56 S	Navigation school	C-47 Dakota	Nîmes-Garons
59 S	Deck-landing school	Etendard/Alizé/Zéphyr	Hyères
SRL	Communications	Paris/Falcon 10	Landivisiau
SSD	Communications	DC-6/N262/Navajo	Dugny
SVS	Air experience	Rallye 100S	Lanvéoc-Poulmic
SES	Air exp/trials	Rallye/Noratlas	St Raphael

Abbreviations: SRL, Section Reacteur Léger; SSD, Section de Soutien de Dugny; SVS, Section de Vol Sportif; SES, Section Experimental et de Servitude

joint project which will also fulfill the latter's PAH-2 specification.

Until the re-organisation of the front line is complete, the ALAT will have old and new formations in existence. In both cases, squadrons are allocated at corps and division level, the latter for observation and liaison work only. Centred on the French 1st Army and its two component Corps, the second of which is stationed in Germany, the original deployment comprised two Groupes d'Aviation Légère de Corps d'Armée, GALCA 1 at Phalsbourg and GALCA 2 at Friedrichshafen, Germany. Each consisted of an Escadrille d'Helicoptères Légers (EHL–light helicopter squadron) for liaison with ten Gazelles or Alouette IIs, one Escadrille d'Hélicoptères Anti-Chars (EHAC–anti-tank helicopter squadron) with 12 Alouette IIIs equipped with SS 11 missiles, and two Escadrilles d'Hélicoptères de Manoeuvre (EHM–assault and heavy-lift helicopter squadrons), each with 11 Pumas.

At divisional level came six Groupes d'Aviation Légère de Division (GALDiv) each with four attached helicopter flights, or *pelotons*, having similar functions to the GALCA units. The operational picture is completed by seven Groupes d'Aviation Légère de la Région (GALRegs), attached to regional army HQs for liaison and communications. These each comprised one Escadrille Avions with two Max Holste MH1521 Broussards and five Cessna O-1 Bird Dogs and one Escadrille Hélicoptères with four Alouette IIs. The unit at Les Mureaux qualifies for an additional Escadrille Hélicoptères equipped with four Alouette IIIs by reason of its proximity to the Paris area and consequently increased communications commitments.

The large-scale re-grouping of front-line formations now nearing completion will deprive the Corps units (GALCAs) of their anti-tank and assault helicopters, establish six enlarged battlefield helicopter units and convert liaison flights to all-helicopter equipment. The six previous GALDivs form the basis for the new Régiments d'Hélicoptères de Combat (RHCs), but with a 300 per cent increase in anti-armour capability. Although the basic framework will be complete by the end of this year, it will not be until 1982–83 that all units reach their authorised complement of Gazelles.

Each RHC consists of a Commandement, or HQ staff, an Escadrille de Soutien et Ravitaillement (supply and provision squadron) and seven *escadrilles* of helicopters designated in the same manner as previously. These are two EHLs each with ten SA 341 Gazelles for liaison, three EHACs for anti-armour work with 12 SA 341M or SA 342M Gazelles and two EHMs of 11 Pumas each. In the interim, the Alouette III will substitute for the awaited HOT Gazelles, although it has been stated that the type may be retained in some small measure after the prescribed strength is met. The early Gazelles now being converted to HOT standard will form the equipment of three RHCs, first deliveries to Etain beginning in September 1978.

A second level of formation, the Groupe d'Hélicoptères Legers will replace both the former GALRegs and the two Corps GALCA units. In the latter case, the existing mixed-function *groupes* will be replaced by two communications GHLs, each with a strength of ten Alouette IIs, ten Alouette IIIs and ten Gazelles and based at Phalsbourg and Friedrichshafen with 1 and 2 Corps respectively. Six more GHLs will comprise only ten Alouette IIs and ten Gazelles.

Second-line army units remain largely unaffected by the current restructuring, and comprise two training schools, one weapons trials unit, one HQ squadron and three attached flights. Pilot training begins at the Ecole de Specialisation at Dax with a one month grading course of 15 hours on 19 Cessna O-1 Bird Dogs, and is followed by a 22 week (115 hour) helicopter course on 25 Alouette IIs and three weeks (20 hours) on the school's ten Gazelles. Instrument training and three weeks' tactical flying follow at the Ecole d'Application, based at Luc/le Canet des Maures. The latter school operates the Gazelle, Alouette II and Puma, together with a detached flight of Pumas designated the Groupe de Manoeuvre de l'Ecole d'Application based at Aix. After co-pilot experience with an operational unit, personnel may return to the Luc school for a seven-week course to achieve Puma captain standard, or to train as an instructor during a 13 month period.

Equipment trials and evaluation are undertaken by the Groupement ALAT de Section Technique de l'Armée de Terre at Valence with three or four examples each of the Puma, Alouette II/III and Gazelle. Headquarters of the French forces in Germany at Baden-Baden, the Commandement

Centrale des Forces Francais Allemande (CCFFA) is served by an *escadrille* consisting of a small number of Alouettes, O-1s, Broussards and Gazelles. In addition, flights of three Alouette IIs are attached to the Artillery College at Chalons-sur-Marne, Cavalry College, Saumur and the Infantry College, Montpellier. The ALAT presence overseas is now minimal, and is confined to a flight of Pumas and Alouette IIIs at Djibouti.

Transport support for all French forces is provided by some 270 aircraft and 120 helicopters of the air force's Commandement du Transport Aérien Militaire (CoTAM) which also controls communications and rescue units. CoTAM currently comprises 17 squadrons and nine flights within five wings or groups.

Top: an Etendard IVP tactical reconnaissance aircraft of 16F, whose shore base is Landivisiau, but which also operates from carriers.

Above: the remaining LTV Crusader carrier fighters in Aéronavale service are being allocated to 12F at Landivisiau, with the current re-equipment of 14F with the Super Etendard.

Above left: an Alouette III of 22S, undertakes search and rescue, offshore delivery and communications duties for aircraft carriers. Based at Lanvéoc-Poulmic, 22S is responsible for 2 (Atlantic) Region and operates both the Alouette II and III

The mainstay of the tactical transport force is 61 Escadre de Transport which operates 48 of the 53 Transall C160 freighters originally delivered from Franco-German production from 1967 onwards. Based at Orleans-Bricy, ET 61 comprises squadrons ET 1/61 'Touraine', ET 2/61 'Franche Comté' and ET 3/61 'Poitou'. At Evreux, the majority of the ageing Nord Noratlas aircraft have been concentrated within 64 Escadre de Transport and its three *escadrons*, ET 1/64 'Bearn', ET 2/64 'Anjou' and ET 3/64 'Bigorre'. The Noratlas is scheduled for replacement between 1981 and 1984 by an additional order for 25 Transalls. With the exception on new avionics and navigational equipment and extra fuel capacity, the new aircraft will be similar to their predecessors.

With the withdrawal of the Douglas DC-6 aircraft of the previous ET 2/64 'Maine' in 1977, the present incumbent was transferred from the disbanded 62 Escadre in June 1978, while ET 1/62 'Vercors' moved to Toulouse minus its aircraft at the same time. Here, it combined with the existing CoTAM operational conversion unit, 340 Centre d'Instruction des Equipages de Transport with its 15 Noratlas and six Nord 262D Frégates to form the basis of 63 Escadre de Transport, which is expected to be the first recipient of the new Transall order.

The principal communications squadron for government agencies and HQ staff at Villacoublay is 60 Escadre de Transport and its two components. First of these, Escadron 1/60, otherwise the Groupe des Liaisons Aériennes Ministérielles (GLAM), has a mixed inventory of the two presidential Caravelles, six Dassault Falcon 20s, two Pumas, three Alouette IIIs and a Falcon 50. At Paris-Roissy Airport (Charles de Gaulle), ET 3/60 'Esterel' has four Douglas DC-8s for VIP and long range transport roles.

ET 2/60, the Groupe d'Entrainement et de Liaison, was raised to wing standard as ET 65 at Villacoublay in 1972, with the responsibility of training communications pilots and operating light transport aircraft on liaison and MedEvac work. ET 1/65 'Vendome' flies eight of the 18 Nord 262s delivered to the Armée de l'Air as Dakota replacements from 1970 onwards, together with a single Falcon 20. ET 2/65 flies 24 MS 760 Paris aircraft and a dozen Broussards. ET 65 is at the disposal of 2e Region Aérienne–2 Air Region (Paris)–but the remaining three regions have independent flights with the Broussard, Paris and

Far left: a Breguet Alizé of 3S based at Hyères. The unit operates several types on liaison and special duties, and is one of 11 Aéronavale escadrilles de servitude or second-line units.

Left: a Breguet Atlantic of 23F, based at Lann-Bihoué. Armed with AS 12 missiles and carrying comprehensive radar and electronic detection equipment, the Atlantic is operated in the ocean patrol and anti-submarine roles. A total of 41 such aircraft was delivered to the Aéronavale.

Below: an Aéronavale Lockheed SP-2H Neptune of 25F maritime reconnaissance unit, based at Lann-Bihoué. The type entered service in 1953 and a total of 60 was delivered, of which 15 remain in service. The type is intended to be replaced by the Atlantic Nouvelle Génération in the 1980s.

N262 Frégate. One further transport squadron is Group Aérienne Mixte 56 'Vaucluse' at Evreux with six Noratlas, two Puma and a newly acquired DHC Twin Otter for the transport of security service personnel.

In May 1975, the helicopters of CoTAM were rationalised within one wing, the 67 Escadre d' Hélicoptères, with five squadrons deployed throughout France for liaison and SAR. EH 1/67 'Pyrénées' at Cazaux was the first to operate two of the 20 Pumas delivered to the Armée de l'Air from December 1974 onwards, and also has four Alouette IIs and three Alouette IIIs used for security patrol at Landes missile test range. EH 2/67 'Valmy' at Metz has eight Alouette IIs and ten Alouette IIIs, while EH 3/67 'Parisis' at Villacoublay operates 13 Alouette IIs and eight Alouette IIIs within the Paris area and is a frequent user of Issy Heliport which is conveniently located beside the Armée de l'Air headquarters at Boulevard Victor, Paris.

EH 4/67 'Durance' has the primary task of logistic support for 1er Groupement Missiles at Apt, with three Pumas and three Alouette IIs, and EH 5/67 'Alpilles' has similar equipment plus three Alouette IIIs at Istres-le Tube. The helicopter OCU is Centre d'Instruction des Equipages d'Hélicoptères 341 at Chambery/Aix-les-Bains, formed in 1975 with nine Alouette IIs, ten Alouette IIIs and two Pumas. CIEH 341 provides *ab initio* helicopter training for pilots arriving at the unit with some 250 hours flying experience. Following a 30-week course consisting of 70 hours on the Alouette II and a further 15 on the Alouette III, trainees qualify as second pilots. After 250–300 hours with an operational squadron, they may return to Chambery for nine weeks (35 hours) training to crew chief standard.

The Aéronavale relies on the Armée de l'Air for the major part of its training requirements, but additionally has specialist schools among the second-line units, or Escadrilles de Servitude, all of which have an 'S' suffix to their title. At Lann-Bihoué, 2S operates the Nord 262, Piper Navajo and Alizé as the liaison squadron for 1 and 2 Maritime Regions while 3 S at Hyères performs a similar function for 3 Region using these three types and two Dassault Falcon 10s. Both Escadrilles have an additional training commitment for radar and instrument-flying instruction.

At Tontouta, in New Caledonia, 9S flies two Douglas C-47 Dakotas and single examples of the Douglas C-54E and P-2H Neptune for maritime patrol and communications work in the Pacific area, but is additionally involved with mapping and survey duties. Similar duties are undertaken by 12S at Faaa, Tahiti with eight Neptunes, some of which are detached to Hao and Mururoa in connection with the Pacific Nuclear Test Centre.

Nearer home, 20S represents the rotary-wing section of the CEPA flight-test establishment operating a mixture of Alouette II, Super Frelon, H-34 and, from May 1978, Lynx helicopters in ongoing service trials. Communications and SAR detachments for aircraft carriers are provided by two similar *escadrilles*, each with a mixed fleet of Alouettes II and III. 22 S at Lanvéoc-Poulmic has responsibilities for 2 (Atlantic) Region and 23 S at St Mandrier for 3 (Mediterranean) Region. Five heavier Super Frelon helicopters of 27 S handle cargo requirements of the Pacific Nuclear Test Centre from their base at Mururoa.

At Hao, Section Alouette du Pacifique has four Alouette IIIs detached from 2 S, similarly assisting the nuclear trials programme, while at Le Bourget-Dugny, the Section de Soutien de Dugny flies three Nord 262s, a Navajo and the Aéronavale's sole DC-6 on VIP duties in support of the Naval Headquarters at Paris. Section Jeanne d'Arc is based on the helicopter of that name to administer detachments of H-34 and Alouette III helicopters periodically embarked for training. Other Aéronavale detachments include single Atlantics or Neptunes based at Dakar and Djibouti in support of the forces of former French colonies.

Among several experimental and trials units, the Centre d'Experiences Aériennes Militaires (CEAM) at Mont-de-Marsan undertakes trials of new combat aircraft under direct control of the Etat Major, or Air Staff. CEAM controls Escadre de Chasse 24/118–equipped with the Mirage III, Mirage F1 and Jaguar responsible for the working-up of the first squadron of new types to enter service–CEAM 332 with similar equipment plus the MS760 Paris, and the associated transport flight, Escadrille de Transport et Liaison 26/118.

Armée de l'Air representation overseas is now greatly diminished, although there has, of late, been some expansion in the Pacific area. Re-inforcement operations have recently included Jaguar detachments to Chad for anti-guerilla operations, and the permanent deployment of the Mirages of EC 3/10 to Djibouti in December 1978. There, Escadron de Transport d'Outre Mer (ETOM) 88 operates six Noratlas, four Alouette IIs and two Broussards, with two of the Noratlas having been transferred to the Republic of Djibouti Air Force late in 1978 as its sole equipment. At St Denis, on the East African island of Réunion, ETOM 50 has two Alouette IIs and a detachment of three Transalls from ET 61. ETOM 52 'Tontouta' flies one Alouette II and three Pumas from Noumea in New Caledonia, while ETOM 55 'Ouessant' at Dakar, Senegal, has Alouette IIs and Noratlas to support French forces in the area. Communications in the Antilles and Guyana are undertaken by the Alouette, Broussard and Puma aircraft of ETOM 58, aided by a small Transall detachment and based at Point-à-Pitre. Lastly, long-range communications and transport for the nuclear test centre are delegated to the three Caravelles of ETOM 82 at Papeete.

CHAPTER SIX

West Germany

Luftwaffe/Heeresflieger/Marineflieger

The Lockheed F-104G Starfighters of the Federal German Luftwaffe have earned undeserved notoriety, due to the number of crashes involving this aircraft. In fact the Starfighter's safety record compares favourably with other advanced combat aircraft

With the exception of the United States, the Federal Republic of Germany now provides the largest single air contribution to NATO, and the achievement of this prodigious feat is even more remarkable when it is remembered that at the inauguration of the North Atlantic Alliance thirty years ago, West Germany possessed not one military aircraft, vehicle or ship. Germany's vital strategic location, in addition to its economic strengths, confirms its position as the cornerstone of NATO, and its armed forces are both highly-trained and well-equipped to counter the threat posed by the massive Warsaw Pact armoury stationed in the Eastern sector of the once unified nation.

· For ten years after World War II, all military activity was prohibited in Germany by the Allied Control Commission and, when the time came for the Luftwaffe to be re-born, many of the aircraft factories which were to provide its equipment had lain untouched since they had last been visited by Lancasters or Flying Fortresses a decade before.

However, not only would an air force and air arms of the two other fighting services be formed, but within five years they would be operating 670 fighters, 1,250 transports and trainers and 180 helicopters. To achieve this formidable total, most orders were placed outside Germany for aircraft types already in production, in the certain knowledge that many would be nearing obsolescence when they entered service and a second round of re-equipment would be required almost immediately.

On 24 September 1956, the Luftwaffe formally came into being with a short ceremony in which it received its first three training aircraft and awarded pilot's 'wings' to ten former World War II pilots. By 1961 only 15 of the projected 20 *geschwader*, or wings, had been formed, but it was a shortage of airmen, not aircraft, which bedevilled the Luftwaffe during its early years. The Heeresflieger–army aviation–began operations in 1957, forming small flights equipped with helicopters and the indigenous Dornier Do 27 light aircraft, each attached to an

army ground unit for liaison and communications. Bases for the Marineflieger – naval aviation – were established in northern Germany where they could co-operate with Danish forces in patrolling the strategically important Baltic, and initially the Navy was trained and equipped by Britain.

As service aircraft changed in the ensuing years, so did patterns of procurement. When the Luftwaffe replaced its North American F-86 Sabres and Republic F-84 Thunderstreaks in the early 1960s, it did so with Lockheed F-104 Starfighters and Fiat G 91s produced largely in German factories, and became increasingly self-sufficient as time progressed. Modification of NATO thinking away from the tripwire strategy towards one of 'flexible response' imposed further changes on Luftwaffe administration towards the end of the decade, but this was more significantly demonstrated by a rapid expansion in the army's fleet of medium-lift helicopters. Flexible response acknowledged that a future war might take the form of a conventional attack by Soviet forces, which would be met by NATO conventional forces, rather than nuclear weapons. Accordingly the deployment and supply of army ground units became of increasing importance. The vulnerability of airfields in a protracted campaign received detailed consideration. However, after brief flirtations with NATO-inspired VTOL strike aircraft and transports, and even a rocket-assisted zero-length launch Starfighter, the Luftwaffe remains firmly wedded to its concrete runways, or specially-prepared strips of Autobahn for emergency use.

The Luftwaffe of today comprises some 1,100

THE LUFTWAFFE

Unit	Aircraft type	Base
JBG 31	F-104G Starfighter	Norvenich
JBG 32	F-104G Starfighter	Lechfeld
JBG 33	F-104G Starfighter	Büchel
JBG 34	F-104G Starfighter	Memmingen
JBG 35	F-4F Phantom	Pferdsfeld
JBG 36	F-4F Phantom	Rheine-Hopsten
LKG 41	Fiat G 91	Husum
LKG 43	Fiat G 91	Oldenburg
JBG 49	Alpha Jet	Fürstenfeldbrück
AKG 51	RF-4E Phantom	Bremgarten
AKG 52	RF-4E Phantom	Leck
JG 71	F-4F Phantom	Wittmundhafen
JG 74	F-4F Phantom	Neuburg
LTG 61	C160 Transall	Landsberg
LTG 63	C160 Transall	Ahlhorn
HTG 64	UH-1 Iroquois	Ahlhorn
WS-10	F-104G/TF-104G	Jever
FFS S	Transall/Skyservant	Wunsdorf
FAR	Piaggio P149	Fürstenfeldbrück
FBS	several	Köln-Bonn
ESt 61	several	Manching
FVS 61	Hansa Jet/Skyservant	Lechfeld
3525PTW	T-37B/T-38A	Williams AFB
4510CCTW	F-104G/TF-104G	Luke AFB

THE MARINEFLIEGER

Unit	Aircraft type	Base
MFG 1	F-104G Starfighter	Schleswig
MFG 2	F-104G Starfighter	Eggebeck
MFG 3	Atlantic	Nordholz
MFG 5	Sea King/Skyservant	Kiel

aircraft and 111,000 personnel including 39,000 conscripts serving a 15-month term. Defence expenditure for all armed forces totalled DM34,200 –£8,964.6 million–in 1978 and, although only about 22 per cent of this was available for equipment purchases, it has enabled the Luftwaffe to continue its policy of re-equiping its entire combat inventory between 1974 and 1983 with Phantoms, Alpha Jets and Tornados. Together with ten of her NATO neighbours, Germany is participating in the Boeing E-3A AWACS project and is providing the second largest financial contribution (after the United States) as well as supplying the main operating base at Geilenkirchen.

Eighteen AWACS aircraft will be delivered to the alliance in Western Europe between 1982 and 1985 to provide early warning of hostile air activity far behind the Iron Curtain and to control interception by NATO air defence forces. At the same time, the Luftwaffe is devoting resources to 'hardening' its bases against the threat of attack by aircraft and surface-to-surface missiles, improving ground-based radars and reporting networks and integrating these with the forthcoming AWACS system. Operational planning has for some years been concerned with requirements of the Luftwaffe for the latter years of this century, and a replacement for the Phantom and ultimately, the Tornado, is now being formulated. Under the project name *Taktische Kampfflugzeug für 1990*, or TKF-90, the Luftwaffe is looking for advanced tactical combat aircraft optimised for air superiority at low and medium levels, and with secondary close-support roles. It is expected that the total Luftwaffe inventory will reduce to 700 aircraft by 1987 and 500 by 1992, and it is therefore vital that the new equipment be as versatile as possible for optimum effectiveness.

The recent British decision to separate VTOL capability from its AST 403 requirement brings this more into line with the TKF-90 project, and discussions are now in hand to assess the possibility of a second-generation Anglo-German combat aircraft as a follow-on to the Tornado. France and the United States present alternative partners for a collaborative venture. Research and development funding of DM1,800 million has been set aside until 1982, and current TKF-90 thinking revolves around an RB 199-powered, fixed-wing-geometry design making extensive use of carbon composite structures. Full all-weather capability, linked to Luftwaffe insistence on multiple tracking and target aquisition systems will produce a highly competent aircraft for the twenty-first century. The TKF-90 will begin replacement of the Phantom from 1992 onwards, by which time the early RF-4Es will have been in service for 20 years–something of a record for front-line aircraft, possibly only equalled by the last few Starfighters withdrawn with the Tornado's entry into service.

Changes brought about by the flexible response concept resulted in a re-structuring of the Luftwaffe chain of command which came into effect on 1 July 1970. The previous system of grouping units according to their geographical location gave way to one of commands dedicated to a single operational role. However, the polarisation of airfields to the

Left: the Lockheed F-104G Starfighter has been the principal West German interceptor and fighter-bomber throughout the 1960s and into the late 1970s. Three F-104G interceptor and strike geschwader have been re-equipped with the McDonnell Douglas F-4F Phantom. An F-104G of Jagdgeschwader 74 is illustrated.
Below left and below: in 1971–72 the McDonnell Douglas RF-4E re-equipped Luftwaffe units AKG 51 and AKG 52, which had previously operated the RF-104G. A multi-sensor reconnaissance aircraft, the RF-4E is a considerable improvement over the RF-104G. RF-4Es of AKG 51 'Immelmann' are illustrated

north and south of the country, with the majority of NATO airfields situated centrally, has perpetuated some regional sub-divisions. Overall control is exercised by the Fuhrungsstab der Luftwaffe (air force high command) and its associated management group – Inspizienten Gruppe – and security group – Flugsicherheit Gruppe – under overall control of the Federal Defence Ministry in Bonn, the Bundesministerium der Verteidigung. Combat units are integrated within the Luftflotten Kommando (air fleet command) and sub-divided into Taktische (tactical) and Luftverteidigungs (air defence) Divisions for the co-ordination of operational planning and training.

Central agencies, such as those for communications, electronics, radar sites, flying safety, transport and training are subordinate to the Luft-

waffenamt, or Air Force Central Office. Third and last element in the command structure is the Unterstutzungskommando (support command) which has no flying units, but is sub-divided into a Materialamt for control of logistics and a Northern and Southern Unterstutzungsgruppe to which supply depots, air force regiment squadrons and military bands are attached. Tactical element of the Luftflotten Kommando (LKdo) comprises the 1 and 3 Luftwaffe Divisionen with some 590 combat aircraft and a further 200 in reserve or stationed in the United States for training purposes, these equipping six fighter-bomber, two tactical reconnaissance, two light ground-attack and two operational conversion wings. Unlike squadrons of the RAF, Luftwaffe *staffeln* do not exist as independent units, but are permanently attached to a *geschwader*, or

wing. As originally formed, the *geschwader* had three squadrons, but this has now been reduced to two. In the Starfighter era, wing complement was 52 aircraft equally divided between the two *staffeln*, each with 18 F-104s operational, five undergoing servicing and three in reserve at any one time. This has now been scaled-down and a present-day Phantom *geschwader* has an overall total of 42 aircraft.

Luftflotten Kommando has two wings of Martin Pershing 1A surface-to-surface missiles, each with four squadrons and 36 launch ramps for nuclear weapons delivery. Flugkorpergeschwader 1 (FKG 1) is stationed at Landsberg and FKG 2 beside the former RAF airfield at Geilenkirchen. Additional nuclear strike capability was previously exercised by five wings of Starfighters, but since 1971 they have relinquished their one megaton tactical nuclear weapons for conventional ground attack in keeping with the flexible response philosophy.

As the spearhead of the German strike force, four Lockheed F-104G Starfighter wings, classified as heavy fighter-bomber units—*jagdbombergeschwader* (JBG)—are stationed on central and southern airfields. At Norvenich is JBG 31 'Boelcke', with JBG 32 at Lechfeld, JBG 33 at Büchel and JBG 34 at Memmingen. Eight two-seat TF-104Gs are attached to each *geschwader* for continuation training and can additionally participate in operational sorties with full armament. In common with all other flying units, each *geschwader* has a flight of four Dornier Do 28D Skyservant communications aircraft.

Since its early days with the Luftwaffe, the Starfighter has been surrounded by controversy. In common with most other NATO countries, Germany decided on the adoption of the F-104 in 1958 in a far-reaching project which envisaged production centres in Canada, Germany, Belgium, Holland and Italy. Based on experience gained during the Korean War, the original design for the Starfighter project was for a high-level interceptor aircraft with good speed and climb rate as its principal features. A prototype was flown in February 1954, followed by a total of 300 F-104As, Bs, Cs and Ds for the US Air Force. In order to meet the requirements of European operation, the design was radically altered to incorporate strengthened fuselage and wing fittings and structure to accommodate the increased stresses imposed by low-level operations. Fin area was increased to permit greater stability for weapon launching at low level and an enlarged braking parachute provided to cater for the shorter West European runways.

By far the greatest innovation was the NASARR (North American Search And Range Radar) housed in the nosecone and incorporating an optional fire-control computer to provide the pilot with positioning data for missile release, and also displaying information in the gunsight for visual attack with Vulcan M61 cannon. These modifications changed the aircraft designation to F-104G Super Starfighter but in daily usage, the 'Super' was soon discarded. The Luftwaffe selected the Starfighter in October 1958 after considering thirteen other types for its requirements, and in February 1959 announced an order for an initial quantity of 66 single-seat aircraft and 30 dual control F-104Fs—the latter to

earlier standards but adequate for the initial pilot training. The first F-104F was handed over the following October and soon afterwards, German officials travelled to Ottawa to discuss joint Starfighter procurement with the Canadians.

Belgium and the Netherlands had similar needs for an advanced interceptor and ground-attack aircraft, and joined the programme soon afterwards. By January 1960, Lockheed was negotiating with Fokker of the Netherlands for a licence production agreement to complement that signed with Messerschmitt a year earlier. The Belgian firm of SABCA was also granted production rights and, in November 1960, Italy entered the programme with a manufacturing licence awarded to Fiat. All three firms produced Starfighters for their own air forces, the Luftwaffe and Marineflieger, Holland contributing 255, Belgium 88 and Italy 50.

To add to Messerschmitt's 210 aircraft, Lockheed eventually produced 30 F-104Fs, 96 F-104Gs and 137 TF-104Gs with the result that only 30 per cent of the German Starfighters were actually assembled and flown in their own country. This offsetting of orders enabled economic production totals to be achieved by all the assembly lines and the cost to Germany was to a large extent retrieved through avionics purchases through German firms. Canadian production went to the RCAF and various smaller air forces, such as those of Denmark, Norway and Greece. Such was the size of the Starfighter enterprise that a management organisation was established in Koblenz to co-ordinate the activities of the 45 companies directly involved. Thanks to an earlier start, Messerschmitt were able to fly their first aircraft in October 1960, followed by the Belgians in August 1961 and the Netherlands the following November. The two latter production lines began deliveries to Germany almost immediately, but arrivals from Italy were delayed whilst Fiat built their first 100 aircraft for the Italian air force.

From an eventual total of 916 Starfighters obtained by Germany, 146 were operated by the Marineflieger despite their earlier preference for the Buccaneer as a strike aircraft for the Baltic area. As deliveries reached their peak between 1963 and 1964, considerable numbers of Starfighters were lost in training and familiarisation exercises, gaining the aircraft notoriety in the German parliament and the world press. Over 200 have now been written off in accidents, but when this high figure is viewed as a percentage of the total purchased, the Starfighter emerges twice as reliable as its immediate predecessor, the Republic F-84F; it also has a better record than similar Lockheed F-104 Starfighters flown by the air forces of the United States, Canada and Italy.

At the height of its service career, the Starfighter was flown by five fighter-bomber, two interceptor, two reconnaissance and two training wings, but the 1979 inventory includes only 300 aircraft, including 30 based in the United States for training and a further 100 TF-104G tandem-seaters of which 25 are also American-based. Critical shortages suffered in the late 1960s were alleviated by the arrival of Phantoms for the reconnaissance squadrons; stripped of their photographic equipment, the surplus RF-104Gs made welcome replacements.

Right: a Dassault-Breguet/Dornier Alpha Jet carries a 30mm Aden cannon belly pack and stores on its wing hardpoints. The Alpha Jet can carry twice the load of the Fiat G 91, which it is replacing in the close air support role. Deliveries of the type began in 1979.
Far right: the Dornier/Bell UH-1D serves with the Heeresflieger and Luftwaffe in the light transport role. HTG 64, the Luftwaffe's helicopter wing, based at Ahlhorn and Landsberg, flies the UH-1D. A re-organisation will result in three wings, each equipped with a mixed complement of Transalls and UH-1Ds. Each of the three Heeresflieger aviation commands (HFK) has two staffeln of UH-1Ds.
Below: a Fiat G 91R-3 light strike fighter pictured in 1979.

All have now been retro-fitted with the J79-MTU-J1K turbojet for increased performance, and their withdrawal is expected to begin in 1981 with the arrival of Tornados at Waffenschule 10, Jever.

The Marineflieger, too, is eagerly awaiting the Tornado, to convert its two Starfighter wings, Marinefliegergeschwader 1 at Schleswig and MFG 2 at Eggebeck, both on the Jutland peninsula. Naval Starfighters are armed with the Aérospatiale AS 30 air-to-surface missile, which is licence-built by MBB in Germany for shipping strike. MFG 1 is receiving the new MBB AS 34 Kormoran of which 350 rounds are on order. MFG 2 aircraft are evenly divided between strike and reconnaissance versions. First Tornado deliveries to the German navy will be to MFG 1 in 1980, whilst MFG 2 will have completed its conversion by November 1984 with the acceptance of the last of 112 examples, including ten training versions.

Unlike the Starfighter, the Tornado is well-suited for low-level operations over the Baltic. This strategically important sea area, like the Mediterranean and Black Sea, restricts Soviet naval movements by reason of its natural bottle-neck through the Skagerrak. As with the Luftwaffe, the naval air arm has lost many Starfighters exercising over the Baltic and has taken over 14 aircraft from surplus air force stocks to maintain its present fleet of 118 fighters and 12 trainer versions.

Luftwaffe plans for the deployment of the Tornado have undergone several revisions since formulation of the original plan to obtain 700 as direct replacements for all Starfighters then in service. Panavia Aircraft GmbH was established in Munich during 1969 with Germany, Britain and Italy as shareholders. Production responsibility was divided between BAC for the nose, rear fuselage and tail surfaces, MBB the centre fuselage, and Aeritalia, the wings, while similar combines were set up to manufacture engine and electronics. Reflecting the varied demands of the three customer air forces, the project became known as the Multi-Role Combat Aircraft or MRCA. For the Luftwaffe, fighter-bomber variants were specified, but anticipated delivery dates in the late 1970s would be too late to begin replacement of the dwindling Starfighter fleet, and so the order was reduced, first to 420 and then 322, including the 112 for the Bundesmarine. The Luftwaffe aircraft will comprise

165 fighter-bombers and 47 training examples, plus two pre-series aircraft. The first German Tornados will enter service with the Tri-National Tornado Training Establishment at RAF Cottesmore in 1980 with the final aircraft scheduled for completion in September 1986, although the project is running several months late. The weapons inventory will include the MBB MW-1 belly-pack dispenser for laterally-ejected anti-tank 'bomblets' and a wide variety of other ordnance.

First of the Luftwaffe wings to retire their Star-fighters was reconnaissance unit Aufklärungsgesch-wader 51 'Immelmann' at Bremgarten and AKG 52 at Leck. They were recipients of a 1968 order for 88 McDonnell RF-4E Phantoms delivered from January 1971 onwards. By May 1972, the last Starfighter had been withdrawn. With two-engined reliability, complex optical and infra-red sensors, the RF-4E represented a significant improvement over its predecessor, 42 being allocated to each *geschwader* with the remaining four employed as trials and maintenance-training aircraft.

Well-satisfied with performance of the Phantom, the Luftwaffe decided on a second batch to replace more Starfighters in the interceptor and strike roles. Half the total of 175 slat-equipped F-4Fs were delivered to the EKdo Tactical Division, where they entered service with JBG 36 at Rheine-Hopsten in April 1975 and the newly-formed JBG 35 (formerly the Fiat G 91 wing LKG 42) at Pferdsfeld. A further ten remained in the United States with the 35 TFW at George Air Force Base where German crews are trained, but were replaced by a supplementary batch of ten F-4Es bought in 1977.

The tactical element of LKdo is completed by two *geschwader* and one training unit equipped with the Fiat G 91R-3 light strike fighter and G 91T-1 tandem-seat trainer. The Luftwaffe took delivery of 394 single-seat 'Ginas' as they are known – of which 50 were the R-4 variant from a cancelled Greek and Turkish order and served only briefly in the training role before being passed on to Portugal. A total of 66 trainers was procured, the final 22 as late additions in 1971–73 to increase output of navigators for the Phantom programme. They were originally segregated in strike and reconnaissance wings, but in 1967 these were pooled to form four Leichten-kampfgeschwader (light strike wings) with one reconnaissance and one strike squadron each, for close battlefield support. These were LKG 41 at Husum, LKG 42 Pferdsfeld, KLG 43 Oldenburg and LKG 44 Leipheim, together with the conversion unit, Waffenschule 50 at Fürstenfeldbrück.

The G 91 was the outcome of the first NATO international fighter specification and is also used by the Italian air force, but despite its two 30 mm internally-mounted cannon and wing-pylons for external stores, its replacement has for some time been a matter of urgency to the Luftwaffe. In July 1975, LKG 44 disbanded, and at the same time LKG 42 converted to the Phantom under the new designation JBG 35, their 'Ginas' entering storage at Leipheim from where 20 were passed on to the Portuguese air force in 1976, leaving 130 in reserve. Selected replacement is the Breguet-Dassault/Dornier Alpha Jet A, a strike-trainer which is already in service with the French and Belgian air forces in an instructional role.

The Alpha Jet, with a low-level speed of 417 knots and twice the load carrying ability of its predecessor, will significantly enhance the support which can be given to units on the battlefield, following arrival of the first 50 production aircraft at Fürstenfeldbrück between February and December 1979. Here, the former Waffenschule 50 has been re-named JBG 49 and will retain a few 'Ginas' for training forward air controllers in battlefield observation. Early in 1980, a further 17 Alpha Jets will be flown out to the Mediterranean island of Sardinia where they will be attached to the Base Flight of the NATO weapons range at Deci-momannu. During 1980, LKG 43 will become

Opposite top: the Luftwaffe acquired three BAC Canberra B Mark 2s in 1966, which are operated by ESt 61, at Manching, for experimental work.
Opposite centre: the multi-role Panavia Tornado is the most important aircraft scheduled to serve with NATO. The Luftwaffe will re-equip JBGs 31, 32, 33 and 34 with the Tornado from 1982, to replace these units F-104Gs.
Opposite lower: since the early 1970s, the C.160 Transall has formed the fixed-wing component of the Luftwaffe's transport command, currently equipping LTGs 61 and 63.
Opposite bottom: the Luftwaffe has used the OV-10B(Z) as a target-tug. It is fitted with a GE J85 jet-booster above the wing.
Below: the first production Dassault-Breguet/Dornier Alpha Jet for the Luftwaffe pictured on display at Erding in May 1978, with a selection of the stores carried by the type.
Bottom: a Fiat G 91R-3 of LKG 43, a light strike unit based at Oldenburg, which has flown the type since 1966. It will shortly re-equip with the Alpha Jet, becoming JBG 43

THE HEERESFLIEGER

LHFTR 10	UH-1 Iroquois	Celle
MHFTR 15	CH-53G	Rheine
LHFTR 20	UH-1 Iroquois	Roth
MHFTR 25	CH-53G	Laupheim
LHFTR 30	UH-1 Iroquois	Fritzlar
HFVS 101	Alouette II	Rheine-Bentlage
HFVS 201	Alouette II	Laupheim
HFVS 301	Alouette II	Niedermendig
HFS 1	Alouette II	Hildesheim
HFS 2	Alouette II	Fritzlar
HFS 3	Alouette II	Rotenburg
HFS 4	Alouette II	Mitterharthausen
HFS 5	Alouette II	Niedermendig
HFB 6	Alouette II/UH-1	Itzehoe
HFS 7	Alouette II	Celle
HFS 8	Alouette II	Oberschlessheim
HFS 10	Alouette II	Neuhausen
HFS 11	Alouette II	Rotenburg
HFS 12	Alouette II	Niederstetten
HFWS	Alouette/UH-1/CH-53	Bückeburg

JBG 43 with 52 Alpha Jets, and the following year will see LKG/JBG 41 follow suit. With four more examples allocated technical training and experimental duties, this will complete delivery of the 175 aircraft so far on order. However, a further 25 are expected to be fitted out for ECM (electronic countermeasures) work. JBG 49 will be tasked with evaluating operational techniques, but it has already been decided that each wing should be divided into three, rather than the normal two, squadrons as 18 aircraft constitutes the optimum size unit for detached operations.

In parallel to the tactical side of LKdo, the air defence division has responsibility for maintaining air superiority over Germany. Those F-4F Phantoms not allocated to strike duties were received by former Starfighter wings Jagdgeschwader 71 'Richthofen' at Wittmundhaven and JG 74 'Mölders' at Neuburg in 1974. Each wing has 42 aircraft now armed with the AIM-9L Sidewinder air-to-air missile, following cancellation of the indigenous Dornier Viper. Further air defence capability is maintained through four Flugabwehrraketenregimenten (anti-aircraft missile regiments), FRR 1, 2, 3

Above: a Westland Sea King Mark 41 operated by MFG 5's first squadron for search and rescue.
Left: the Dassault-Breguet Atlantic is operated by MFG 3 primarily in the ASW role in the Baltic.
Below left: one squadron of MFG 5 operates the Dornier Do 28 Skyservant on communications and utility duties for Marineflieger HQ.
Opposite far left: the Dornier/Bell UH-1D equips LHFTRs 10, 20, and 30 for combat transport of infantry

and 4, each with six batteries of nine launch ramps and 144 Nike-Hercules missiles on establishment. Two more units, FRR 13 and 14 have a total of 36 batteries, each with six launch ramps for the Hawk surface-to-air missile. Germany is participating in the Hawk Improvement Programme which will extend the useful lives of these missiles by 10–15 years and at the same time improve performance, but has also been looking into the question of a Nike-Hercules replacement and considering introduction of the Raytheon Patriot in 1983. Aircraft and missiles are linked to a network of defence radars operated by eight Fernmelderregimenten comprising two Type A control units, FmRgt 11 and 12; four subsidiary, Type B units FmRgt 31, 32, 33 and 37; and two ECM control stations, FmRgt 71 and 72.

Overriding priority so far as the army is concerned is anti-armour capability. The Warsaw Pact countries now have a superiority in numbers of battle tanks in the ratio of 1:2.8 to NATO's disadvantage and both on the ground and in the air, anti-tank missiles such as the Euromissile Hot, Hughes BGM-71 Tow, MBB Cobra and Mamba and Euromissile Milan are being supplied in increasing numbers, while specialist combat helicopters are on the drawing board for delivery in six or seven years' time. The army's battle plans are based on the use of highly-mobile helicopter units with a primary anti-tank and transport role and Heeresflieger squadrons are also tasked with providing heavy assault lift for other allied forces, notably elements of the US and British armies. Now entirely a rotary-wing force, army aviation has some 530 helicopters and a further 300 on order for imminent delivery. Reorganisation in 1970 placed greater emphasis on air mobility through the use of helicopters, and modified the former system whereby each of the three army corps and 12 divisions had *flieger-battalionen*, (air battalions) attached at their respective levels. Further changes in 1972 established the present deployment in which there are now 16 armoured brigades, 12 armoured infantry brigades, three light infantry brigades, two mountain brigades and three airborne brigades with an overall total of 336,200 troops. Of these, about 3,000, including 750 pilots, serve the Heeresflieger. Plans for further reinforcement were announced late in 1978, involving the formation of three brigades and five tank battalions.

Three types of helicopter serve the Heeresflieger, of which the VFW-Fokker/Sikorsky CH-53G is the largest. From July 1972 onwards, 110 of these medium-lift aircraft were distributed between the three army corps which received 32 each, with the remainder used for training. They gave each corps the ability to transport a mobile battalion of up to 1,000 men in a single airlift. Light transport is the domain of the Dornier/Bell UH-1D Iroquois, the army receiving 204 licence-built examples between 1968 and 1970. Like their larger brothers the 'Hueys', as they are colloquially known, they are attached to

the corps and are capable of supplying the daily needs of a brigade in the field, again in a single airlift.

Most numerous of the Army's helicopter trio is the Aérospatiale Alouette II and its higher-powered equivalent, the Alouette Astazou. Entering service in 1958, 226 Artouste-engined Alouette IIs were distributed to units at corps and divisional level, and were later joined by the survivors of an additional 21 used as trainers by the Luftwaffe. From 1968, they were augmented by a further 54 examples with the Astazou engine, but about half will be retired in the next few years with the arrival of the MBB Bo 105.

Attachment of flying units to the Army order of battle begins at corps level, each having a Heeres-fliegerkommando (HFK) or army aviation command, comprising HFK 1 at Handorf, HFK 2 at Laupheim and HFK 3 at Neidermendig. Each HFK comprises an HQ company and supporting and administrative elements plus three main aviation formations. The smallest of these is the single Verbindungsstaffel (HFVS–army aviation communications squadron) attached to each HFK and comprising an establishment of 12 Alouette IIs. As with all army aviation units, their designation denotes their attachment to I, II or III Korps, and thus HFVS 101 at Rheine-Bentlage is in HFK 1, HFVS 201 at Laupheim in HFK 2 and HFVS 301, Neidermendig, in HFK 3.

The main Heeresflieger battlefield elements are two helicopter regiments within each HFK, these comprising two light helicopter squadrons (*leichtes heeresfliegertransportstaffeln*) which make up an LHFT Regiment of 40 UH-1D Iroquois and two *mittleres heeresfliegertransportstaffeln*, or medium helicopter squadrons, combining to form an MHFT Regiment with a total of 32 CH-53Gs. In addition to these three formations, the corps aviation element also includes associated training and repair units. Ten of the 12 divisions attached to the three corps have a flight of ten Alouette IIs in a *heeresflieger-staffel* (HFS) for communications, general staff duties, observation and liaison. All HFS units were re-designated from battalions in 1972 when their complement was reduced from 15 to 10 light observation helicopters, but HFB 6 at Hungriger Wolf (Itzehoe) retains its former title by reason of an oversize fleet of 20 UH-1s.

Experimental flying is undertaken by the Heeres-flieger Versuchsstaffel at Celle which evaluated the MBB Bo 105 helicopter in the anti-tank role between September 1973 and January 1977. Two versions of the Bo 105 will enter service from September 1979: the Bo 105P Panzerabwehrhub-schrauber 1, or first anti-armour helicopter, with deliveries at the rate of six per month from a total of 212 on order; and the Bo 105M communications and observation helicopter at the lower priority of two per month from 100 on order for partial replace-

flight of seven for detachment to forward areas as demanded by a tactical situation. Availability of the Bo 105M will permit the attachment of small light-helicopter flights to most units, and in future, each corps will have at its disposal a new HQ flight of three VBH (communication/observation) helicopters to be increased to 15 on mobilisation. The HFVS will have 12 VBHs, the PAR 56 PAH-1s and five VBHs, the LHFTR 48 UH-1Ds and five VBHs, and the MHFTR 32 CH-53Gs and five VBHs.

HFS units at divisional level will have one or two light helicopters for battlefield observation and liaison, two reconnaissance helicopters linked to airmobile operation, four to five with missiles for anti-tank work, two or three carrying anti-tank ground teams, and one or two with infantry and ground radio teams. The upper-limit figures will be achieved by converting 127 PAH-1s to normal standard when the PAH-2 becomes available. The independent HFB 6 will comprise 15 communications, 21 PAH-1 and 24 UH-1 helicopters.

Agreement has been reached between the German and French governments for development of their respective aircraft industries, with design leadership vested in Germany. The helicopter will be specifically tailored for battlefield operation, but some points of conflict between the PAH-2 specification and the French Hélicoptère Anti-Char requirement still remain to be resolved. France favours a more sophisticated approach, incorporat-

The F-104Gs of the Marineflieger's MFGs 1 and 2 form the front-line of defence in the crucial Baltic area. In war it would be essential to prevent the Soviet fleet breaking out of the Baltic into the North Sea. Based on the Jutland peninsula, MFG 1 comprises two squadrons of F-104Gs operating in the attack role, but MFG 2 comprises one attack squadron with F-104Gs, two of which are illustrated and one squadron of reconnaissance RF-104Gs. MFG 1's F-104Gs are armed with the MBB/Aérospatiale AS 30, but MFG 2's F-104Gs have the more effective MBB AS 34 Kormoran anti-shipping missile, specifically designed for high-performance aircraft

ment of the Alouette II with corps and certain divisional units.

Intended as an interim counter to the Warsaw Pact armoured superiority, the PAH-1 will be replaced in service from 1986 onwards by the purpose-built PAH-2 helicopter gunship. Armed with six Euromissile Hot anti-tank missiles on outriggers, the PAH-1 is restricted to day and visual operations, and by October 1982 will be in service with three corps *panzerabwehrregimenten* (PAR) located at the bases of the present UH-1D regiments which will then be re-deployed to Fassberg (LHFTR 10), Neuhausen (LHFTR 20) and Niederstetten (LHFTR 30).

The PAR will comprise 56 PAH-1s and an observation and communications flight of five Bo 105Ms, these to be deployed in a *schwarm*, or

ing a retractable undercarriage for optimum high-speed performance, whilst Germany would prefer a more sturdy fixed gear. Additionally, there is disagreement over the FLIR night-vision system; Germany would prefer off-the-shelf US equipment despite French belief in the ability of European industry to develop a comparative system before the PAH-2/HAC enters service.

Both countries will arm their helicopters with eight advanced versions of the Euromissile Hot anti-tank missile with a new smokeless propellant, and France has additionally specified a large calibre gun, which is not presently required by Germany. With twin turbine power plants and integrated dynamic systems, the PAH-2, of which 210 are to be ordered, will have a tandem seating arrangement within a fuselage of minimal cross-section.

Within Germany itself, there are disagreements over replacement of the UH-1Ds of the Heeresflieger and the Luftwaffe. The 'Hueys' will be due for retirement by the late 1980s, and the army currently favours a replacement type with double the payload, placing it in the 'Ten Tonne' class. Faced with this upgrading of army requirements, the Luftwaffe has expressed a preference for about 25 'Ten Tonne' helicopters with the balance of the order directed towards purchase of a 'Six Tonne' type, as originally specified by the army, of which some 65 would be needed. The new Westland WG 30 may be a contender for this latter requirement, but NATO agreements on new helicopter projects have allocated this weight bracket to Aérospatiale, which has recently completed design studies for such an aircraft.

Army missile forces include 11 battalions with 65 Honest John launchers and four battalions in the process of receiving 175 Vought MGM-52C Lance and 26 launchers in the role of corps artillery. Target pinpointing is undertaken by Canadair AN/USD 501 (CL-89) drones with infra-red line-scan, operated at divisional level. Surface-to-air missiles, in the form of 1,400 General Dynamics Redeyes, are issued to infantry units, and airfield defence will be greatly strengthened with the delivery of 5,240 Euromissile Roland 2s now on order.

Further helicopter procurement for the Bundesmarine centres round the recently concluded contract for 12 Westland Lynx for deployment aboard six new Type 122 frigates presently under construction. Traditionally, Germany has not operated aircraft or helicopters at sea, and its last aircraft-carrier, the *Graf Spee* spent the greater part of World War II in various dockyards. The German navy's Type 122 vessels will each have eight McDonnell Douglas RGM-84A Harpoon ship-to-ship missile launchers, for which a total of 142 rounds is on order, whilst for air defence, 96 Raytheon RIM-7H Sea Sparrows will be included in the armament. Three destroyers have a single General Dynamics RIM-24 Tartar anti-aircraft missile launcher, and 34 patrol boats have four MM-38 Exocet anti-ship missiles each.

Currently the only helicopter in use with the Marineflieger, the Westland/Sikorsky Sea King HAS Mark 41 is used in its search and rescue (SAR) role, rather than as an anti-submarine aircraft. The first of 22 Sea Kings was delivered to Marinefliegergeschwader 5 at Kiel-Holtenau in March 1974, after initial crew training had been completed by the Royal Navy at Culdrose, Cornwall. At Kiel the Sea King replaced eight Grumman Albatross amphibians and a flight of Sikorsky SH-34Gs, and assumed responsibility for SAR over the German area of the Baltic, in conjunction with similar helicopters of the Danish forces. MFG 5 is a dual-function wing, its second squadron operating 20 Dornier Do 28D-2 Skyservants as communications and utility transports in support of the Marine-Flieger HQ, which is also at Kiel. Under the operational command of Flotillenadmiral Deckert, the Marine-fliegerdivision is responsible for the administration of the four wings, comprising 6,000 men and 190 aircraft, which together with the Lehrganggruppe (ground training school) at Westerland-Sylt constitute the German naval aviation element.

As the remaining naval air wing, the Atlantic-equipped MFG 3 *Graf Zeppelin* at Nordholz replaced its Gannets with two squadrons totalling 20 of the Franco-German anti-submarine patrol aircraft in 1965–66. One further prototype example was later obtained from France for static instruction of technicians at the Lehrganggruppe. In 1977, five Atlantics were modified by Ling-Temco-Vought for electronic countermeasures and intelligence gathering over the Baltic, eavesdropping on communications and radar signals behind the Iron Curtain from international waters. One of the latter was lost in 1978, but the remaining 15, more conventionally-equipped aircraft continue to operate on NATO ASW tasks armed with AS 20 air-to-surface missiles, although the older AS 12 is used for training. Proposals to replace the Atlantic force with 15 Lockheed S-3A Vikings in 1979 were abandoned due to the high cost of the project. Instead, Dornier will upgrade the aircraft's reconnaissance and detection systems with avionics from the United States.

The vitally important functions of transport and training for the Luftwaffe (and to a lesser extent, the Marineflieger and Heeresflieger) are allocated to the Luftwaffenamt, the second major air force command. The Luftwaffenant also administers the Sammdienstelle, or service records branch at Cologne, the Amt für Wehrgeophysik (cartographic section) at Porz-Wahn and the Luftwaffe Fuhrungsdienstkommando (planning research command). The last-mentioned has one attached flying unit, the Fernmelde Lehr und Versuchsstaffel 61, otherwise referred to as the Flugvermessungsstaffel and based at Lechfeld for the calibration of ground radio-aids. Previously maintaining a fleet of 20 Pembrokes and Dakotas, the FmLVs now operates Hansa Jets and Do 28D Skyservants.

The Marineflieger is scheduled to replace its F-104G Starfighters with Panavia Tornadoes from 1980. Marinefliegergeschwader 1 will be the first front-line unit of any air arm to operate the Tornado and its sister geschwader, MFG 2, is due to complete its conversion onto the type before the end of 1984

Transport aircraft for the Luftwaffe are provided by the Lufttransportkommando, established at Munster in 1968, with a current complement of some 200 aircraft and 160 helicopters. Air Transport Command underwent a complete re-equipment programme between 1968 and 1971 involving the purchase of 110 Transall C160 aircraft and 132 Dornier/Bell UH-1D helicopters, although revised estimates rendered 20 Transalls surplus to requirements. These were transferred to Turkey in 1972.

The remaining Transalls are allocated to two 32-aircraft wings: LTG 61 at Landsberg and LTG 63 at Hohn, each of which is sub-divided into two *staffeln*. HTG 64 at Ahlhorn has four *staffeln* of UH-1Ds together with flights of three helicopters at Bremgarten, Hopsten, Jever, Neuberg, Norvenich and Pferdsfeld for SAR duties.

The Luftwaffe is shortly to implement a revised plan of Transall and UH-1D allocations in which three wings will each have a mixture of aircraft and helicopters to provide a better-integrated transport force. A third Transall wing will be formed at Ahlhorn, previous base of the former Noratlas wing LTG 62, to receive 12 aircraft each from LTG 61 and LTG 63. This new wing will absorb two *staffeln* of HTG 64, amounting to some 50 UH-1Ds, whilst the remaining two *staffeln* will be distributed between the other two wings. The result of this revision will be LTG 61 and LTG 63 with 24 Transall C160s and 25 UH-1Ds each, and the re-formed LTG 62

with 4 aircraft and 50 helicopters.

The new plan also envisages an expansion in light transport capability and the consequent purchase of more Skyservants, the majority of the Luftwaffe fleet of 101 being distributed in flights of four to each major flying wing. A further requirement exists for 20 light helicopters, for which the MBB Bo 105 would be an obvious choice, but in view of the already divergent plans for a UH-1D replacement, some rationalisation of procurement may be imposed on the Luftwaffe. Lufttransportkommando maintains a VIP flight, the Flugbereitschaftstaffel, or special air missions squadron at Cologne/Bonn Airport with a mixed fleet of aircraft and helicopters. Lufttransportkommando operates its own OCU, Flugzeugführerschule 'S', at Wunsdorf with 15 Transall C160s and four Skyservants as the remaining element of a much larger training organisation with a light aircraft section at Diepholz and helicopter school at Fassberg.

The Luftwaffe Ausbildungskommando, training command, has similarly undergone contraction, principally as a result of the decision to undertake basic and most advanced flying training in the United States. Under the previous system, pilots were graded at Flugzeugführerschule 'C' (FFS 'C') at Diepholz on the Piper L-18C Super Cub, then receiving basic instruction on North American T-6 Harvards and Fouga Magisters at FFS 'A', based at Landsberg, and advanced training at

Although more commonly employed in an anti-submarine role, the Westland/Sikorsky Sea King fulfils the search and rescue requirement in Bundesmarine service. The type operates from Kiel and, flown by pilots of Marinefliegergeschwader 5, combines with Danish aircraft to safeguard Baltic shipping

Fürstenfeldbrück on T-33As of FFS 'B'. Restrictions imposed by weather and controlled airspace caused FFSs 'A' and 'B' to be disbanded.

FFS 'C' became the Fluganwaerterregiment at Uetersen, and later Neubiberg, and is now a section of JBG 49 (the former Waffenschule 50) at Fürstenfeldbrück with 45 Piaggio P.149Ds for grading and primary training.

With primary flying skill obtained in Germany, pilot cadets intended for high-performance flying progress to 1 Ausbildungsgruppe USA or the 3525 Pilot Training Wing, USAF, at Williams Air Force Base, Arizona. Here they undergo 132 hours basic training on the T-37B, followed by 130 hours advanced instruction on the T-38A Talon. Thereafter, Starfighter pilots transfer to 2 Ausbildungsgruppe, the 4510 Combat Crew Training Wing CCTW, USAF, at Luke Air Force Base, Arizona to complete their course with 125 hours.

Opened in April 1964, the 4510 CCTW reached a peak level of 75 F-104G and 37 TF-104G aircraft in use, one course passing-out every six weeks. This has now been scaled-down to 30 and 25 aircraft respectively. While in the United States, the Starfighters fly in USAF colours. Returning to Germany, newly qualified pilots undergo 30–40 hours flying at Waffenschule 10, Jever, on some 50 TF-104G and 12 F-104G Starfighters for acclimatisation and familiarisation with European operational procedures before joining their first squadron.

Pilots destined for the Alpha Jet are posted directly from Williams AFB to JBG 49 at Fürstenfeldbrück for 40 hours acclimitisation and a further 80 hours tactical training. The unit also trains navigators for the Alpha Jet and Phantom, and will continue to fly a small number of G 91s for the training of forward air controllers when it is fully-equipped with the newer aircraft. Conversion of Phantom pilots is undertaken by the USAF within the 35th Tactical Fighter Wing at George Air Force Base on ten German-owned F-4Es.

The tri-national Panavia Tornado is scheduled to play a major part in the inventories of both the Luftwaffe and Bundesmarine in the 1980s. The latter service will employ the type as a platform for the MBB Kormoran anti-ship missile, four of which are carried by the pictured Tornado.
Designed for use in the Baltic, where islands abound, the Kormoran has necessarily a short range, but is nonetheless a potent weapon. It was previously employed by the Lockheed F-104 Starfighters of Marineflieger-geschwader 1, which will convert to the Tornado in 1980

Transport pilots are selected at the primary stage of training, and remain in Europe with the Lufthansa civilian airline school at Bremen, after which pilots complete their training at FFS 'S' with 65 hours Transall C160 flying to reach co-pilot standard. Marineflieger flying training is integrated with the Luftwaffe courses, with the exception of prospective Atlantic pilots, who transfer from the Lufthansa school to MFG 3 to complete type conversion within the operational unit.

Army helicopter trainees undergo 15 hours selection at Fürstenfeldbrück, rather than the normal 25 before being posted to the Heeresflieger-waffenschule at Buckeburg for an initial 100 hours conversion to the Alouette II. The HFWS took over the 19 remaining Luftwaffe Alouettes of the Hubschrauberführerschule when this closed in 1974 and now provides basic helicopter training for the three services. Light-helicopter pilots complete their course with a further 65 hours tactical training on the same type, also at the HFWS, but Marine-flieger candidates transfer to MFG 5 for 50 hours conversion to the Sea King. With selection of the Lynx for the Type 122 frigates, conversion of the first crews, at least, will probably be undertaken with the Royal Navy at Yeovilton. Pilots for the UH-1D progress from the 100 hour Alouette course to 50 hours UH-1D conversion, followed by 75 hours of tactical training and a further 75 hours instrument instruction on the same type prior to qualification. Conversion to the CH-53G may also follow at the HFWS after experience in an operational unit.

Combined services target facilities are provided from Lübeck by the civilian-manned Schiess-platzstaffel (firing-range squadron) with 18 OV-10 Broncos from the United States, and 12 ex-Luftwaffe Fiat G 91s. This unit additionally operates the last three surviving German Noratlas transports and the three Canberras obtained in 1966 for trials and experimental work.

The second major armament practice camp is situated in an area blessed with more favourable weather conditions than the north of Germany, at the Sardinian base of Decimomannu. Universally referred to as 'Decci', the aerodrome is regularly visited by several NATO air forces and has a small flight of T-33As maintained by the resident Italian air force. As principal user, the Luftwaffe had a permanent detachment of Starfighters for range observation duties attached to the permanent German headquarters at 'Decci', the Ubungsplatz-kommando. The allocation was changed in 1977 to ten Fiat G 91s and two Starfighters, and these are shortly to be exchanged for Alpha Jets.

In the foreseeable future, the Luftwaffe and the air arms of the Army and Navy will continue to maintain a strong defensive posture against the Warsaw Pact forces in close co-operation with NATO. In times of peace, the finances available for armaments are not unlimited, but the strong economic position of Germany has allowed both first- and second-line units to be equipped with modern and effective aircraft types. Only the Fiat G 91 is overdue for replacement, but the production of six Alpha Jets per month from the German assembly lines will rapidly transform the situation and, by 1986, the front line will consist of Tornadoes, Alpha Jets and the slightly older, but still effective, Phantom.

CHAPTER SEVEN

NATO
Belgium/Canada/Denmark/Netherlands/Norway

After two decades of service, the ubiquitous Lockheed Starfighter will shortly be replaced by a new strike and interceptor aircraft to maintain NATO's front-line combat force. Four European nations, Belgium, Denmark, the Netherlands and Norway have already decided upon the General Dynamics F-16, which is now beginning to roll off the production lines in Holland and Belgium. However, their transatlantic partner, Canada, is still engaged in lengthy deliberations to select an aircraft to fulfill the NFA (New Fighter Aircraft) specification, although the F-16 is one of the two types which have reached the competition finals.

When the question of a replacement for the Starfighter arose, it was clear that a suitable American aircraft planned for large-scale production would be economical, insofar as development costs could be shared with the US forces. The USAF was at that time sponsoring the F-16 and its rival from the Northrop stable, the F-17, and prototypes of each were flown in January and June 1974 respectively, ready to embark on a programme of competitive evaluation.

By January 1975, however, the F-16 had emerged as the victor for the USAF, work beginning on a batch of six single-seat F-16As and two dual-control 'B' models, followed shortly afterwards by the announcement that the American air force would buy an initial batch of 650. Despite fierce opposition from the Viggen, Mirage F1 and F-17, the four European nations signed a Memorandum of Understanding with the United States in June 1975, calling for the licence-production of 348

Left: the two-seat Mirage Dassault 5BD trainer was one of three licence-built variants of the type to be delivered to the Belgian air force, equipping two wings. The Mirage 5 was a low-cost, clear-weather development of the earlier Mirage III and, despite its limitations, has performed well.
Below: in common with so many European air arms, the Belgian air force purchased the Lockheed F-104 Starfighter in quantity in the early 1960s

BELGIUM

Unit	Aircraft type	Base
1 Sq	Mirage 5BA	Bierset
2 Sq	Mirage 5BA	Florennes
7 Sq	Magister	Brustem
8 Sq	Mirage 5BD/BA	Bierset
9 Sq	Magister	Brustem
11 Sq	T-33A	Brustem
16 Sq	Alouette/Islander	Butzweilerhof
17 Sq	Alouette II	Werl
18 Sq	Alouette II	Merzbruck
20 Sq	Hercules	Melsbroek
21 Sq	Merlin/BAe748/Mystère 20	Melsbroek
23 Sq	Starfighter	Kleine Brogel
31 Sq	Starfighter	Kleine Brogel
40 Sq	Sea King/Alouette III	Koksijde
42 Sq	Mirage 5BR	Florennes
349 Sq	F-16A/B	Beauvechain
350 Sq	Starfighter	Beauvechain
EAL/SLV	Alouette/Islander	Brasschaat
EPE/EVS	SF-260	Gossoncourt

F-16s and offset agreements to produce equipment and components for all aircraft built on both sides of the Atlantic.

At Gosselies, in Belgium, the first European aircraft, an F-16B for the Belgian air force, made its initial flight in December 1978, and SABCA is now building a total of 174 to be shared with Denmark. A further 174 will come from Fokker at Schiphol to meet Dutch and Norwegian contracts. As with most high-cost projects, the F-16 has already been criticised on estimates that Norway and Denmark will not receive their promised share of offset contracts. Further doubt has been expressed concerning the aircraft's all-weather performance and fighting capability, particularly with regard to its avionics. Fitted with the appropriate 'black boxes'

to accommodate all eventualities, the perfect fighter would be inordinately expensive and, to quote the hyperbole of the exasperated designer, too heavy to leave the ground.

In Canada, the NFA deliberations, which must also take account of the need to replace the McDonnell Douglas Voodoo, continue to drag on. The ultimate selection will be influenced by the number of aircraft which can be obtained within the ceiling project cost of $C 2,340 million, rather than ability or industrial offset agreements. This figure will be corrected for inflation since November 1977, and will allow the Canadians to obtain 142 F-16s or 127 F-18s, with overall requirements quoted as 130-150 examples, of which 54 will serve in Europe with a further 24 allocated to reinforcement of NATO's northern flank in Norway, 36 for home air defence and the balance as reserves and attrition replacements.

A decision date has been postponed on several occasions and is currently fixed for December 1979 with deliveries to begin in 1982/83 for completion in 1987. This has forced the CAF to embark reluctantly on a costly programme of life-extension for their Starfighters, while further consternation resulted from a proposal by the newly-elected government to buy Iran's 78 surplus F-14s, plus an additional quantity of F-18s for use in Europe, as a mixed-purchase alternative.

Arab objections to Canadian relationships with Israel brought F-14A negotiations to an abrupt conclusion, and the Ottawa government now appears to favour the F-18A despite its higher cost. Whichever type is eventually chosen, it will take its place alongside the aircraft of the smaller NATO nations, Belgium, Denmark, the Netherlands and Norway, in maintaining the strength of the alliance into the twenty-first century.

With headquarters at Caserne Prince Baudouin in Brussels, the Belgian air force operates some 270 aircraft and five helicopters supported by an army

aviation component of 70 helicopters and 12 fixed-wing types, and a flight of three naval helicopters. Personnel strength amounts to 18,500 of all ranks manning four combat wings, a transport wing, plus training and support units. The last-mentioned include two surface-to-air missile wings with seven squadrons of Nike-Hercules, assisted by two army Hawk battalions with 24 launchers. With the exception of one transport squadron, all units are assigned to NATO.

The dual language situation existing in Belgium requires most military formations to operate under joint names. A squadron may be referred to as a *smaldeel* (Flemish) or an *escadrille* (French), but the English term 'wing' is used to designate groups of two or more squadrons in both languages. The air force is known as the Force Aérienne Belge/Belgische Luchtmacht (FAB/BLu).

Belgium replaced a considerable proportion of its air force inventory in the first half of the decade, re-equipping the entire transport force and converting reconnaissance and some strike squadrons to the Mirage 5. Remaining strike and interceptor units are preparing to accept the F-16 during the next few years, while the training programme is currently being revised following the recent introduction of the Alpha Jet.

Combat formations are administered by the Tactical Air Force Command (Force Aérienne Tactique/Groepering Tactische Luchtmacht) from headquarters at Evere, and form part of the Anglo-Belgian-Dutch-German Second Allied Tactical Air Force under control of No 1 Tactical Operations Centre. For national and regional air defence, 1 All-Weather Fighter Wing (1ère Wing de Chasse Tous-Temps/1 Wing Alle-Weder) at Beauvechain comprises two squadrons, (Nos 349 and 350) in the early stages of converting from the F-104G Starfighter to the F-16.

Between 1963 and 1965, Belgium received 100 single-seat Starfighters and 12 TF-104G dual-

control trainers, divided between air defence and tactical strike roles, of which 33 have been lost in accidents to date. During 1978, No 1 Wing was considerably reduced in strength in anticipation of F-16 deliveries, the first of which, a two-seat F-16B, was received by No 349 Squadron on 29 January 1979. The air force will acquire 104 F-16A and 12 F-16B aircraft, and has options on a further 14, all of which will be produced by SABCA. By November 1980, No 349 Squadron will be fully operational on its new mounts, and during the period July 1980 to July 1981, No 350 Squadron will undergo a similar transformation.

Following the re-equipment of No 1 Wing, remaining F-16 deliveries will be directed to the second Starfighter unit, 10 Wing de Chasseur-Bombardier

Above: the Dassault Mirage 5BA is flown in the fighter-bomber role by No 1 Squadron and No 2 Squadron of the Belgian air force. The two units form part of NATO's Second Allied Tactical Air Force, which will operate over the northern part of the Central Front in time of war. The Mirage 5BA, of which 49 of the original buy of 63 remained in service in 1979, replaced the venerable Republic F-84F in Belgian service at the beginning of the 1970s. Below: the Mirages of No 2 Squadron carry the famous comet insignia, which dates back to World War I

Left: Les Diables Rouges aerobatic team fly the Fouga Magister, which has served as the Belgian air force's basic trainer since 1960. By the spring of 1980, however, the Magisters will have been replaced by Alpha Jets, which are also supplanting the Lockheed T-33A advanced trainers. Below left: the Lockheed C-130H is the standard tactical transport in Belgian service, 12 serving with No 20 Squadron at Melsbroek. Right: three Aérospatiale SA330H Puma transport helicopters serve with Belgium's Gendarmerie. They are based at Brasschaat, where they are maintained by the army air corps. Below: The Sikorsky S-58 helicopters of No 40 Squadron were superseded by the Westland Sea King for SAR

based at Kleine Brogel in the fighter-bomber role. Comprising Nos 23 and 31 Squadrons, the wing is equipped to fulfill NATO nuclear strike commitments in addition to conventional close-support work, and these capabilities will be perpetuated by its successor. Final F-16 deliveries are scheduled for early 1984, each of the four squadrons having 18 aircraft, with the remainder allocated to reserve or routine servicing.

Further strike missions are undertaken by the Mirage 5 which entered service in 1971 as a successor to the Republic F-84F and RF-84F. Despite Dutch hopes of a joint agreement to buy the F-5 and thus perpetuate the combined training syllabus, Belgium signed an initial order for 88 Mirages in August 1968, and subsequently increased this total to 106, all except three of which were licence-produced by the home industry. Principal of the three versions delivered, 63 Mirage 5BA strike aircraft were issued to two squadrons, together with 27 Mirage 5BR tactical reconnaissance variants and 16 Mirage 5BD twin-seat trainers to equip a further two units.

Mirages are flown by two dual-purpose wings. At Florennes, No 2 Wing Tactique consists of No 2 Squadron with the Mirage 5BA and No 42 Squadron in the reconnaissance role flying the Mirage 5BR. No 3 Tactical Wing at Bierset comprises No 1 Squadron (Mirage 5BA) and No 8 Squadron, an operational conversion unit having 12 Mirage 5BRs and six BAs, and providing 75 hour transition courses for newly-qualified pilots. More experienced aircrew may subsequently convert to the Starfighter via the TF-104G conversion flight at Kleine Brogel. Pending the arrival of the F-16, 60 Lorel Rapport 2 ECM (electronic countermeasures) pods have been fitted to Mirages to improve their survivability in a hostile environment.

Day interception is performed by two wings of Nike-Hercules surface-to-air missiles, totalling seven squadrons, each with 16 launchers. Established in February 1962, No 9 Wing d'Engins Téléguides Sol-Air/9 Wing Telegeleide Tuigen Grond-Lucht has its headquarters at Grefrath and administers No 53 Squadron at Kaster, No 55 Squadron at Kapellen-Erft, and Nos 54 and 56 Squadron at Grefrath. First to form, however, was No 13 Wing in May 1959, this controlling No 50 Squadron at the headquarters base of Düren, No 51 Squadron at Blankenheim and No 52 Squadron at

Euskirchen. Belgium additionally operates two warning and control stations in the NADGE chain, at Glons and Zemerzaeke, the latter recently-equipped with a new 3-D GE592 solid-state phased-array early warning radar as part of a wide-ranging NATO programme of updating radar and control installations.

Tactical transport and communications aircraft are concentrated within No 15 Wing at Melsbroek, the military portion of Brussels-Zaventem international airport. Between July 1972 and March of the following year, 12 C-130H Hercules transports were received by No 20 Squadron to replace C-119G Packets operated by Nos 20 and 40 Squadrons. These were followed by the progressive withdrawal of the older Dakotas, Pembrokes and DC-6s of No 21 Squadron in favour of more modern equipment. First to arrive, in the early spring of 1973, were two Dassault Falcon 20s for VIP communications, later augmented by six Swearingen Merlin 3As in March-October 1976 and three HS748 tactical transports during the following summer.

Acting as Pembroke replacements, the two final Merlins were equipped for photographic survey, but returned temporarily to the manufacturers with

Above and right: all the Lockheed CF-104 Starfighters which survive in Canadian service are stationed in Europe, with the exception of those of No 417 Squadron, which acts as an OCU

stability problems. At the same time, two Boeing 727s were obtained from the national airline SABENA for transportation of the Belgian contingent of the Allied Mobile Force of NATO to Norway or Turkey, and support detachments at the weapons ranges of Solenzara (Corsica) and Namfi (Crete). In mid-1978, however, both were leased to a civilián operator, but remain on call for air force operations. After precisely 30 years of service, the last Dakota was withdrawn from service late in 1976, completing a re-equipment programme totalling $25 million exclusive of the Falcon purchase.

Last of the units under Tactical Air Force control is the joint-service SAR flight at the coastal base of Koksijde, which adopted the title No 40 Squadron in 1974. The final survivors of 11 Sikorsky S-58 helicopters were retired in 1978 and, of these, two were attached to the navy (Force Navale/Zeemacht), and two allocated to the support of Nos 9 and 13 Missile Wings in Germany. No 40 Squadron has now standardised on the Westland Sea King Mk 48, five of which were delivered to the Royal Navy training flight at Culdrose, England, in mid-1976 prior to taking up station at Koksijde for a significant reinforcement of search and rescue capability. A second component of the unit is the Naval Flight, which received three Alouette IIIs in 1971 to operate from the support ships *Zinnia* and *Godetia*.

Personnel training is managed by the Groupement de l'Instruction et de l'Entrainement at Evere, incorporating flying schools, the Technical School at Saffraanberg and Air Force Regiment School at Koksijde. Flying training begins with 125 hours on the SIAI-Marchetti SF 260MB high performance primary trainers of the Ecole de Pilotage Elementaire at Gossoncourt. Between late 1969 and early 1971, 36 SF-260s were delivered to Belgium as replacements for the SV-4 biplane, and three – one

of which is now a simulator – have so far been lost in accidents. The aerobatic tradition of the SV-4 'Penguins' team, renowned for their mirror-flying in the 1960s, has been maintained through the formation of the 'Swallows' team of three SF-260s in 1972 and their subsequent appearance at many international air displays.

From Gossoncourt, student pilots progress to Brustem, home of the Basic and Advanced Flying Schools. The former course, lasting 125 hours, is now transferring to the Alpha Jet, of which Belguim has 33 on order as replacements for the Fouga Magister. A total of 45 Magisters was delivered in 1960–61, followed by five attrition replacements from ex-Luftwaffe stocks in 1969. In June 1971, these adopted the 'shadow' identities of Nos 7 and 9 Squadrons, which had previously been operational Hunter units. The first indigenously-produced Alpha Jet was delivered on 14 December 1978, the day following the arrival of the French-built first production model. Five more had been accepted by the middle of 1979, with production at Gosselies now increasing to two aircraft per month. Alpha Jets will also replace the T-33As of the Advanced Flying School, otherwise No 11 Squadron, on which 100 hours are currently flown prior to qualification. Of 38 T-33s received from 1952 onwards, some 24 remained in mid-1979, although only half were in active service.

Aviation elements of the army (Force Terrestre/Landmacht) gained independence from the air force in May 1954 under the Army Light Aviation (Aviation Légère/Licht Vliegwezen) and are employed for liaison and artillery observation with mechanised and armoured brigades assigned to NATO. From 1959 onwards, Belgium received an eventual total of 90 Alouette II helicopters, the final 48 of which are powered by the uprated Astazou engine. Six of the latter model were

Above: although long retired from first-line service by the United States, the McDonnell Voodoo will continue to provide a substantial portion of Canada's interceptor capability into the 1980s. The type is the only Canadian-operated aircraft to carry nuclear missiles, these being the Douglas MB-1 Genie; Hughes AIM-4D conventional weapons may also be used.

Below left: based at Bagotville, Quebec, the 433 Escadrille Tactique de Combat Aérienne flies the Northrop CF-5A fighter as half of the strike force of No 10 Tactical Air Group. Together with No 434 Squadron at Cold Lake, the unit is constantly at readiness to reinforce the Canadian presence in Europe; transatlantic refuelling in-flight exercises are frequently practised for that purpose with Boeing CC-137 aircraft of No 437 Squadron

CANADA

Unit	Aircraft type	Base
VU 32	CT-129/CT-133/CH-135	Halifax, NS
VU 33	CP-121/CT-133	Comox, BC
400 Sq	CSR-123	Toronto, Ont
401 Sq	CSR-123	Montreal, Que
402 Sq	—	Winnipeg, Man
403 Sq	CH-135/CH-136	Gagetown, NB
VP 404	CP-107	Greenwood, NS
VP 405	CP-107	Greenwood, NS
VT 406	CH-124/CP-121	Halifax, NS
VP 407	CP-107	Comox, BC
408 Sq	CH-135/CH-136	Edmonton, Alta
409 Sq	CF-101F	Comox, BC
410 Sq	CF-101F/CT-133	Bagotville, Que
411 Sq	CSR-123	Toronto, Que
412 Sq	CC-109/CC-117/CC-132	Ottawa & Trenton
413 Sq	CH-113/CH-115	Summerside, PEI
414 Sq	CF-100/CT-133/Falcon	North Bay, Ont
VP 415	CP-107	Summerside, PEI
416 Sq	CF-101F	Chatham, NB
417 Sq	CF-104	Cold Lake, Alta
418 Sq	—	Edmonton, Alta
419 Sq	CF-5A/D	Cold Lake, Alta
420 Sq	—	Shearwater, NS
421 Sq	CF-104	Sollingen, Germany
422 Sq	CH-135/CH-136	Gagetown, NB
HS 423	CH-124	Halifax, NS

diverted to the Gendarmerie (which also operates three Pumas), but the five survivors are now allocated to Army units. Three army aviation squadrons in Germany fly the Alouette, the first of which is a corps liaison unit, No 16 Squadron, at Butzweilerhof. The 16th Armoured Division is allocated No 17 Squadron at Werl and the 1st Infantry Division has call on No 18 Squadron based at Merzbruck.

Each unit has 18 aircraft, and No 17 is additionally known for its demonstration team, the 'Blue Bees'. Twenty further Alouettes serve the Light Aviation School (formerly No 15 Squadron) at army aviation headquarters, Braaschaat. The School also operates six of the 12 Britten-Norman Islanders received as Dornier Do 27 replacements in 1975–76, the remainder being allocated to No 16 Squadron. The initial five Islanders are to normal transport and liaison standards, but were followed by three equipped for paratrooping and four fitted with Vinten 360 survey cameras. Examples of all versions serve with both squadrons.

No 1 Army Corps is equipped with MBLE Epervier battlefield reconnaissance drones at divisional level and is in the process of receiving 40 examples for deployment in Germany. For anti-aircraft defence, two battalions equipped with MIM-23A Hawks (43rd and 62nd) totalling 59 launch units are to benefit from the European Hawk Improvement Program through conversion of their

missiles to MIM-23B standard. Two further surface-to-surface missile battalions have recently completed the change-over from Honest John to the LTV MGM 52C Lance. In common with most NATO countries, Belgium is increasing its anti-tank capability through purchases of Swingfire and Milan missiles to augment the earlier SS 11s and ENTACs. Future army requirements include 80 new light helicopters for Alouette replacement. These are to be delivered in two batches, initially for observation and anti-armour, followed by the assault role.

Canada's forces have in the past been the subject of far-reaching governmental changes which in 1968 resulted in the replacement of the three services by a single organisation known as the Canadian Armed Forces, or Forces Armées Canadiennes. A steadily decreasing proportion of the defence budget available for capital expenditure, which reached the low figure of 10·7 per cent in 1974–75, resulted in the withdrawal of aviation units administered by Maritime Command (Navy) and Mobile Command (Army) for which responsibility was allocated to the new Air Command, formed in September 1975. Air Command now controls all Canadian military aviation through one senior commander with the aim of improving flexibility, although the service integration previously imposed has undergone some modification in recent years and the forces have re-gained some of their individuality.

424 Sq	CC-115/CH-113A	Trenton, Ont
425 Sq	CF-101F	Bagotville, Que
426 Sq	CC-130	Trenton, Ont
427 Sq	CH-135/CH-136	Petawawa, Ont
429 Sq	CT-129/CC-130/CT-133	Winnipeg, Man
430 Sq	CH-135/CH-136	Valcartier, Que
433 Esc	CF-5A	Bagotville, Que
434 Sq	CF-5A	Cold Lake, Alta
435 Sq	CC-130	Edmonton, Alta
436 Sq	CC-130	Trenton, Ont
437 Sq	CC-137	Ottawa South, Ont
438 Sq	CSR-123	Montreal, Que
439 Sq	CF-104	Sollingen, Germany
440 Sq	CC-113/CC-138	Edmonton & Yellowknife
441 Sq	CF-104	Sollingen, Germany
442 Sq	CH-113/CC-115	Comox, BC
HS 443	CH-124	Halifax, NS
444 Sq	CH-136	Lahr, Germany
447 Sq	CH-147	Ottawa, Ont
450 Sq	CH-147/CH-135	Ottawa, Ont
VS 880	CP-121	Halifax, NS
1 FTS	CF-5A/D	Cold Lake, Alta
2 FTS	CT-114	Moose Jaw, Sas
3 FTS	CT-134/CH-136	Portage, Man
AN & IRS	CC-130	Winnipeg, Man
FIS	CT-114	Moose Jaw, Sas
TU	CP-107	Summerside, PEI
MPEU	CP-107	Greenwood, NS
ASU	CC-117	Ottawa South, Ont

Air Command tasks include the provision of operationally ready regular and reserve forces to meet Canada's sovereignty requirements, together with defence of North America in conjunction with US forces, and contributing to the strength of NATO. In order to meet these commitments, a five year plan was instituted in 1977 to raise equipment funding to 20 per cent of the annual budget, which for the fiscal year 1978–79 stands at $C 4·13 billion. Personnel strength of Air Command amounts to 23,000 with some 900 aircraft, although 150 of these are in storage.

Following the 1975 re-organisation, the command now comprises four operational groups, re-designated from the former commands: Air Defence Group, Air Transport Group, Maritime

Top left: the Canadair-built CF-5 variant of Northrop's Freedom Fighter is due to serve on in the advanced training role once supplanted by the NFA.
Top: the McDonnell Voodoo is operated by three interceptor squadrons, including No 425 Sqn.
Left: a Lockheed CF-104 Starfighter of the Europe-based 1 Canadian Air Group touches down

Below: designated CT-133 in Canadian service, the Lockheed trainer supplements some 22 two-seat Starfighters with 1 CAG. The pictured aircraft is based at Sollingen.
Centre: the Canadair Argus was developed from the Bristol Britannia for long-range maritime reconnaissance.
Bottom: VS 880 flies fishery protection patrols over east coast waters. Aircraft operated are Grumman Trackers

Air Group and 10 Tactical Air Group. Air Command also exercises control over air training schools and the Air Reserve, and has training and reinforcement responsibilities for 1 Canadian Air Group (1 CAG) in Europe. Of 40 squadrons in Air Command, 11 (including OCUs) operate jet fighters in combat roles, while the first-line element also includes seven maritime air and eight tactical helicopter squadrons. Air Command also has five air transport and four SAR squadrons, plus seven additional Air Reserve squadrons.

Canada's contribution to NATO takes the form of three squadrons (Nos 421, 439 and 441) based at Bad Soellingen, Germany, and equipped with the CF-104 Starfighter for tactical ground attack with conventional weapons. Together with the 14 Bell CH-136 Kiowa light observation helicopters of No 444 Squadron at Lahr operating in co-operation with the ground forces of 4 Mechanised Brigade Group and the home-based operational training unit, No 417 Squadron at Cold Lake, they constitute No 1 Canadian Air Group, with headquarters at Lahr within NATO's 4th Allied Tactical Air Force.

From 200 single-seat and 38 dual-control Starfighters received by the CAF, attrition and overseas disposals have reduced No 1 CAG to an effective total of 78 strike versions, plus a further 22 trainers supported by a small number of Canadair CT-133s (Lockheed T-33) and a single, radar-nosed Dakota at Cold Lake. Starfighters have been converted to F-104G standard by the installation of 20mm M-61-A1 Vulcan cannon, underwing stores and Sidewinder air-to-air missiles, and are receiving new tail assemblies and other modifications to extend their service lives into the 1980s in the face of continued delays in the NFA programme.

Mobile Command, formed at St Hubert in October 1965, is supported by No 10 Tactical Air Group which transferred to Air Command in September 1975, although still remaining under operational control of the Commander in Chief, Land Forces. No 10 TAG controls ten squadrons for fire-support, reconnaissance and tactical transport over battle areas, and has a strike element comprising two squadrons of Canadair-built CF-5A tactical fighters: No 434 Squadron at Cold Lake and the French-speaking 433 Escadrille at Bagotville.

In support of Canada's commitments to Allied Command Europe's Mobile Force, CF-5As of No 10 TAG complete monthly flight-refuelling exercises with Boeing CC-137 tanker/transports of No 437 Squadrons in readiness for transatlantic reinforcement. The two units have been operating 27 CF-5As from the original procurement of 89 in 1966, of which 44 were delivered straight into storage. A further 16 were sold to Venezuela

together with two of the 26 two-seat CF-5Ds, and the resulting funds financed a further 18 CF-5Ds to replace the CT-133 Silver Stars of 1 FTS at Cold Lake.

From January 1976, the operational training role formerly held by No 434 Squadron was passed to a unit–No 419 Squadron–which formed as a component of 1 FTS at Cold Lake with CF-5As and -5Ds withdrawn from storage. Although some 20 F-5s remain in storage, the type will be relegated to the advanced training role when the NFA eventually appears, and is therefore likely to remain in the CAF inventory for some considerable time.

Remaining tactical support units in 10 CAG include eight helicopter squadrons, of which four are attached to home-based Battle Groups and the training garrison centre, with a mixture of 6–7 Bell CH-135 Twin Hueys and 8–10 Bell CH-136 Kiowas for transport and observation duties respectively. Canada ordered 50 Bell CUH-1N Twin Hueys in late 1969 together with 74 COH-58As in early 1970 to replace smaller helicopters and fixed-wing aircraft. The Hiller UH-12s used in Europe were then sold while UH-1H Iroquois were converted for base SAR flights at Cold Lake, Moose Jaw, Bagotville and Chatham under their Canadian designation, CH-118. Battle Group units comprise No 408 Squadron at Edmonton, No 427 Squadron, Petawawa and No 430 Squadron, Valcartier, together with No 422 Squadron at the garrison centre of Gagetown.

Similar rotorcraft are operated by No 403 Operational Training Squadron at Gagetown, while remaining 10 TAG helicopter units comprise No 447 Squadron at Edmonton and No 450 Squadron, Ottawa. The latter unit replaced its Boeing-Vertol CH-113 Voyageurs with eight CH-147 Chinooks in 1975, although the first of these was destroyed in an accident whilst on its delivery flight, and also maintains a VIP/utility flight of three CH-135 Twin Hueys. A second flight, detached to Edmonton with four Chinooks, was re-designated No 447 Squadron in January 1979.

Naval air elements, previously part of the Royal Canadian Navy, became Maritime Command in January 1966, adopting their present title of Maritime Air Group (MAG) in September 1975. MAG is a component of Air Command managing all air resources engaged in sea patrol, surveillance and anti-submarine warfare (ASW). With the disposal of the sole light fleet aircraft carrier HMCS Bonaventure in 1970, all maritime units became shore-based, with the exception of two helicopter squadrons deployed to destroyer platforms for ASW duties.

For the principal task of coastal and anti-pollution patrols, fishery protection and arctic surveillance, the existing force of 26 Canadair CP-107 Argus long-range reconnaissance aircraft now

Below: the indigenous Canadair CT-114 Tutor provides basic jet training for students at No 2 FTS, Moose Jaw. Centre: prospective helicopter pilots undergo instruction on the Bell CH-136 Kiowa, military version of the Jet Ranger. The type also serves as an observation platform, as pictured here. Bottom: No 442 Squadron RCAF fly the de Havilland Canada CC-115 Buffalo from Comox on the west coast

Above left: the Lockheed Starfighter equips two interceptor units of the Royal Danish Air Force, 723 and 726 Eskadrilles, both based at Aalborg. Left: despite a high attrition rate in Danish service, the F-100 Super Sabre has served with the RDAF in the fighter-bomber role since 1959 and it will remain in service until the advent of the General Dynamics F-16. Below: two-seat Starfighters undertake operational conversion and electronic-countermeasures duties

overdue for replacement will give way to 18 Lockheed Orions, to be designated CP-140 Aurora in CAF service. Fitted with the navigational, acoustic and anti-submarine avionics employed by the S-3A Viking, the first Aurora was flown in March 1979, and all 18 examples will be delivered between May 1980 and March 1981.

Initial recipients will be Nos 404 and 405 Squadrons at Greenwood, currently operating six Argus each, followed by No 415 Squadron at Summerside, on the East Coast, and No 407 Squadron at Comox. The Argus is additionally employed by the Trials Unit at Summerside and the Maritime Proving and Evaluation Unit at Greenwood, the latter a component of No 404 Squadron, the maritime OCU. MAG units have now adopted US Navy role prefixes to their numbers, all patrol squadrons being designated 'VP'. For shorter-range surveillance of Bay of Fundy and Gulf of St Lawrence, 17 CS2F-3 (CP-121) Trackers are operated by VS 880 Squadron from CFB Shearwater near Dartmouth, and a further ten remain in storage at Saskatoon. Stripped of their ASW equipment, the Trackers are predominantly flown on fishery protection sorties following the extension of territorial waters to 200 miles in January 1977.

Three Trackers are also operated on similar patrol duties along the West Coast by VU 33 from Comox, where this utility squadron is additionally equipped with three CT-133 (T-33) Silver Stars for fleet support and target towing. The East Coast equivalent is VU 32 at Halifax with six Silver Stars, two CC-129 Dakotas and two CH-135 Twin Huey helicopters. Halifax is also the home base of fleet helicopter squadrons HS 423 and HS 443, each with a complement of 16 CH-124 Sea Kings detached to the four DDH 280 and eight older helicopter destroyers of the Atlantic Fleet, and certain replenishment ships. Previously intended for retirement in 1980, the Sea Kings, equipped with Fairey Beartrap haul-down systems for recovery in high seas, have been re-furbished to extend their useful lives until 1985. Operational training is assigned to VT 406 at Halifax with seven Sea Kings and two Trackers. Two of the former are to CH-124U standard with ASW equipment removed for operation in the transport and utility role.

Once the largest of the original RCAF formations, the Air Defence Group has now been reduced in strength, although it still remains integrated with USAF units through the NORAD joint air defence agreement. The previous NORAD boundaries between the United States and Canada have been re-aligned in order that each country may assume entire responsibility for its own air space, and a seven-year programme of modernisation completed in 1978 has extended the capabilities of the command post inside Cheyenne Mountain for a further ten years of operation without resort to replacement of major equipment. Two NORAD regions with control centres at Edmonton and North Bay sub-divide Canada into two equal areas in the east and west.

Integrated with NORAD through the Semi-Automatic Ground Environment (SAGE) air defence system are four main and 18 auxiliary radar sites of the Distant Early Warning (DEW) lines; 25 long-range radar sites of the Pinetree Line extending across the entire continent, from Gander in Newfoundland to Holberg, British Columbia, together with satellite tracking stations at Cold Lake and St Margarets.

Interceptor element of ADG comprises three squadrons of McDonnell CF-101 Voodoos: Nos 409 at Comox, 416 at Chatham and 425 at Bagotville. During the early 1970s, 58 CF-101Bs were exchanged for 66 refurbished CF-101Fs, of which nine have been lost in accidents, and a further 15 are in storage. Principal armament of the Voodoo is the Hughes AIM-4D Falcon and Douglas MB-1 Genie air-to-air missiles, the latter with a nuclear warhead, representing the one exception to Canada's non-nuclear principles.

Given a final decision in the NFA competition by the promised date of December 1979, the Voodoos will be required at least until 1983, and to maintain an effective air defence network, the Starfighters of No 417 Operational Training Squadron at Cold Lake have been assigned interception tasks in the Prairie Air Defence Zone, although nominally a component of No 1 CAG. Voodoo operational training is conducted from Bagotville by No 410 Squadron which also maintains a detachment of CT-133s at Chatham.

As an integral component of air defence, electronic warfare capabilities have been recently enhanced by the delivery of three modified Dassault Falcons, surplus to communications requirements, to No 414 Squadron at North Bay. The unit also flies 16 CT-133s and the 14 remaining CF-100 all-weather fighters in the CAF inventory, similarly equipped for ECM and target facilities work from detachments at Comox and Uplands.

Primary role of the Air Transport Group, which has some 80 fixed-wing aircraft and helicopters, has remained virtually unchanged since unification, although some increase in troop-carrier and mobility provisioning has been made. Headquartered at Trenton, the Group also supports Canada's UN commitments and provides SAR services through five airlift and four support/SAR squadrons. Mainstay of ATG logistic and army support forces is a fleet of 19 C-130E Hercules, five newer C-130Hs and a similar number of Boeing CC-137 (707-320C) tanker/transports. Despite an ATG request for an additional two CC-137s early in 1978, both types are currently scheduled for retirement in 1985–86.

All 24 Hercules are pooled between three squadrons; No 435 at Edmonton, and Nos 436 and 426 (also the operational training unit) at Trenton. Since 1973, all C-130Es have undergone refurbishing and incorporation of new centre-wing sections. CC-137s replaced the 12 CC-106 Yukons in No 437 Squadron at Ottawa South during 1970, and two have been retro-fitted with Beech flight-refuelling equipment in wingtip installations for the simultaneous replenishment of two CF-5s.

For VIP and utility transport, No 412 Squadron operates from Uplands and Trenton with seven Convair CC-109 (Model 580) Cosmopolitans fitted with cargo doors and reinforced freight-loading floors with a detachment to NORAD headquarters at Colorado Springs. A similar flight at Lahr for CFE headquarters began re-equipment with the first of two DH Canada CC-132 Dash 7 STOL transports in March 1978. The original Dassault CC-117

Falcons, mostly employed for governmental communications, have been reduced to three following transfers to ECM duties.

Canada provides SAR facilities for large areas of the Atlantic and Pacific under ICAO agreements, and while almost all Air Command units have a secondary SAR function, the primary force consists of four transport and rescue squadrons and four base rescue flights. Two transport and rescue squadrons—No 413 at Summerside and No 442 at Comox on the East and West coasts respectively—are equipped almost solely for SAR with three CC-115 Buffalo transports incorporating special search equipment and three Boeing-Vertol CH-113 (CH-46A) Labrador helicopters.

Two similar inland units comprise No 424 Squadron at Trenton with the Buffalo and Labrador and No 440 Squadron at Edmonton. The latter is concerned with both transport and SAR, and flies the Buffalo (of which two are assigned to UN forces in the Middle East) and four CC-138 Twin Otters. Fitted for operations from ground, water or snow, the Twin Otters form the equipment of the two-aircraft detachment to Yellowknife.

Following deliveries of Chinooks to No 450 Squadron, Boeing-Vertol Voyageur helicopters previously used for medium-lift support of Mobile Command now supplement Twin Otters within No 424 Squadron, for SAR in Ontario and parts of Quebec and North West Territories from Trenton. Further Voyageurs have joined the Buffaloes of No 442 Squadron at Comox, while SAR detachments at Cold Lake, Moose Jaw, Bagotville and Chatham operate Bell CH-118 Iroquois, formerly with No 406 Squadron, and now fitted with a searchlight, hoist and radar altimeter. ATC units are completed by the Airborne Sensing unit at Ottawa South, with single examples of the Falcon and CF-100 Canuck for atmospheric analysis, particularly in conjunction with automatic fall-out sampling.

Following the 1975 re-organisation, former Training Command air units transferred to Air Command and the training system headquarters moved from Winnipeg to Trenton to supervise the three flying training schools and one composite training squadron. Initial selection and grading of prospective pilots totalling 27 hours is undertaken by No 3 FTS, Portage la Prairie, on 18 of the 25 Beech CT-134 Musketeers obtained as Chipmunk replacements in 1971, two similar aircraft providing communications facilities for the MU at Trenton.

Basic training at No 2 FTS, Moose Jaw extends for 200 hours on the CT-114 Tutor, the previous CT-113/CT-114 courses having been combined with the phase-out of the former aircraft in mid-1974. Sales and attrition have reduced the original 190 Tutors to 145, but although only two-thirds of the fleet is in service at any one time, the additional burden placed on the aircraft following CT-133 requirement is now reducing the anticipated service life by an unacceptable degree.

A revision of the training syllabus is being considered, whereby the Musketeers would be replaced by some 50 slightly more advanced lightplanes in the Beech T-34C, Pilatus PC-7 or BAe Bullfinch class, enabling an extension of the 3 FTS course, and subsequent reduction in Tutor workload. Twelve modified Tutors are used by the 'Snowbirds' aerobatic team, and a further five serve the Flying Instructors' School, also at Moose Jaw.

After receiving their 'wings' at No 2 FTS, students are then streamed, with combat jet pilots continuing to No 1 FTS, Cold Lake for 100 hours' advanced flying on 12 CF-5As and 19 CF-5Ds. Operational training and tactical instruction is then completed by 92 hours with No 419 Squadron, an OCU unit attached to No 1 FTS, although subordinate to No 10 TAG.

Helicopter students return to No 3 FTS, which also operates 18 Bell CH-136 Kiowas, for 70 hours of basic and advanced training, followed by appropriate operational conversion in VT 406 for Sea King pilots of No 403 Squadron for tactical training. Pilots trained on multi-engined aircraft leave Moose Jaw for either operational training with the Hercules OTU, No 426 Squadron, or a regular unit for 'on-job' conversion.

For crew instruction, No 429 Squadron at Winnipeg provides instrument, observer and navigator training and has a secondary role as an SAR support unit. From an original 29 CT-129 Dakotas with the Air Navigation and Instrument Rating School component of No 429, only five remain in service, having been largely replaced by four C-130Es which have been specially modified with palletised interiors as flying classrooms. The unit is completed by an allocation of CT-133 Silver Stars, used as instrument trainers.

Air Reserve Group received increasing attention

DENMARK		
Unit	Aircraft type	Base
721 Esk	C-47/C-130/T-17	Vaerløse
722 Esk	S-61/Alouette III	Vaerløse
723 Esk	F-104 Starfighter	Aalborg
725 Esk	A35XD Draken	Karup
726 Esk	F-104 Starfighter	Aalborg
727 Esk	F-100 Super Sabre	Skrydstrup
729 Esk	S35XD Draken	Karup
730 Esk	F-100 Super Sabre	Skrydstrup
FSk	T-17 Supporter	Avnø
HF	T-17/Hughes 500M	Vandel

from 1975 onwards when consideration was given to an eventual doubling of the six squadrons then extant. In 1968, these units had been transferred from Air Transport to Mobile Command, but the new 'total force' concept brought about the establishment of the Air Reserve Group.

Under the present system, seven squadrons operate either the CSR-123 Otter or have a pooling arrangement with a regular unit. In the first category are Nos 401 and 438 Squadrons at Montreal and Nos 400 and 411 Squadrons at Toronto, while paired squadrons are No 402 Squadron at Winnipeg sharing No 429 Squadron's Dakotas, No 418 Squadron at Winnipeg using No 440's Twin Otters, and No 420 Squadron at Shearwater with the Trackers of CS 880.

Opposite above: the pilot of an F-100D of Eskadrille 730 RDAF is pictured during an exchange visit to the United Kingdom. Such visits are intended to give crews experience of an unfamiliar operating environment.

Opposite: the nose-section of the Saab S35XD slides forward for access to its cameras. This tactical recce version of the Draken is flown by Eskadrille 729 based at Karup.

Above: eleven Sk35XD operational conversion trainers were bought by the RDAF. All flying training up to this stage, apart from initial grading, is undertaken in the United States

Denmark has a small tactical air force, the Kongelige Danske Flyvevabnet (KDF), with a statutory strength of 116 first-line aircraft in six operational squadrons as decreed by the 1973 Defence Act, and supporting Allied Air Forces Baltic Approaches. A second Defence Act for the years 1977–81 will maintain annual expenditure at the unchanged level of $910 million at April 1976 price levels. The 1973 plan switched expenditure emphasis from personnel to equipment, but with the consequent fall in manpower from 9,500 to 6,900 in the KDF, the air force is seriously under-strength.

This deficiency is unlikely to improve in the next few years, with the major share of appropriations up until April 1981 devoted towards the F-16 programme of 58 aircraft, including 12 trainers. In 1970, the Danish forces underwent a degree of unification in command, rather than structure, intended to simplify joint control and administration of all three services without abandoning their separate identities. Under the revised scheme, army, navy and air force commands were abolished as separate organisations and merged within a unified Defence High Command under the Chief of Defence. Like Canada, Denmark is opposed to the employment of nuclear weapons, and is thus restricted to the use of conventional armament in undertaking its NATO duties.

The Royal Danish Air Force operational organisation now comprises two main headquarters for Tactical Air and Air Material Commands, together with a missile-equipped Air Defence Group. From Karup, Air Tactical Command

(Flyvertaktisk Kommando) exercises control of all squadrons and missile units, including three fighter-bomber, two interceptor, two transport and one fighter-reconnaissance squadron, eight surface-to-air-missile batteries and supporting units. All combat squadrons have a normal establishment of 20 aircraft (less attrition), apart from the reconnaissance squadron, which has 16.

Long-term mainstay of the tactical fighter force is the F-100D/F Super Sabre, of which 48 single-seat and 10 tandem F-100F models were supplied in 1959, for 725,727 and 730 Eskadrilles. The first of these squadrons disbanded in 1971, distributing its aircraft between the remaining two units, although the high rate of attrition common to many Super Sabre operators soon reduced the force to 26 F-100Ds and four F-100Fs. Squadron establishments were restored by the delivery of 14 replacements from USAF Air National Guard squadrons, all of which were two-seat models refurbished to the unique TF-100F standard during 1974.

A continued and unacceptable loss rate resulted in the Super Sabre being grounded for several months in 1977–78 whilst modified afterburners were fitted and crews returned to the .USA for continuation training. By July 1978, both squadrons had resumed operations from their base at Skrydstrup with a diminished fleet of 22 F-100Ds and 14 F-100Fs, destined to remain in service until arrival of the F-16 in 1980–82. Both units will then be re-designated Esk 727-100 and 730-100 and will not disband until their newly-formed parallels, Esk 727-16 and 730-16, become operational.

NETHERLANDS		
Unit	**Aircraft type**	**Base**
298 Sq	Alouette III/Bö 105	Soesterberg
299 Sq	Alouette III/Bö 105	Deelen
300 Sq	Alouette III	Deelen
306 Sq	RF-104G Starfighter	Volkel
311 Sq	F-104G Starfighter	Volkel
312 Sq	F-104G Starfighter	Volkel
313 Sq	NF-5A/B	Twenthe
314 Sq	NF-5A	Eindhoven
315 Sq	NF-5A	Twenthe
316 Sq	NF-5A	Gilze-Rijen
322 Sq	F-104G Starfighter	Leeuwarden
323 Sq	F-104G Starfighter	Leeuwarden
334 Sq	Fokker F.27	Soesterberg
CAV	TF-104G Starfighter	Volkel
2 Sq	—	Valkenburg
7 Sq	Lynx UH-14A	de Kooij
320 Sq	P2V-7 Neptune	Valkenburg
321 Sq	Atlantic	Valkenburg
860 Sq	Wasp AH-12A	de Kooij

From 1970 onwards, the Super Sabres were supplemented by an initial order for 46 SAAB-Scania Drakens which had been ordered two years previously. These comprised 20 A 35XDs (F 35s) armed with Bullpup ASMs and Sidewinder AAMs for strike/interceptor squadron Esk 725 at Karup; 20 S 35XDs (RF 35s) for tactical reconnaissance unit Esk 729 at the same base, and six two-seat Sk 35XDs

Opposite top: a CH-130H of Eskadrille 721 pictured at Keflavik, Iceland. Opposite centre: the Danish army flies Hughes 500M light helicopters. Opposite: float-equipped Alouette III helicopters equip the naval air arm. Above: No 313 Squadron R Neth AF is the NF-5 operational conversion unit

to be divided between the two eskadrilles. Draken procurement was subsequently increased to 51 when a second batch of Sk 35s replaced the remaining T-33s at Karup.

Main air defence role is undertaken by F-104 Starfighters of 723 and 726 Eskadrilles at Aalborg. As originally constituted in 1964, the squadrons shared 25 Canadian-built F-104Gs and four TF-104G trainers, but these were augmented by 15 CF-104Gs and seven two-seat CF-104Ds from surplus Canadian stocks in 1972–73 to increase establishments to 20 per squadron. Some CF-104Ds have been fitted with ECM equipment to replace eight similarly equipped T-35s at Karup, the last of which was withdrawn in May 1977. The 41 remaining aircraft will continue in KDF service until the mid-1980s.

Air Defence Group under Tactical Command is solely-missile-armed, controlling four MIM-14B Nike Hercules squadrons, each with nine launchers, transferred from the army in 1962, and four squadrons, each with six launchers, formed on permanent sites around Copenhagen in 1965. Denmark is a participant in the Hawk Improvement Program, four sites being designated for conversion to MIM-23B standard.

One transport squadron, Esk 721 at Vaerløse, employs three C-130H Hercules for general airlift duties and air supply to Greenland, assisted by eight Dakotas obtained in 1953–56. Plans to replace the Dakota by Friendships as long ago as 1970 were thwarted by financial constraints, but Denmark now has a requirement for three dual-purpose fishery patrol and medium transport aircraft, which the Friendship would meet.

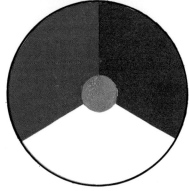

Top: the Lockheed F-104G Starfighter operates in the close support, air defence and reconnaissance roles with the Royal Netherlands Air Force.
Above: the Dutch air force operates three helicopter squadrons on behalf of the army, Alouette IIIs of No 300 Squadron being pictured.
Left: the Lockheed Neptune maritime patrol aircraft of No 320 Squadron at Valkenburg are due to be replaced by Lockheed Orions in 1982

Consideration is being given to combining this with a second requirement for two more Hercules and two VIP jets, with the Gulfstream 3, Dassault Falcon Guardian, Boeing 737 and Hercules as contenders for a single-type purchase. Announcement of a final decision has twice been postponed, and is now scheduled for the end of 1979. One aircraft would be based in Greenland to replace the Dakota detachment from Esk 721 which undertakes internal supply work. SAR duties are performed by Esk 722, also at Vaerløse, with detachments to Aalborg, Skrydstrup and Greenland, using eight Sikorsky S-61A helicopters.

Aircrew training is undertaken principally in the United States following an initial 25–30 hours grading on 14 SAAB T-17 Supporters of the Flyveskolen at Avnø. Between October 1975 and December 1976, 32 Supporters replaced DHC Chipmunks and KZ VII trainers of the air force and army; in addition to those at Avnø, the aircraft is used by Base Flights at Aalborg, Skrydstrup, Karup and No 721 Squadron.

On arrival in the United States, students fly a further 25 hours on Cessna T-41s at a civilian flying school under USAF contract before continuing to Vance AFB for basic training on the Cessna T-37B. This is followed by an advanced stage at Laughlin AFB on the Northrop T-38A, to complete in excess of 200 hours' flying time. Output totals some 40 pilots per year and, with retirement of the Lockheed T-33, newly qualified aircrew undertake European familiarisation in two-seat versions of the Lockheed Starfighter, Saab Draken and North American Super Sabre before becoming fully operational. Helicopter training is provided by the US Army

at Fort Rucker, Alabama.

Army aviation gained partial autonomy in July 1971 under the title of Haerens Flyvetjeneste, although the KDF continues to be responsible for maintenance and servicing of liaison aircraft and helicopters. Seriously weakened by the loss of 11 of its 12 Piper Super Cubs in a hangar fire during April 1968, the Army Flying Service received 12 Hughes 500M light observation helicopters by way of replacement in 1971, a further three being delivered in June 1974. Nine Supporters completed re-equipment of the Vandel-based air component early in 1976, although one was subsequently lost in a flying accident.

Eight float-equipped Alouette III helicopters constitute the naval air unit, Søvaernets Flyvevaesen which, although flown by naval officers from five

Right: the badge of No 860 Squadron painted on the nose of one of the unit's Westland Wasp helicopters. Wasps operate from the Royal Netherlands Navy's frigates.
Below: the Westland Lynx is to replace the Wasp as the main Dutch shipboard helicopter, but the first Lynx deliveries are to the shore-based No 7 Squadron for search and rescue and utility duties

frigates and patrol vessels, are otherwise attached to No 722 Squadron. Naval requirements for additional helicopters for fishery patrol duties are shortly to be satisfied by delivery of seven Westland Lynx Mark 80s.

After suffering long-term defence economies, the Dutch forces have benefited from a change of government and a subsequent increase in defence spending to meet NATO-specified targets. Reassessment of operational tasks during the General Dynamics F-16 selection process had, however, removed high-altitude air defence and deep penetration from the responsibilities of the air force (Koninklijke Luchtmacht). Requirements for limited air superiority and offensive air capability remain, following the rejection of nuclear weapons.

Attached to 2 ATAF of the Allied Air Forces Central Europe, the KLu comprises nine first-line squadrons, of which five have the Lockheed F-104 Starfighter and the remainder the Northrop F-5. Eight Hawk and four Nike-Hercules surface-to-air missile squadrons, a transport unit and three helicopter squadrons operated on behalf of the Army made up the service's strength in 1979. A ceiling on Starfighter replacement costs imposed in 1975 reduced the proposed purchase from 112 to 84 aircraft, but options on a further 18 will provide 102 General Dynamics F-16s, including 22 examples of the tandem-seat F-16B, which will be sufficient for the conversion of five squadrons.

Augmenting the 160 combat aircraft and 17,700 personnel of the KLu are 20 patrol aircraft and 18 helicopters of the Marine Luchtvaartdienst (MLD or naval flying service) operating from shore bases and frigates of the Netherlands Navy with the primary role of submarine detection. Progressive reductions since 1973, and disposal of the sole aircraft carrier, Hr Ms *Karel Doorman*, have reduced the MLD to a personnel strength of 1,900 within the navy's total of 17,000.

The organisation of the KLu concentrates all flying units within Tactical Air Command (Commando Tactische Luchtstrijdkrachten or CTL) with supply and training support from Commando Logisteik en Opleidingen, with headquarters at Zeelst and Gilze-Rijen respectively.

Nuclear strike commitments within NATO were formerly undertaken by two F-104G squadrons, Nos 311 and 312 at Volkel, operating alongside No 306 Squadron equipped with infra-red camera-carrying RF-104G aircraft in the tactical reconnaissance role. Each unit has a complement of 18 aircraft from an original procurement of 138, including 18 trainer versions. Leeuwarden will be the first operator of the F-16, and conversion of No 322 Squadron will begin with the establishment of a training flight of seven F-16As and five F-16Bs. Both units are scheduled to be fully equipped with the new fighter during 1981.

Scheduled to be the third and fourth recipients of the F-16, Nos 311 and 312 Squadrons will convert to the new type in 1982–83 for close-support operations in tactical roles with conventional weapons and will be followed by No 306 Squadron in 1984 with a continued reconnaissance commitment. Volkel also houses the Starfighter OCU, Conversie Afdeling Volkel, equipped with the TF-104G for instrument training and conversion

NORWAY		
Unit	Aircraft type	Base
Skv 330	Sea King Mk 43	Sola
Skv 331	F-104 Starfighter	Bødo
Skv 332	Northrop F-5A	Rygge
Skv 333	P-3B Orion	Andøya
Skv 334	F-104 Starfighter	Bødo
Skv 335	Hercules/Mystère 20	Gardermoen
Skv 336	Northrop F-5A	Rygge
Skv 338	Northrop F-5A	Ørland
Skv 339	UH-1B Iroquois	Bardufoss
Skv 717	Northrop RF-5A	Rygge
Skv 718	Northrop F-5A/F-5B	Sola
Skv 719	Iroquois/Twin Otter	Bardufoss
Skv 720	UH-1B Iroquois	Gardemoen

of pilots posted from F-5 units.

Between 1969 and 1972, the Netherlands received 75 Canadair-built NF-5As and 30 NF-5B tandem seat trainers for three squadrons and one conversion unit, as replacements for the Republic F-84F Thunderstreak in the tactical strike role. At Twenthe, as the F-5 OCU, No 313 Squadron is predominantly equipped with the two-seat F-5B alongside an operational unit, No 315 Squadron. Remaining aircraft are flown by No 314 Squadron at Eindhoven and No 316 Squadron from Gilze-Rijen, although Eindhoven is scheduled for closure and the two will thenceforward operate from Gilze-Rijen.

Air Defence Command (Commando Luchtverdediging) became a component of CTL in January 1973, controlling two Starfighter squadrons, Nos 322 and 323 at Leeuwarden, and the unique 32 TFS at Soesterberg (Camp New Amsterdam) equipped with the McDonnell Douglas F-15 Eagle as the only USAF squadron under foreign command. Semi-automatic computerised intercept control for air defence is provided by the control centre at den Helder employing an Elliott Firebrigade system based on MCS 900 computers.

Missile defence forces have been reduced to three groups of surface-to-air weapons (Groepen Geleide Wapens) each with four squadrons, based in Germany and Holland and equipped with Raytheon MIM-23B Improved Hawks and Nike-Hercules missiles. In addition to participating in the NATO Hawk Improvement Program, Holland is also interested in the new SHORAD missile to replace both Nikes and Hawks, while the army is evaluating several SAMs, including the Rapier, ROLAND and Chaparral, for tank installation. 1st Army Corps is also replacing eight Honest John SSM installations with six LTV MGM-52C Lance units beginning in late 1978. The latter are equipped with nuclear warheads, Holland having failed to persuade West Germany to take over its entire atomic role in NATO. Army units are receiving Raytheon/Kollsman FGM-77A Dragon anti-tank missiles.

Transport facilities are provided by a single squadron, No 334 at Soesterberg, equipped with nine Fokker F-27M Troopships and three F-27 Friendships, although three of the former were converted to navigation trainers in 1973-74 to replace withdrawn Beech TC-45J twins. Long-term funding for the period 1984-88 includes provision for a small number of medium-range transports, for which the Hercules appears suited, and some two or three will be obtained for airlifting the Dutch contingent of the NATO Mobile Force and supplying detachments at the weapons range in Crete.

Army aviation requirements are furnished through the Light Aircraft Group (Groep Lichte Vliegtuigen) of the KLu, headquartered at Deelen with three squadrons of light helicopters. Two mixed-complement units, No 298 Squadron at Soesterberg and No 299 Squadron at Deelen comprise two flights of Aérospatiale Alouette IIIs and one of MBB Bö 105s, the latter equipped with TOW anti-tank missiles. Also at Deelen is No 300 Squadron with three Alouette flights including one assigned to operational conversion (Helicopter Vliegopleiding).

GpLV received 72 Alouettes between 1964 and 1969, of which five have so far been lost. An extra five were bought by the KLu for the SAR Flight, now at Leeuwarden, this unit providing helicopters for Royal transport when required. Six of the 30 Bö 105 helicopters received during 1975-76 were placed in immediate storage against attrition, and the remainder equally divided between Nos 298 and 299 Squadrons.

Helicopter pilot training procedures were revised in October 1977 and now consists of 13 weeks at the National Flying School (civilian operated for the fixed-wing stage) at Eelde, followed by 30 weeks flying the Hughes TH-55A and Bell UH-1 at Fort Rucker, USA, in a joint Dutch/German/Norwegian and Danish programme. Training is completed by 38 weeks with No 300 Squadron. KLu aircrew under Logistics and Training Command train in Canada on the Beech Musketeer, Canadair Tutor and Northrop F-5. Operational conversion is completed with No 313 Squadron at Twenthe.

The continued effectiveness of the Naval Air Service in its primary role of submarine detection was assured by the decision to buy 13 Lockheed Orions in December 1978. After delays caused by Government investigation of Lockheed payments to Prince Bernhard and consideration of the British Aerospace Nimrod and Dassault-Breguet Atlantic ANG, Orions will replace the Lockheed Neptunes of No 320 Squadron at the main naval air base of Valkenburg in 1982, some six years after the planned date for Neptune phase-out. Since 1974, the squadron has additionally provided a detachment of four aircraft to Curaçao in the Dutch Antilles for naval patrol and SAR duties.

Operating as the second NATO-committed fixed-wing ASW unit for long-range surveillance of the North Atlantic and Channel areas, supplementing Dutch seaborne escort groups, No 321 Squadron operates seven Dassault/Breguet SP-13A Atlantics from nine received between 1969 and 1972 as substitutes for the withdrawn aircraft carrier, but plans for an Atlantic replacement by 1983 seem unlikely to be fulfilled before the last half of the next decade. Both Neptunes and Atlantics may be armed with the Aérospatiale AS 12 anti-shipping missile.

Further ASW commitments to SACLANT Command are undertaken by No 860 Squadron, shore-based at de Kooij, with detachments of single Westland AH-12A Wasps on each of the six *Van Speijk* class frigates from 12 examples delivered in 1966-67 and one further attrition replacement in 1974. No 7 Squadron, also at de Kooij is the sole second-line helicopter units. It withdrew seven Agusta-Bell UH-1 Iroquois in 1977 following the delivery of the first of six Westland UH-14A Lynx helicopters in May of that year.

Defence plans published in 1974 foresaw the procurement of 24 new helicopters for the MLD in two four-year periods until 1983 for both SAR and ASW duties, and the Lynx was chosen late 1974 as the standard type. An initial order for utility versions for No 7 Squadron was followed by a further requirement for ten SH-14Bs with uprated power plants for Wasp replacement and an option, now taken up, for an additional eight SH-14Cs.

Eventual Lynx total will be some 30 helicopters for deployment aboard two new guided missile frigates and twelve ASW frigates on order for the

Left: five Lockheed P-3B Orion maritime patrol aircraft equip No 333 Squadron, whose primary task is monitoring Soviet submarines entering and returning to Murmansk from patrols in the Atlantic. Below left: the Northrop F-5A serves in the fighter-bomber role, some 62 remaining in service in 1979. Below: Starfighters undertake both air defence and attack roles with the Royal Norwegian Air Force

Koninklijke Marine. Two SH-14Bs will be used for training by No 860 Squadron in addition to facilities provided by No 7 Squadron, which also supports operations by the Netherlands Marine Corps. Fixed-wing instruction is allocated to No 2 Squadron at Valkenburg, which has no aircraft of its own but borrows Atlantics and Neptunes from combat squadrons as required. Basic and advanced pilot training for helicopter pilots follows the KLu/GpLV courses at Eelde, Canada and Deelen.

In partnership with Denmark in NATO's Northern Command, Norway occupies an important strategic position on the northern flank of the Alliance and regularly hosts exercises and deployments from other NATO countries perfecting winter battle techniques.

Norway's air force, the Kongelige Norske Luftforsvaret, is divided into two commands, Luftkommando Nord-Norge and Sør-Norge, each with its own headquarters and operational control, integrated within the NADGE air defence system. Luftforsvaret establishment, including anti-aircraft components, approximates to 10,000 personnel—half of whom are conscripts—and a dozen squadrons with an inventory of 130 first-line aircraft. A significant proportion of defence expenditure for 1978 was devoted to the purchase of 60 General Dynamics F-16A and 12 F-16B fighters and the US version of the Euromissile ROLAND 2 surface-to-air missile. Extension of national waters has increased the responsibilities of the maritime component, and an unfulfilled requirement for three long-range patrol aircraft has existed since 1975.

Manned interceptor capability is confined to the Lockheed F-104G Starfighters of No 331 Squadron (Skv 331) at Bødo which received 21 aircraft, including two TF-104G trainers in 1963–66. Additional CF-104s were obtained in 1973–74 from surplus Canadian stocks, entering service with Skv 334 at Bødo after modification (by Scottish Aviation at Prestwick) to F-104G standard with 20mm Vulcan cannon. To these 19 single-seat and three CF-104D trainers were added a further two TF-104Gs from Germany in 1975, raising overall deliveries to 45 aircraft, of which five were lost.

Skv 334 operates in the strike role alongside the Northrop F-5s of Skv 332 and Skv 336 at Rygge and Skv 338, Ørland, each with 16 aircraft. A further 13 RF-5As equip Skv 717 flying tactical reconnaissance missions from Rygge. Norway received a total of 108 F-5s, including 14 F-5B tandem-seat trainers, the majority of the latter serving Skv 718 at Sola, the operational tactical school. Originally the sixth F-5 squadron, Skv 334 converted to the Starfighter after heavy F-5 losses, which now total 18 or more aircraft.

Air defence Starfighters are augmented by four Nike-Hercules batteries situated around Oslo and Eastern Norway and from 1980, ROLAND 2 SHORAD systems will significantly increase SAM capability by a further 900 units. Delivery of F-16s from 1981 onwards will allow eventual replacement of the Starfighter; some F-16s will be equipped with Penguin or Harpoon anti-ship missiles.

Offensive naval capability is also possessed by the five Lockheed P-3B Orions of Skv 333 at Andøya, whose vital peacetime role is the reporting of Soviet submarines leaving Murmansk for the

Atlantic. On leaving Norwegian areas of responsibility, the vessels continue to be tracked by RAF and, ultimately, US Navy squadrons.

Transport support is provided by Skv 335 at Gardemoen with six C-130H Hercules, received in 1969 as replacements for eight Fairchild C-119s and three Douglas C-47s. The squadron also operates three Dassault Falcon 20s for VIP transport, calibration and ECM. Skv 719 at Bødo has four DHC Twin Otters and two Bell UH-1B Iroquois for additional SAR and liaison roles.

A total of 29 Iroquois helicopters was delivered

between 1965 and 1970, being followed by a further half-dozen early in 1976. Fifteen are operated by Skv 339 at Bardufoss for army support, together with a further nine by Skv 720 at Rygge on similar duties and SAR. The principal rescue role is, however, undertaken by the Sea Kings of Skv 330, which accepted 10 Mk 43 machines from Westland Helicopters in 1972, augmented by a further attrition replacement example in 1978. Squadron headquarters are at Bødo, with other permanent detachments to Banak, Ørland and Sola.

Norwegian defence forces' requirements for up to 40 new rotorcraft for transport, assault and ASW have been partially satisfied by an order for six Westland Lynx to be based at Sola and Stavanger. Although intended primarily for operation from airfields, shipborne capability was an essential prerequisite of the Lynx purchase of an initial four helicopters, together with two options taken up late in 1978.

Most KNL aircrew are NATO-trained, although primary instruction is given on SAAB Safirs at Vaernes. Since being delivered in 1956–57, four Safirs have been lost, although a further two were issued to No 720 Squadron. The anticipated replacement has not yet been selected, and as an interim measure, five more examples are being delivered from the Swedish air force to extend the service career of this rapidly obsolescent aircraft.

The army Field Artillery Observer Service has two squadrons equipped with the nine Piper L-18C Super Cubs which remain from 16 received in 1955 and 24 of the 27 Cessna O-1A/L-19A Bird Dogs delivered between 1961 and 1968. Both types are due for replacement, in all probability by helicopters, in the early 1980s.

No 330 Squadron, with headquarters at Bødo, flies the Westland Sea King Mark 43 on air/sea rescue duties. Detached flights cover most of Norway's extensive coastline, with 'A' Flight operating from Bødo, 'B' Flight from Banak, 'C' Flight from Ørland and 'D' Flight from Sola

CHAPTER EIGHT

NATO/Southern Front
Portugal/Greece/Turkey/Italy

Portugal's strategy and tactics were for over forty years dictated by a dual national policy, which had remained unaltered since the emergence of the New State concept in 1928. This called for the preservation of the territorial integrity and political sovereignty of Portugal itself, and the defence of the country's Overseas Provinces in Africa and Asia as an integral part of the Portuguese Nation. The Portuguese air force (Força Aérea Portuguesa) was from its very beginning an instrument of national policy – the first sovereignty operations in Africa took place as early as 1917 – and, despite limited funds and political difficulties, maintained a degree of success in adapting itself to changes as they took place. The far-reaching reorganisation of the 1960s, brought into being by the escalation of guerilla warfare in the three major overseas provinces (Guinea-Bissau, Angola and Mozambique), trans-formed it into a tactical, Africa-oriented force, its contribution towards NATO being limited to a squadron of Neptune maritime patrol aircraft and the availability of a number of bases when required.

Following the revolution in April 1974, the main aim of the new, left-wing government was total withdrawal from the overseas provinces and the transfer of their government to the leading guerilla factions. FAP operational sorties were suspended (an obvious measure, as the FAP was never won over to the pro-communist side), personnel gradually evacuated, and large numbers of aircraft abandoned or handed over to the territories' new governments. The FAP became a shadow of its former self, and the consequent loss of morale, together with growing political instability and actual cases of sabotage, almost brought flying to a standstill by mid-1975. However, in November of that year, an anti-communist counter-coup staged by the armed forces brought a degree of stability to the country. The FAP then decided to begin a much-needed reorganisation, to be spread over the next few years, which would adapt it to the new situation.

However desirable it would have been to make radical changes, these had to be ruled out due to a desperate lack of funds and an overall strength of a mere 264 aircraft, many of which were unserviceable and only two dozen having a true offensive capability. Short-term re-equipment was out of the question; on the contrary, total strength had to be further reduced to improve efficiency. It was therefore decided that the FAP should retain a small operational nucleus, mainly for home defence, internal policing and coastal patrol, concentrating on transport and training duties at the same time; external help was to be requested for re-equipment. The FAP was to become a volunteer force, with periodic recruiting drives. This led to the formation of the Asas de Portugal (Wings of Portugal) aerobatic team, whose standards are surprisingly high, considering the absence of such a team since 1961 and the type of aircraft used.

The basic FAP formation remains the *base aérea* (air base), a military establishment housing training units or up to three *esquadras* (squadrons), as well as small rescue/communications detachments. The main transport base is still Lisbon's international airport, Portela de Sacavém, which, not being exclusively military, is designated *aeródromo-base* (airfield-base). Small airstrips without permanent FAP units are called *aeródromos de manobra* (auxiliary airfields). Continental Portugal has six *bases aéreas*, four of which specialise in training – BA 2 and BA 7 for primary/basic training, BA 1 for advanced training and BA 3 for conversion

PORTUGUESE AIR FORCE – FORÇA AEREA PORTUGUESA, 1979		
Base Aérea 1 (BA 1)	Sintra	(Advanced training)
		Cessna T-37C
		CASA 212 Aviocar
Base Aérea 2 (BA 2)	Ota	(Primary/basic training)
		de Havilland DHC-1 Chipmunk
Base Aérea 3 (BA 3)	Tancos	(Operational training unit)
		Nord 2501 Noratlas
		CASA 212 Aviocar
		Northrop T-38A
		Aérospatiale Alouette III
Base Aérea 4 (BA 4)	Lajes (Azores)	(Transport/Rescue)
		Aérospatiale SA 330C Puma
		CASA 212 Aviocar
Base Aérea 5 (BA 5)	Monte Real	(Interception/ground attack)
		Grupo Operacional 51 (GOp 51)
		Esquadra 103 Lockheed T-33A
		Esquadra 201 North American F-86F
		Esquadra Fiat G 91R/G 91T
Base Aérea 6 (BA 6)	Montijo	(NATO/tactical support)
		Grupo Operacional 62 (GOp 62)
		Esquadra Lockheed P2V-5F
		Esquadra 301 Fiat G 91R/G 91T
		Esquadra Aérospatiale Alouette III and SA 330C Puma
Base Aérea 7 (BA 7)	São Jacinto	(Primary/basic training)
		de Havilland DHC-1 Chipmunk
		North American T-6 Texan/Harvard
		Cessna F337G
Aeródromo-Base 1 (AB 1)	Portela de Sacavém	(Transport)
		Douglas DC-6, Lockheed C-130H

Right: the North American Sabres of
Portugal's Esquadra 201 operate from
Base Aérea 5 at Monte Real. In
addition to its American-built F-86F
Sabres, the unit also flies a number
of Canadair Sabre Mark 6 fighters.
Re-equipment with the Northrop F-5E
is scheduled for 1980, the more-modern
fighters being supplied and funded
by the United States.
Inset above: six Northrop T-38A
Talon supersonic jet trainers have
been loaned to the Portuguese by
the USAF to prepare pilots for the
transition from the Sabre to the F-5E.
Inset below: advanced training is on
the Cessna T-37C, which also equips
the Asas de Portugal aerobatic team

training; the remaining two, BA 5 and BA 6, house Grupos Operacionais (Operational Groups), respectively for home defence (interception and ground attack) and NATO duties (coastal reconnaissance, internal security).

Montijo is fast becoming the FAP's major base; its Neptune *esquadra* maintains periodical surveillance within Portuguese territorial waters, and Esquadra 301 specialises in tactical support, developing the techniques successfully applied in the African campaigns. Montijo's helicopter element currently flies the Puma and Alouette III, but may receive a small number of Agusta A109A armed helicopters in 1980–81. The Lockheed P2V-5 Neptunes are to be temporarily superseded by ASW-equipped Lockheed C-130H Hercules, pending a probable delivery of a couple of US Navy surplus early model P-3 Orions.

Esquadra 201 at Monte Real still operates the venerable North American F-86F Sabre, as pre-revolution plans to re-equip with the Dassault Mirage were abandoned, but is scheduled to receive an initial batch of Northrop F-5E Tiger IIs under MDAP agreement in 1980. Conversion training began in the summer of 1977 with six Northrop T-38As loaned by the USAF, which were the first FAP aircraft in fifty years not to be allotted Portuguese serial numbers. Plans exist for the gradual replacement of the Fiat G 91R with additional deliveries of F-5E Tiger IIs, and there is a possibility that a modern multi-role jet may be procured (with NATO or US funds) by 1982–83.

The FAP's transport force was streamlined by the decision to re-equip with the Lockheed C-130H Hercules, five of which were initially ordered; a few Douglas DC-6s still in service are to be grounded, and such types as the Douglas C-47 and the Nord 2501 Noratlas were disposed of–although a few Noratlases remain for operational training and paradropping. Training units are concentrating on the DHC-1 Chipmunk, due for replacement in the near future, and the Reims-Cessna F337G, some of which have provision for light armament and are occasionally used for internal policing duties. The faithful North American T-6 Texan (Harvard), which was used to such good effect in Africa as a ground-attack machine, is currently being phased out, and there is a requirement for a basic piston-engined trainer.

The only *base aérea* outside metropolitan Portugal is BA 4 (Lajes) in the Azores, which is shared with the USAF and houses two detachments–a transport flight with Spanish-built CASA 212 Aviocar light twins, and a search-and-rescue flight with SA 330 Pumas. The Portuguese Atlantic Islands (now the Autonomous Regions of the Azores and Madeira) have become increasingly important for Portugal, particularly because of the existence of separatist movements–a thing unheard of before the 1974 revolution. Consequently, measures have been taken to integrate the islands in the overall FAP strategy; in August 1978, Esquadra 301's Fiat G 91Rs from Montijo took part in Operação Atlântida 1978 (Operation Atlantis), to evaluate the possibility of fast tactical intervention against enemy landings or local insurgency. At the same time, the FAP began consolidating its position in the hearts and minds of the islanders.

On 6 January 1978, two Alouette IIIs were airlifted to Madeira in a Hercules and were attached to the Funchal Infantry Regiment for a week, to study the feasibility of a permanent helicopter detachment in the islands, mainly for rescue and communications duties; the findings were quite favourable.

There has been no need to form a separate army air force, because the FAP has always maintained a high degree of co-operation with the ground forces, and no such need arose during the African campaigns; but future changes in the peninsula – such as the uncertain political situation in neighbouring Spain, or the bitter political rivalry within the country – may dictate otherwise. On the other hand,

Top: the Fiat G 91R ground-attack fighter currently serves with two esquadras and previously saw action against guerrillas in Angola and Mozambique.
Above left: an esquadra of Lockheed P2V-5F Neptune maritime patrol bombers are based at Montijo for surveillance of Portuguese waters.
Left: the SA 330C Puma helicopter equips an esquadra based at Montijo, which also flies the smaller Aérospatiale Alouette III.
Below: this F-86F serves with Esquadra 201.
Right: a section of Greek air force Northrop F-5A fighters pictured at Nea Ankhialos, where the 111ᵃ Pterigha operates three squadrons

there are plans, formulated shortly after the revolution, to form a naval air arm basically because the Navy and the Marines were instrumental in the success of the changeover. However, the present political and economic situation suggests that no such move will take place for the time being.

All things considered, the FAP's recovery from the post-revolutionary chaos has been remarkable. Although no long-term planning can yet be made, positive steps have been taken towards increasing its efficiency, and it seems certain that the FAP will once again become an essential component of NATO.

Greece is a curious mixture of strength and weakness. Its strategic importance is undeniable–the country's position in the European map, together with a long coastline and a large number of

HELLENIC AIR FORCE–HELLINIKI AEROPORIA, 1979			
28th Tactical Air Command			
110ª Pterigha	Larissa	345ª Mira	Vought A-7H
		348ª Mira	Republic RF-84F
		349ª Mira	Northrop F-5A
		Base Flight	Lockheed T-33A
111ª Pterigha	Nea Ankhialos	337ª Mira	Northrop F-5A/B, RF-5A
		341ª Mira	Northrop F-5A/B, RF-5A
		343ª Mira	Northrop F-5A/B, RF-5A
		Base Flight	Lockheed T-33A
114ª Pterigha	Tanagra	336ª Mira	Dassault Mirage F1CG
		342ª Mira	Dassault Mirage F1CG
		Base Flight	Lockheed T-33A
115ª Pterigha	Soudha (Kriti)	338ª Mira	Vought A-7H

		340ª Mira	Vought A-7H
		Base Flight	Lockheed T-33A
116ª Pterigha	Araxos	335ª Mira	Lockheed F/TF-104G
		Base Flight	Lockheed T-33A
117ª Pterigha	Andravida	339ª Mira	McDonnell F-4E
		344ª Mira	McDonnell F-4E
		Base Flight	Lockheed T-33A
363ª	Elevsis		Grumman HU-16B (ASW)
30th Air Material Command			
355ª Mira	Elevsis		Lockheed C-130H, Nord 2501D Noratlas, Canadair CL-215, Douglas C-47
356ª Mira	Elevsis		
357ª Mira	Elevsis		Bell 47G
358ª Mira	Elevsis		Agusta-Bell 204B/205A
359ª Mira	Elevsis		Sikorsky UH-19D
362ª Mira	Tatoi-Dekelia		Agusta-Bell 205A/206A
31st Training Command Ethniki Aeroporiki			
Akademia	Tatoi-Dekelia		Cessna T-41D
360ª Mira	Kalamata		Rockwell T-2E
361ª Mira	Kalamata		Cessna T-37C

dependent islands, assures the effective control of the Eastern Mediterranean. On the other hand, there are obvious weaknesses–Greece has land frontiers with four countries, three of which (Albania, Bulgaria and Yugoslavia) are communist, and the fourth the Greeks' traditional enemy, Turkey, currently a NATO ally but nevertheless a traditional enemy; another serious weakness is the unstable internal political situation. This being so, it is not surprising that Greece's armed forces are very large in relation to the country's population; besides being supported by conscription, defence expenditure accounts for a large share of the national budget.

The Greek air force (Helliniki Aeroporia) is therefore a very important part of the country's defence network. Re-formed by the British in 1944–45 with surplus RAF equipment, mainly to oppose a massive communist uprising supported by Soviet-backed governments across the border, it was gradually modernised, with substantial assist-

NATO/SOUTHERN FRONT

Inset right: a Greek order for 60 Vought A-7H attack aircraft was completed in 1977 and the type now equips three squadrons. The two-seat TA-7H is due to enter service in 1980. Below: a McDonnell Douglas Phantom touches down at its base at Andravida, where the 117ª Pterigha is equipped with the F-4E variant. In 1979 the Greek air force was reported to have 64 aircraft of this type on charge. Inset below: two squadrons of Dassault Mirage F1CGs of the 114ª Pterigha fly from Tanagra. France supplied 40 of these aircraft to Greece in the mid-1970s to replace Convair F-102A Delta Daggers. They operate in the air defence role, covering Athens

Inset left: a single squadron of the Greek air force flies the Lockheed Starfighter, this being the 335^a Mira of 116^a Pterigha at Araxos with some 30 aircraft on strength. A two-seat TF-104G, which serves with this unit as a conversion trainer, is shown. Inset below: in addition to the tactical reconnaissance RF-5As of the 111^a Pterigha, a squadron of elderly Republic RF-84F Thunderflashes of the 110^a Pterigha operate in this role. The Thunderflashes are being replaced by eight McDonnell Douglas RF-4Es

ance from the United States, as part of the 6th Allied Tactical Air Force (6th ATAF), until it reached a high degree of efficiency. The air force has suffered from occasional embargoes, due to Greece's conflicts with Turkey over Cyprus, but it may be said that the only serious current threat to the Aeroporia's efficiency lies in political interference from within the country.

The Helliniki Aeroporia comprises three Commands – Tactical, Air Material and Training. The 28th Tactical Air Command (28 Taktiki Aeroporiki Dynamis), the force's contribution to the 6th ATAF, was formed in June 1952 and presently comprises six wings or *pterighe*, with a total of thirteen *mire* (squadrons), and an independent anti-submarine warfare *mira*. Each *pterigha* also includes a base flight flying a couple of Lockheed T-33As. The 110[a] Pterigha, based at Larissa in Thessalia (north-eastern Greece), consists of two close-support *mire*, the 345[a] with Vought A-7H Corsair IIs and the 349[a] with Northrop F-5As, and a tactical reconnaissance *mira*, the 348[a] Mira Anagnoriseos Taktikou, with the few remaining Republic RF-84F Thunderflashes, currently being re-equipped with the McDonnell RF-4E Phantom II. There is another close-support *pterigha*, the 115[a] at Soudha Bay near Iraklion, in Crete; this has two Vought A-7H-equipped *mire*, the 338[a] and the 340[a]. The 111[a] Pterigha at Nea Ankhialos, near Volos (south of Larissa), comprising the 337[a], 341[a] and 343[a] Mire Dioxeos Taktikou or Tactical Fighter Squadrons, operates the lightweight Northrop

F-5A and its photo-reconnaissance variant, the RF-5A, with a few tandem-seat F-5Bs for operational training.

There are three interceptor *pterighe*, all of them located in central Greece. The 114ᵃ Pterigha, based at Tanagra, north-west of Athens, consists of the 336ᵃ and 342ᵃ Mire flying the 40 Dassault Mirage F1CGs received in 1975 to replace the Convair F-102A Delta Daggers supplied as interim equipment by the USA, as well as a dozen Lockheed F-104G Starfighters. The Starfighter still equips the only Mira (the 335ᵃ Mira Dioxeos) of the 116ᵃ Pterigha, based at Araxos in northern Peloponnesus. The 117ᵃ Pterigha, at Andravida, south of Araxos, operates the McDonnell Douglas F-4E Phantom II, 58 of which were delivered between 1974 and 1978; the first Mira to re-equip was the 339ᵃ, and a second squadron was subsequently formed, this being the 344ᵃ Mira. (The actual squadron number remains to be confirmed.) The 28th Tactical Air Command also includes an Operational Conversion Unit, formed in mid-1978 with eight Northrop F-5Bs and four Lockheed TF-104Gs, and an independent *mira*, the 363ᵃ, based at Elevsis, flying the survivors of thirteen ASW-equipped Grumman HU-16B Albatross amphibians obtained from Norway, and an ex-US Navy HU-16D.

The 30th Air Material Command combines all the Aeroporia's transport, communications and helicopter units. It comprises six *mire*, all but the 362ᵃ being based at Elevsis, west of Athens. The 355ᵃ and 356ᵃ Mire, the force's fixed-wing transport units, operate a mixture of Lockheed C-130H Hercules, ex-West German Nord 2501D Noratlases,

with Rockwell T-2E Buckeye jets (forty of which were delivered in 1976–77 to replace Lockheed T-33As and the 361ᵃ with eighteen of the twenty Cessna T-37Cs delivered in 1964.

Future procurements will be limited by available funds, but it is believed that further Northrop F-5s and McDonnell Douglas F-4Es and RF-4Es will be supplied by the USA to make up for service attrition. Transport aircraft will probably be obtained from West Germany or other NATO member countries.

The Hellenic Army, reorganised in 1952 along American lines, has an air component based on US Army Aviation organisation and equipped with US-supplied aircraft. The Army Aviation (Aeroporia Stratou) flies both light aircraft (Cessna U-17s, de Havilland Canada U-6A Beavers, Piper L-21B Super Cubs) and helicopters (Bell 47Gs and UH-1Ds, Agusta-Bell 204Bs and 205As) for artillery spotting, close support, communications and casualty evacuation. Headquarters is at Athinai-Megara, where a major Army airfield has been built; Megara also houses a staff transport flight with two Aero Commander 680FL twins and a Beechcraft Super King Air 200. Aviation Companies, fourteen of which have been formed so far, are attached to Army field units of battalion and division strength. Serial numbers of Army aircraft are allocated in blocks, according to type and primary role, and are normally prefixed Epsilon-Sigma (EΣ) for Hellenic Army (Helliniki Strateia).

The Hellenic navy's recently-formed Naval Wing (Pterigha Navtika) operates an embryo flight of four Aérospatiale Alouette III helicopters equipped for air-sea rescue and anti-submarine duties;

Top: the 360ᵃ Mira based at Kalamata is an advanced training squadron. operating the Rockwell T-2E Buckeye. Students finish their flying training with this unit after initial instruction on the Cessna T-41, followed by basic jet training with the 361ᵃ Mira on Cessna T-37Cs. Above left: the Bell 47G is operated by both the Greek air force and by the army, with some 15 in service. Left: an Augusta-Bell 205A search and rescue helicopter pictured at Larissa. Above right: the Grumman HU-16B Albatross flies anti-submarine patrols from Elevsis with the 363ᵃ Mira

former USAF and RAF Douglas C-47 Dakotas, and the survivors of eight Canadair CL-215 twin-engined amphibians. The 357ᵃ Mira flies Bell 47G light helicopters on liaison and casualty evacuation duties; utility helicopters are operated by the 358ᵃ Mira (Agusta-Bell 204B and 205A), 359ᵃ Mira (Sikorsky UH-19D) and 362ᵃ Mira (Agusta-Bell 205A and 206A). The 362ᵃ Mira, based at Tatoi-Dekelia, north of Athens, also provides transport for Armed Forces executives. Both the 357ᵃ and the 358ᵃ Mire regularly deploy to Kalamata airfield for operational training purposes. The 31st Training Command consists of the National Air Academy (Ethniki Aeroporiki Akademia) at Tatoi-Dekelia with Cessna T-41D primary trainers, and two advanced training *mire*, both at Kalamata–the 360ᵃ

they are normally deployed aboard destroyer escorts (frigates). Serial numbers, allocated in sequence, are prefixed Pi-Nu (πN). Greece's main naval bases are Piraeus, Athens' harbour; Thessaloniki in the north; Volos on the central Aegean coast; and Mitilini on the island of Lesvos, off the coast of Turkey.

Turkey's unique strategic importance lies in its position between Europe and Asia, and between the Mediterranean and the Black Sea. It may be said that Turkey's control over the Dardanelles has always prevented Russian expansion towards southern Europe and the Middle East. This together with the Turkish people's opposition to communism makes the country's role in NATO an important one. This is why the whole of the Turkish

TURKISH AIR FORCE–TÜRK HAVA KUVVETLERI, 1979

1st Tactical Air Force (Birinci Taktik Hava Kuvveti)

1st Air Base	111 Squadron	North American F-100D/F
Eskişehir	112 Squadron	Northrop F-5A/B, RF-5A
	113 Squadron	McDonnell Douglas F-4E
3rd Air Base	131 Squadron	North American F-100C/D/F
Konya	132 Squadron	North American F-100C/D/F
	Station Flight	Lockheed T-33A, Bell UH-1H
4th Air Base	141 Squadron	Lockheed F/TF-104G
Murted	142 Squadron	Convair F/TF-102A
	143 Squadron	Aeritalia F-104S
6th Air Base	161 Squadron	Northrop F-5A/B
Bandirma	162 Squadron	McDonnell Douglas F-4E
9th Air Base	191 Squadron	Lockheed F/TF-104G
Balikesir	192 Squadron	Northrop F-5A/B

3rd Tactical Air Force (Üçüncü Taktik Hava Kuvveti)

5th Air Base	181 Squadron	Convair F/TF-102A

Air Force (Türk Hava Kuvvetleri) is assigned to NATO's 6th Allied Tactical Air Force (ATAF), a component of Allied Forces Southern Europe.

The THK comprises two Tactical Air Forces, an Air Transport Force and an Air Training Command. Squadrons (Filo) are grouped in Air Bases (Üs), which are the exact equivalent of the USAF Wings and consist of two or three squadrons. An impressive force of 50,000 personnel, and 80,000 trained reserves and about 600 aircraft, over half of which have an offensive capability, the THK has been suffering from spares shortage and periodical US arms embargoes due to Turkey's opposition to Greece over Cyprus. Previous US governments have tried to maintain a delicate balance between the two countries by supplying them with equivalent amounts of weaponry, but the Carter Administration, influenced by the Greek lobby in Washington, has tended to support Greece rather than Turkey. The effects of this policy on THK operations have been staggering, the force's efficiency being drastically reduced to such a level that the US Pentagon had to warn Congress about its consequences. The recent embargo has been lifted (but not before the THK threatened to procure aircraft from the Soviet Union) and new aircraft promised. Nevertheless, not much has been accomplished so far to improve US-Turkish relations.

The First Tactical Air Force covers Western Turkey, Cyprus, the Greek Islands of the Aegean Sea, and the frontiers with Greece and Bulgaria. It comprises five air bases: Eskişehir, Konya, Murted; Bandirma on the Marmara Sea, and Balikesir to the south of Bandirma, the last two bases controlling the air space over the Dardanelles. The Third Tactical Air Force keeps watch over the frontiers with Syria, Iraq, Iran and the Soviet Union. It comprises three air bases: Merzifon, Erhac near the city of Malatya, and Diyarbakir north of the Syrian border.

Logically enough, the THK transport force is concentrated in the centre of the country. The two heavy transport squadrons are based at Erkilet,

north of Kayseri, the other squadrons flying from Etimesġut Airport, west of Ankara. A staff transport squadron, whose aircraft regularly fly to most West European countries, is based at Istanbul's Yeşilköy Airport. Both training bases are in Western Turkey, basic training taking place at Gaziemir and advanced training at Cigli near Izmir.

THK aircraft procurement has always been limited by a chronic shortage of funds; many aircraft were delivered under Defence Aid agreements with the USA and West Germany because of Turkey's strategic importance and military reliability. US-supplied aircraft included substantial numbers of North American F-100 Super Sabres, now in need of replacement, as well as Northrop F-5 Freedom Fighters and McDonnell F-4 Phantoms, and two units of Convair F-102A Delta Dagger interceptors. The latter were apparently delivered for the sole reason that Greece received a similar number of the type, and have suffered from questionable serviceability and are now believed to be grounded through lack of spares. Other aircraft were purchased directly from the manufacturers, including a number of Aeritalia F-104Ss to supplement the earlier, MDAP-funded F-104G Starfighters. Recent agreements with the USA provide for the supply of thirty second-hand Northrop T-38A supersonic trainers to replace the elderly Lockheed T-33A still in service, as well as eight tactical reconnaissance McDonnell RF-4E Phantoms, which will be the THK's most modern aircraft.

Merzifon	182 Squadron	Northrop F-5A/B
	183 Squadron	Northrop F-5A/B
7th Air Base	171 Squadron	Northrop F-5A/B
Malatya-Erhac	172 Squadron	Northrop F-5A/B
10th Air Base	101 Squadron	Northrop F-5A/B
Diyarbakir	102 Squadron	Northrop RF-5A, F-5B
Air Transport Force (Hava Nakil Kuvveti)		
12th Air Base	221 Squadron	Lockheed C-130E
Erkilet-Kayseri	222 Squadron	Transall C160
2nd Air Base	121 Squadron	Douglas C-47
Ankara -Etimesġut	122 Squadron	Douglas C-47
	VIP Flight	Cessna 421B, Bell UH-1H
Air Training Command (Hava Eğitim Komütanliği)		
Primary Flying School		Beechcraft T-34A, Cessna T-41D,
Gaziemir		Douglas C-47
Operational Training Base		Lockheed T-33A, Cessna T-37C
Izmir-Cigli		

The USAF was granted the use of two airfields in Turkey: Esenboga near Ankara and Adana-Incirlik in the south. Although the former was handed over to the US Army (which also operates from Izmir-Cigli), the latter has been used for quite some time as a staging post for strategic reconnaissance Lockheed U-2 'spy planes'. As recently as June 1979, the Soviet Union was trying to persuade Turkey to refuse the USAF permission to use its airspace for monitoring Soviet compliance with the Strategic Arms Limitation Treaty (SALT II).

The Lockheed F-104G Starfighter is flown by the Turkish air force's 191 Squadron, which is based at Balikesir alongside a Northrop F-5 unit. The Turkish air force has suffered in recent years from the effects of United States' disapproval of Turkey's action in Cyprus. This has delayed the supply of modern warplanes and spare parts. However, the situation slowly improved in the late 1970s

Left: the Turkish air transport force includes much elderly equipment, such as the Douglas C-54 illustrated, three of which remain in service.
Below left: a North American F-100D fighter-bomber of 111 Squadron.
Lower left: numerically the most important warplane in Turkish service, the Northrop F-5 serves with nine squadrons. An F-5A of 192 Squadron, based at Balikesir, is illustrated.
Bottom left: a Lockheed T-33A advanced trainer

The Turkish army, currently well below strength due to severe financial cutbacks, maintains an Army Air Corps (Kara Ordusu Havaciliği) of some 60 fixed-wing aircraft and 180 helicopters, which operate in flights attached to ground force units of battalion strength or larger. KOH headquarters are located at Ankara-Güverncinlik. Army aircraft have been operating alongside ground forces both in Western Turkey and the critically important eastern Anatolia, where there have been communist attempts to stage an uprising of the Armenian minority against the Ankara government. Internal policing duties are the responsibility of the Gendarmerie (Jandarma Teşkilâti), which operates helicopters attached to its three mobile brigades.

The small Turkish navy includes a naval air arm (Donanma Havaciliği), responsible for the security of the country's territorial waters against intruding submarines. Fixed-wing aircraft operate from Karamürsel, near the city of Izmit and the main naval base of Golcuk; detachments regularly deploy to Sinop in the Black Sea, Izmir (Smyrna) in the West Coast, and Antalya in the south. Anti-submarine helicopters operate in flights attached to naval fleet districts, and are often found in such naval stations as Iskenderun near the Syrian border, Çanakkale in the Dardanelles, and Taskizak near Instanbul. Izmir is also used on occasions by the US Sixth Fleet.

Turkey faces three major political conflicts: internal unrest, either promoted by left-wing organisations or sparked by the Armenian and Kurdish minorities; confrontation with Greece over Cyprus, where the Turkish Cypriots are in minority; or invasion by hostile forces. Despite the difficulties in receiving new equipment and the ever-present threat of further sanctions, there is no doubt that Turkish military aviation will fulfill its tasks as can be expected from a country with Turkey's proud military past–thoroughly and efficiently.

The reason for the strength of Italian military aviation lies in the country's peculiar strategic problems, which became painfully clear during World War II. Most important population and industrial centres are concentrated on the coastal areas, thus being very vulnerable to air attacks; and the long coastline is particularly suited to enemy landings. It is therefore understandable that the Italian air force (Aeronautica Militare Italiana) places a marked emphasis on two roles–air superiority and tactical support. All military flying is co-ordinated by the Air Department of the Defence Ministry (Ministero della Difesa–Aeronautica).

The present-day Italian air force was developed from the Italian Co-belligerent Air Force, which fought alongside the Allies between 1943 and 1945. Although it had been limited to training and trans-

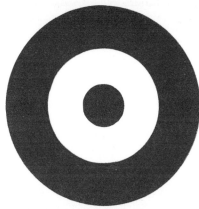

Top: Convair F-102A Delta Dagger interceptors equip two squadrons, although serviceability is poor. Below: a Northrop RF-5A tactical reconnaissance fighter of 102 Squadron based at Diyarbakir

port duties by an agreement signed in September 1947, the changing political scene leading to the formation of NATO in April 1949 caused it to initiate an expansion programme, with extensive aid from the United States. Although reorganisation plans suffered a number of political setbacks, the Italian air force had become a viable component of NATO by the mid-1950s, and successive re-equipment plus the expansion of the Italian aeronautical industry finally made it virtually self-sufficient.

The AMI's first-line formation is the National Air Defence Command (Comando Nazionale della Difesa Aerea), which comprises an *aerobrigata* (air brigade) of MIM-14A Nike-Hercules surface-air missiles; eleven *stormi* (wings), consisting of a total of sixteen *gruppi* (squadrons)–six of interceptor-fighters, four of fighter-bombers, four of fighter-bombers with reconnaissance capability, and two of reconnaissance fighters–and nine *squadriglie collegamenti* (liaison flights) and two *squadriglie collegamento e traino bersagli* (liaison and target towing flights); and an aerobatic *gruppo*, the famous 'Frecce Tricolori' (tricolour arrows) which are well-known throughout Europe due to their participation in air shows. They fly early-production Fiat G 91s converted to G 91PANs (Parruglia Acrobatica Nazionale, or national aerobatic team).

The interceptor *gruppi* fly the Italian-built Aeritalia F-104S, a modernised version of the Lockheed F-104G Starfighter which entered service with nine NATO members and the four fighter bomber *gruppi* are now completely re-equipped with the Aeritalia F-104S. Two fighter-bomber-reconnaissance *gruppi* fly the Fiat G 91R: 14° Gruppo CBR and 103° Gruppo CBR (2^O Stormo CBR 'Mario d'Agostini') at Treviso-San Angelo; one of them is to be transferred to the Udine-Rivolto airfield in the near future. The other two fighter-bomber-reconnaissance *gruppi* operate the much-improved, twin-jet Fiat G 91Y.

The *squadriglie collegamenti* are the 602^a, 608^a and 632^a with the tandem-seat Fiat G 91T, the 603^a, 604^a, 605^a, 606^a, 609^a and 653^a with the Lockheed T-33A and the 636^a and 651^a (with target towing capability) with Lockheed T-33As and a few RT-33As. The last two digits of the *squadriglia* number indicate the *stormo* or *aerobrigata* to which they are attached. Finally, the 1^a Aerobrigata Intercettori Teleguidati, the AMI's anti-aircraft missile force, comprises three *reparti* (units) of four *gruppi* each. The Aerobrigata headquarters are at Padova (Padua), and the *gruppi* are dispersed at several sites. The *reparti* have liaison flights flying Agusta-Bell 204B helicopters.

Support units are concentrated in the Transport and Air Rescue Command (Comando Trasporti e Soccorso Aereo). Its major transport unit is the 46^a Aerobrigata Trasporti Medi (medium transport air brigade), based at San Giusto airfield in Pisa on the west coast, which consists of three *gruppi* TM: the 2^O and the 98° with the twin-engined Aeritalia

G222, which has now completely replaced the elderly Fairchild C-119G and J in the AMI, and the 50° with the Lockheed C-130H Hercules. The 31° Stormo Trasporti Speziali (special transport wing) consists of three *gruppi*. The 93° Gruppo Elicotteri operates most of the AMI's transport and communications helicopters (Sikorsky SH-3D, Agusta-Bell 47J and 204B). The 302° and 306° Gruppi form the Reparto Volo Stato Maggiore (general staff flying unit), responsible for the transport of the country's VIPs. All the *gruppi* of the 31° Stormo Trasporti Speziali 'Franco Lucchini' are based at Rome's Ciampino airport.

The 14° Stormo is responsible for electronic warfare and navaids checking duties. Its two *gruppi*, based at Pratica di Mare near Rome, are the 8° Gruppo Radiomisure equipped with the Lockheed T-33A the PD808RM calibration version of the Piaggio biz-jet, and the venerable Douglas C-47, and the 71° Gruppo Guerra Elettronica with the ECM-equipped Piaggio P166 and PD808, a couple of Douglas EC-47s and a Lockheed EC-130H. The last *stormo* of the Transport Command is the 15° Stormo Soccorso Aereo 'Stefano Cagna', the AMI's air rescue force, consisting of the 84° Gruppo SAR with Grumman HU-16 Albatrosses, and the helicopter-equipped 85° Gruppo SAR. Both *gruppi* are based at Ciampino, but deploy *sezioni* (detachments) to Linate airport in Milan and Grottaglia, near Taranto, respectively.

The general training command (Comando Generale della Scuola) consists of six flying schools and an operational training *gruppo*, as well as other minor maintenance and air traffic control establishments. *Ab initio* flying takes place at Latina-Comani, near Anzio and Nettuno of World War II fame, where the 207° Gruppo SVBIE operates the SIAI-Marchetti SF260AM and the twin-engined Piaggio P166M. Basic jet training is the responsibility of the Lecce-Galatina-based Scuola Volo Basico Iniziale Aviogetti, which consists of three *gruppi* (the 212°, 213° and 214°) equipped with the tandem two-seat Macchi MB326. The 201° and 204° Gruppi SVBAA at Amendola near Foggia provide conversion training on the Fiat G 91T. Helicopter training is given at the Scuola Volo Elicotteri, which consists of a single *gruppo*, the 208°, operating various Agusta-Bell types from Frosinone, east of Latina. There are also the Scuola Centrale Istruttori Volo or central flying instructors' school at Grot-

ITALIAN AIR FORCE–AERONAUTICA MILITARE ITALIANA, 1979		
National Air Defence Command–Comando Nazionale della Difesa Aérea		
1^a Aerobrigata Intercettori Teleguidati		
7° Reparto IT (dispersed)		MIM-14A Nike-Hercules
16° Reparto IT (dispersed)		MIM-14A Nike-Hercules
17° Reparto IT (dispersed)		MIM-14A Nike-Hercules
2^O Stormo Caccia-Bombardieri-Ricognitori		
14° Gruppo CBR	Treviso-San Angelo	Fiat G 91R
103° Gruppo CBR	Treviso-San Angelo	Fiat G 91R
602^a Squadriglia Collegamenti		Fiat G 91T
3° Stormo Caccia Ricognitori		
28° Gruppo CR	Verona-Villafranca	Lockheed F/RF-104G
132° Gruppo CR	Verona-Villafranca	Lockheed F/RF-104G
603^a Squadriglia Collegamenti		Lockheed T-33A
4° Stormo Caccia-Intercettori		
9° Gruppo CI	Grosseto	Aeritalia F-104S
604^a Squadriglia Collegamenti		Lockheed T-33A
5° Stormo Caccia		
23° Gruppo CI	Rimini-Miramare	Aeritalia F-104S
102° Gruppo CB	Rimini-Miramare	Aeritalia F-104S
605^a Squadriglia Collegamenti		Lockheed T-33A
6° Stormo Caccia-Bombardieri		
154° Gruppo CB	Brescia-Ghedi	Aeritalia F-104S
606^a Squadriglia Collegamenti		Lockheed T-33A
8° Stormo Caccia-Bombardieri-Ricognitori		
101° Gruppo CBR	Cervia	Fiat G 91Y
608^a Squadriglia Collegamenti		Fiat G 91T
9° Stormo Caccia-Intercettori		
10° Gruppo CI	Capua-Grazzanise	Aeritalia F-104S
609^a Squadriglia Collegamenti		Lockheed T-33A
32° Stormo Caccia-Bombardieri-Ricognitori		
13° Gruppo CBR	Brindisi	Fiat G 91Y
632^a Squadriglia Collegamenti		Fiat G 91T
36° Stormo Caccia		
12° Gruppo CI	Bari-Gioia del Colle	Aeritalia F-104S
156° Gruppo CB	Bari-Gioia del Colle	Aeritalia F-104S
636^a Squadriglia Collegamenti e Traino Bersagli		Lockheed T-33A, RT-33A
51° Stormo Caccia		
22° Gruppo CI	Treviso-Istrana	Aeritalia F-104S
155° Gruppo CB	Treviso-Istrana	Aeritalia F-104S
651^a Squadriglia Collegamenti e Traino Bersagli		Lockheed T-33A, RT-33A
53° Stormo Caccia-Intercettori		
21° Gruppo CI	Novara-Cameri	Aeritalia F-104S
653^a Squadriglia Collegamenti		Lockheed T-33A
313° Addestramento Aerobatico PAN		Fiat G 91PAN, G 91T

Left: one of the 13 Lockheed C-130Hs operated by the 50° Gruppo, 46^a Aerobrigata TM, the Italian air force's major transport brigade. It is based at San Guisto, near Pisa

taglia, whose 200° Gruppo SCIV has on charge a variety of aircraft and helicopters, and the Scuola Militare Volo a Vela (military gliding school) at Rome's Guidonia airfield with LET L.13 Blanik and CVV Canguro sailplanes and SIAI-Marchetti S 208M and SF 260AM tugs. The operational training unit is the 20° Gruppo Addestramento Operativo, which shares the Grosseto base with the 4° Stormo CI and has Lockheed F-104G and TF-104G Starfighters.

Italy's three air regions (Regioni Aerei) have their own communications squadrons or Reparti

Abbreviations: IT, Intercettori Teleguidati, Interceptor missiles; CBR, Caccia-Bombardieri-Ricognitori, Fighter-bomber-reconnaissance; CR, Caccia-Ricognitori, Fighter-reconnaissance; CI, Caccia-Intercettori, Fighter-interceptor; CB, Caccia-Bombardieri, Fighter-bomber; PAN, Pattuglia Acrobatica Nazionale, National Aerobatic Team.

Transport and Air Rescue Command – Comando Trasporti e Soccorso Aereo

14⁰ Stormo

8⁰ Gruppo RM	Pratica di Mare	Douglas C-47, Piaggio PD808RM, Lockheed T-33A
71⁰ Gruppo GE	Pratica di Mare	Piaggio PD808ECM and P166, Douglas EC-47, Lockheed EC-130H

15⁰ Stormo Soccorso Aereo

84⁰ Gruppo SAR	Roma-Ciampino	Grumman HU-16A
85⁰ Gruppo SAR	Roma-Ciampino	Agusta-Bell 47J/204B, Sikorsky HH-3F

31⁰ Stormo Trasporti Speziali

93⁰ Gruppo Elicotteri	Roma-Ciampino	Agusta-Bell 47J/204B, Sikorsky SH-3D
302⁰ Gruppo RVSM	Roma-Ciampino	Convair 440, Douglas DC-6 and DC-9-32
306⁰ Gruppo RVSM	Roma-Ciampino	Piaggio PD808TA

46ᵃ Aerobrigata Trasporti Medi

2⁰ Gruppo TM	Pisa-San Giusto	Aeritalia G 222
50⁰ Gruppo TM	Pisa-San Giusto	Lockheed C-130H
98⁰ Gruppo TM	Pisa-San Giusto	Aeritalia G 222
646ᵃ Squadriglia Collegamenti		Piaggio P166

Abbreviations: RM, Radiomisure, Navaids checking; GE, Guerre Elettronica, Electronic countermeasures; SAR, Soccorso Aereo, Air rescue; RVSM, Reparto Volo Stato Maggiore, General Staff Flying Unit; TM, Trasporto Medio, Medium transport.

General Training Command – Comando Generale della Scuola

20⁰ Gruppo Addestramento

Operativo	Grosseto	Lockheed F/TF-104G

Central Flying Instructors' School – Scuola Centrale Istruttori Volo

200⁰ Gruppo SCIV	Taranto-Grottaglia	Macchi MB 326, Piaggio P166M, North American T-6, Agusta Bell 47G/J

Primary and Basic Propeller-Driven Aircraft School – Scuola Volo Basico Iniziale Elica

207⁰ Gruppo SVBIE	Latina-Comani	Piaggio P166M, SIAI-Marchetti SF 260AM

Primary and Basic Jet Aircraft School – Scuola Volo Basico Iniziale Aviogetti

212⁰ Gruppo SVBIA	Lecce-Galatina	Macchi MB 326
213⁰ Gruppo SVBIA	Lecce-Galatina	Macchi MB 326
214⁰ Gruppo SVBIA	Lecce-Galatina	Macchi MB 326

Basic and Advanced Jet Aircraft School – Scuola Volo Basico Avanzato Aviogetti

201⁰ Gruppo SVBAA	Foggia-Amendola	Fiat G 91T
204⁰ Gruppo SVBAA	Foggia-Amendola	Fiat G 91T

Helicopter Flying School – Scuola Volo Elicotteri

208⁰ Gruppo SVE	Frosinone	Agusta-Bell 47G/J and 204B

Military Gliding School – Scuola Militare Volo a Vela

	Roma-Guidonia	SIAI-Marchetti S 208M and SF 260AM; LET L.13 Blanik, CVV Canguro (sailplanes)

flight with Piaggio P166Ms and helicopters, which operates from Pratica di Mare, and a gunnery training establishment (Sezione Standardizzazione Tiro) at Decimomannu in Sardinia, one of NATO's most important gunnery ranges – with Lockheed T-33A target tugs.

The AMI intends to standardise on a few modern aircraft types of Italian manufacture, so as to become self-sufficient and free from external political interference. The veteran Lockheed F-104G and Fiat G 91R have been complemented by the improved Aeritalia F-104S and Fiat G 91Y, pending the introduction of the Aeritalia-built version of the Panavia Tornado. There are two medium transport types, the Lockheed C-130H and the twin-engined Aeritalia G 222; the last of the ancient Fairchild C-119G/Js was withdrawn from use in January 1979. Training and communications aircraft, as well as helicopters, will be supplied by the Italian aerospace industry and an urgent requirement for a light electronic countermeasures type is expected to be met by a Piaggio type. It appears that the AMI will fufill its duties within NATO's 5th ATAF quite efficiently, although this depends upon the country's internal situation, which is still far from stable.

The Italian army (Esercito Italiano) has a sizeable air corps, the army light aviation (Aviazione Leggera dell'Esercito), which is currently going through a vast reorganisation initiated in 1976–77. Until that time, ALE units were deployed on a territorial basis, but changes in the structure of the ground forces reflected themselves on its organisation. Some units are now attached to Army High Command Headquarters at Viterbo, others to field units of corps, division and brigade strength. Still others, responsible for training and maintenance, were given semi-autonomy.

Two major units are attached to Army Headquarters. The first is the army light aviation centre (Centro Aviazione Leggera dell'Esercito), which combines the functions of an OCU and a trials unit, and comprises a Reparto Mezzi Aerei (fixed-wing aircraft unit), a Reparto Elicotteri (helicopter unit) and a Campo Sperimentale (Experimental Unit). The second is the 1⁰ Raggrupamento ALE 'Antares' (raggrupamento is the army name for a group of squadrons), with three gruppi. They share the Viterbo base with the Centro Addestramento dell'ALE (training centre) which operates Agusta-Bell helicopters, the CH-47C and SM 1019.

Right: the Aeritalia F-104S is the Italian air force's main interceptor and fighter-bomber, 205 having been delivered. An F-104S interceptor of the 21⁰ Gruppo, 53⁰ Stormo is pictured

Volo Regionali (regional air units), flying a miscellany of light aircraft, some of which – such as the Beechcraft C-45 and North American T-6 Texan – are in need of replacement. Each Reparto has its own Gruppo RVR: the 1⁰ Reparto has the 302⁰ Gruppo RVR at Bergamo's Orio al Serio airfield, the 2⁰ has the 303⁰ Gruppo RVR at Guidonia, and the 3⁰ has the 304⁰ Gruppo RVR at Bari-Palese on the south-western coast. Other AMI units comprise the Reparto Sperimentale Volo or Experimental Flying Unit, comprising the 311⁰ Gruppo RSV with a variety of aircraft types and a communications

Three Raggrupamenti ALE are attached to three army corps (Corpi d'Armata): the 3° at Padova (Padua), the 4° 'Altair' at Bolzano near the Austrian border, and the 5° 'Rigel'. In addition, two Gruppi Squadroni ERI (reconnaissance groups) operate alongside army divisions, with a complement of Agusta-Bell 206 helicopters and another four Gruppi Squadroni are attached to brigades.

It was intended from the outset that the ALE should provide observation and liaison duties to the ground forces to which its units were attached, as well as flying casualty evacuation sorties as required. The delivery of the Vertol CH-47C medium transport helicopter added a new dimension to the ALE, allowing it to ferry troops to the battlefields. The service's high standards of proficiency were confirmed during the very successful Esercitazione Ippogriffo 78 (Exercise Seahorse 78), which took place in October 1978 in Sardinia with a total of 44 ALE helicopters. The next stage in the development of the ALE is the introduction of armed helicopters, which is now under way and comprises the Italian-designed Agusta A.109A and its TOW missile-equipped variants.

Above left: the Macchi MB326 is the Italian's primary and basic jet trainer. It is also operated by the Reparto Sperimentale Volo. Left: a Fiat G 91PAN of the 313° Gruppo Addestramento Acrobatico – the Frecce Tricolori. Below: the Fiat/Aeritalia G.222 serves with the 2° and 98° Gruppo, 46ª Aerobrigata TM. The G 222 has outstanding STOL capabilities

The reason for an anti-tank helicopter force lies in the possibility of major political upheavals in Yugoslavia within the next few years, following the demise of Marshal Tito's present regime. Such a situation could lead to Soviet military intervention in the country as a means of bringing it into the Warsaw Pact as another Soviet satellite. This would present a serious strategic problem, because one of three possible invasion routes into Italy is the Ljubljana Gap on the Yugoslav border–a fairly stable area at present, apart from sporadic civil unrest, but likely to be less so if Yugoslav conditions changed. There would even be a possibility of a Soviet armoured strike against Italy in the Trieste area, which shows how important ALE units stationed in northern bases may become. Whatever the future events, it may be assumed that the ALE would be prepared to play its part

The Naval air arm (Marinavia, short for Aviazione per la Marina Militare) actually comprises two separate formations. The first one is a purely naval force and consists of three Gruppi AMM, with headquarters at Catania-Fontanarossa: the 1º Gruppo AMM with the Agusta-Bell 212AS,

Right: the Aviazione Leggera dell' Escercito (ALE) mainly operates helicopters. The Agusta-Bell 204A (illustrated), 205A and 206B equip one unit of the OCU/trials ALE centre attached to Army HQ. Below: the centre also has ALE's only fixed-wing unit, operating Cessna O-1Es (illustrated) and Piper L-18/L-21s. The O-1E is designated AL-1 under the ALE system introduced in 1977

the 2º with Agusta-Bell 204B(AS)s and the Agusta A.106, and the 3º with Sikorsky SH-3Ds. These helicopters, which have code numbers according to their types rather than their units (for example, SH-3D codes are prefixed '6-', and AB.204B(AS) codes are prefixed '3-'), are normally deployed aboard naval vessels with helicopter provision, such as the cruisers *Vittorio Veneto*, *Andrea Doria* and *Caio Duilio*, the ASW destroyers *Impavido* and *Intrepido*, and a number of frigates.

The second component of the Marinavia, flying fixed-wing aircraft, consists of two air force *stormi antisommergibili* (anti-submarine wings), with air force aircraft and personnel under naval control. These are the 30º Stormo AS and its 86º Gruppo AS, based at Cagliari-Elmas with Breguet Atlantics, and the 41º Stormo AS, consisting of the 87º Gruppo AS at Sigonella in Sicily with the Grumman S-2 and the 88º Gruppo AS at Catania-Fontanarossa, north of Sigonella, with the Breguet Atlantic; the Stormo's communications unit is the 641ª

Miscellaneous units		
Experimental Flying Unit–Reparto Sperimentale Volo		
311º Gruppo RSV	Pratica di Mare	Aeritalia F-104S and G 222, Fiat G 91R/T/Y, Macchi MB 326, SIAI-Marchetti SF260AM and SM 1019 (Agusta-Bell 47G/J and 204B and Piaggio P166M for liaison)
Gunnery Training Establishment–Sezione Standardizzazione Tiro		
	Decimomannu	Lockheed T-33A
Regional Air Units		
1º Reparto Volo Regionale (1ª Regione Aerea)		
302º Gruppo RVR	Bergamo-Orio al Serio	Agusta-Bell 47J and 204B, Piaggio P166M, SIAI-Marchetti S 208M, Beechcraft C-45, North American T-6
2º Reparto Volo Regionale (2ª Regione Aerea)		
303º Gruppo RVR	Roma-Guidonia	P166M, SIAI-Marchetti S 208M, Beechcraft C-45, Douglas C-47, North American T-6
3º Reparto Volo Regionale (3ª Regione Aerea)		
304º Gruppo RVR	Bari-Palese	Agusta-Bell 47J and 204B, Piaggio P166M, SIAI-Marchetti S 208M, Beechcraft C-45, North American T-6

Left: an Agusta-Bell 204B of 2° Gruppo AMM, which also flies the Agusta A 106 and is re-equipping with the Agusta-Bell 212. The helicopter-equipped anti-submarine warfare 1°, 2° and 3° Gruppo AMM, are naval units which are normally deployed aboard frigates and cruisers for operations.
Lower: an Agusta-built Sikorsky SH-3D of the 3° Gruppo AMM, whose helicopters will equip the carrier Garibaldi when it enters service.
Below: the anti-submarine Breguet Atlantic is operated by the 86° Gruppo, 30° Stormo one of whose aircraft is illustrated, and by the 88° Gruppo, 41° Stormo. Naval units are under the command of the naval aviation department but are fully integrated with the air force, and the fixed-wing 30° and 41° Stormi are in fact within the air force establishment

ARMY LIGHT AVIATION–AVIAZIONE LEGGERA DELL'ESERCITO, 1979

Operational units–attached to Army Headquarters
Centro Aviazione Leggera dell'Esercito

Reparto Mezzi Aerei (fixed-wing)	Viterbo	**Piper L-18/L-21 Super Cub, Cessna O-1E**
Reparto Elicotteri (helicopters)	Viterbo	**Agusta-Bell 204B/205A/206**
Campo Sperimentale (trials unit)	Viterbo	various types

1° Raggruppamento ALE 'Antares'

11° Gruppo Squadroni ETM 'Erco'	Viterbo	**111ª Squadriglia ETM (Vertol CH-47C)**
12° Gruppo Squadroni ETM 'Gru'	Viterbo	**121ª Squadriglia ETM (Vertol CH-47C)**
51° Gruppo Squadroni EM 'Leone'	Viterbo	**511ª Squadriglia EM 512ª Squadriglia EM 513ª Squadriglia EM (Agusta-Bell 205A)**

Operational units–attached to Army Corps
3° Raggruppamento ALE

23° Gruppo Squadroni EM	Padova	**231ª Squadriglia EM (Agusta-Bell 205A)**
53° Gruppo Squadroni EM 'Cassiopea'	Padova	**531ª Squadriglia EM 532ª Squadriglia EM 533ª Squadriglia EM (Agusta-Bell 205A)**

4° Raggruppamento ALE 'Altair'

24° Gruppo Squadroni ALE 'Orione'	Bolzano	**241ª Squadriglia ERI 242ª Squadriglia ERI (Agusta-Bell 206) 243ª Squadriglia AL (SIAI-Marchetti SM 1019)**
44° Gruppo Squadroni ALE 'La Fenice'	Bolzano*	**441ª Squadriglia AL (SIAI-Marchetti SM 1019) 442ª Squadriglia ERI (Agusta-Bell 206)**
54° Gruppo Squadroni EM 'Cefeo'	Bolzano*	**541ª Squadriglia EM 542ª Squadriglia EM 543ª Squadriglia EM 544ª Squadriglia EM 545ª Squadriglia EM (Agusta-Bell 205A)**

(*gruppo* headquarters only; *squadriglie* are dispersed)

5° Raggruppamento ALE 'Rigel'

25° Gruppo Squadroni ALE 'Cigno'	Vittorio Veneto	**251ª Squadriglia AL (SIAI-Marchetti SM 1019) 252ª Squadriglia ERI (Agusta-Bell 206)**

Squadriglia Collegamento.

Aircraft are also operated by other Italian security organisations, controlled by the Ministero dell'Interno (internal affairs ministry). These are the 83,500-strong Corpo di Carabinieri (Carabineer Corps), which has thirty Agusta-Bell 47Js and about as many AB.204Bs, 205As and 206As with code numbers prefixed CC; the Servizio Aereo Polizia, the Public Security Police's flying unit, with a dozen Partenàvia P.64B Oscar light aircraft and over thirty Agusta-Bell 47Js, 206As and 212s; and the Guardia di Finanza or Treasury Guard, whose air service has about a hundred Agusta-Bell 47J and Nardi-Hughes 500M helicopters with GdiF codes. A fourth paramilitary unit, the Corpo Nazionale dei Vigili del Fuoco or national fire service has over a score of Agusta-Bell 47G-2s and 47G-3B-1s, 47J 3s, 205A 1s and 206As.

Unit	Base	Squadriglia
55º Gruppo Squadroni EM 'Dragone'	Casarsa	551ª Squadriglia EM 552ª Squadriglia EM 553ª Squadriglia EM 554ª Squadriglia EM (Agusta-Bell 205A)
Operational units–attached to Divisions		
48º Gruppo Squadroni ERI 'Pavone' (Divisione Fanteria 'Mantova')	Mantova	481ª Squadriglia ERI 482ª Squadriglia ERI (Agusta-Bell 206)
49º Gruppo Squadroni ERI 'Capricorno' (Divisione Corazzata 'Ariete')	Casarsa	491ª Squadriglia ERI 492ª Squadriglia ERI (Agusta-Bell 206)
Operational units–attached to Brigades		
47º Gruppo Squadroni ERI 'Levriero' (Brigata Paracadutisti 'Folgore')	Treviso	471ª Squadriglia ERI 472ª Squadriglia ERI (Agusta-Bell 206)
20º Gruppo Squadroni ALE 'Andromeda'	Salerno-Pontecagnano	201ª Squadriglia AL (SIAI-Marchetti SM 1019) 202ª Squadriglia ERI (Agusta-Bell 206)
21º Gruppo Squadroni ALE 'Orsa Maggiore'	Cagliari	211ª Squadriglia AL (SIAI-Marchetti SM 1019) 212ª Squadriglia EM (Agusta-Bell 205A)
26º Gruppo Squadroni EM	Pisa-San Giusto	261ª Squadriglia EM 262ª Squadriglia EM (Agusta-Bell 205A)
28º Gruppo Squadroni ALE 'Tucano'	Roma-Urbe	281ª Squadriglia AL (SIAI-Marchetti SM 1019) 282ª Squadriglia ERI (Agusta-Bell 206)

Abbreviations: ETM, Elicotteri Trasporto Medio, Medium transport helicopter; EM, Elicotteri Multiruolo, Multi-purpose helicopter; ERI, Elicotteri Ricognizione, Reconnaissance helicopter; ALE, Aviazione Leggera dell'Esercito, Army Light Aviation.

Support units		
Army Light Aviation Training Centre-Centro Addestramento dell'ALE (CAALE)	Viterbo	Agusta-Bell 47G, 205A and 206; Vertol CH-47C; SIAI-Marchetti SM 1019
Army Aviation Repair Units–Reparti Riparazione ALE		
1º RRALE	Bracciano-Monte dell'Oro	
2º RRALE	Bologna	
3º RRALE	Bergamo-Orio al Serio	
4º RRALE	Viterbo	

NAVAL AVIATION-AVIAZIONE PER LA MARINA MILITARE (MARINAVIA)

1º Gruppo AMM	Catania-Fontanarossa	Agusta-Bell 212AS
2º Gruppo AMM	Catania-Fontanarossa	Agusta A.106, Agusta-Bell 204B
3º Gruppo AMM	Catania-Fontanarossa	Sikorsky SH-3D
30º Stormo Antisommergibili		
86º Gruppo AS	Cagliari-Elmas	Breguet Atlantic
41º Stormo Antisommergibili		
87º Gruppo AS	Sigonella	Grumman S-2
88º Gruppo AS	Catania-Fontanarossa	Breguet Atlantic
641ª Squadriglia Collegamento		various types

CHAPTER NINE

Non-aligned Europe
Albania/Austria/Finland/Spain
Sweden/Switzerland/Yugoslavia

Right: three Ilyushin Il-14Ms are operated by the Albanian air force's sole transport squadron alongside Lisunov Li-2s and ten Antonov An-2s. Mainly supplied by Communist China, Albania's front-line aircraft comprise two squadrons of MiG-15, two of MiG-17s, three of MiG-19s and two of Mil Mi-4 utility helicopters. They use MiG-15UTIs, Yak-11s and Yak-18s for training purposes.
Opposite: the Saab-105Ö is the Austrian air force's main combat type. Operated as an interceptor and fighter-bomber, it is overdue for replacement. A Karo As aerobatic team Saab 105Ö is pictured before flight

Albania's first attempt to form an air force in early 1914 was frustrated by the outbreak of war later that year, when the three Lohner biplanes it had ordered from Austro-Hungary were taken over by the latter country's air arm. Too poor between the wars, and occupied by Italian forces during most of World War II, it did not acquire an air force of its own until 1946–47, by which time it had become aligned with the Soviet Union.

Albania became a member country of the Warsaw Pact in May 1955, and for the next few years the development and equipment of its air force was undertaken by the USSR. The country's total defence forces, including paramilitary, internal security and frontier guard, were estimated in 1979 at around 67,000 personnel, with another 65,000 in reserve, representing in all nearly 5 per cent of the total population of some 2·7 million. Of the three major armed services, more than half of the person-

of Soviet design, but Albania's membership of the Warsaw Pact was for all practical purposes terminated in 1962 after the severance of diplomatic relations with the USSR in late 1961. Since then its supplies of aircraft have been drawn from China, where numerous Soviet aircraft types are built by the State Aircraft Factories.

A German-Austrian 'Flying Troop' was formed in December 1918, after the dissolution of the Austro-Hungarian empire, which had had a substantial air force during World War I. It was disbanded within a year, and it was 1936 before an air force was again established in the republic, with aircraft imported primarily from Germany and Italy. The Austrian Air Force was absorbed into the Luftwaffe after the 1938 Anschlüss, and Austria did not regain its national sovereignty until May 1955, after a decade of Allied occupation.

The postwar Österreichische Heeresfliegerkrafte

nel are conscripts, most of them in the Army. Some 1,500 conscripts serve in the People's Army Air Force, the personnel strength of which has virtually doubled since 19 2 to a figure in 1979 of about 8,000, a manpower level nearly three times that of the Albanian Navy. As its name indicates, the APAAF is an offshoot of, and subordinate to, the land forces.

The principal air bases in Albania are the centrally-located Tiranë–where the APAAF has its headquarters–and Durazzo/Shijak, further south at Berati/Qytet Stalin and at the south-western seaport of Valona. Front-line combat aircraft are relatively few, numbering about 100 in all; these are supplemented by a handful of obsolete fixed-wing transports, and about 30 Soviet Mi-4 Hound helicopters. All of the country's military aircraft are

was reconstituted shortly afterwards, its first aircraft being four Yak-11 and four Yak-18 piston-engined trainers donated by the Soviet Union. A quartet of Czechoslovak Zlin 126 trainers was acquired during 1957, but since that time the Austrian Air Force has drawn its equipment from western or non-aligned sources.

Out of a total armed forces manpower of about 37,000, the Air Force accounts for approximately 4,000, of whom some 2,000 are conscripts called up for six months' full-time service plus another two months' reservist training a year for 12 years. The Air Force is an integral component of the Austrian Army, the current organisation dating from 1973 when a Fliegerbrigade (now Fliegerdivision) was established under control of the Armeekommando to administer all aspects of air defence except that of anti-aircraft artillery. Commandant of the Flieger-division, whose headquarters is in Vienna, is Oberst Joseph Haiböck.

As might be expected from the statistics, the Austrian Air Force receives a comparatively small percentage of the country's national defence budget, and consequently is in no position to buy or maintain costly aircraft of an advanced nature or such modern defence aids as a sophisticated radar defence network. In fact, the country's total defence expenditure, per head of the population, is the smallest of any European country. The last major updating of Luftstreitkräfte front-line equipment took place in the late 1960s when, abandoning over-ambitious plans to purchase Saab 35 Draken super-sonic fighters from Sweden, it settled instead for 40 examples of the less-costly Saab 105Ö for the assorted roles of training, light strike and recon-naissance. At about the same time it took delivery of two Short Skyvan 3M military transports – Austria was the first customer for the military Skyvan – and two Sikorsky S-65Ö heavy-lift helicopters.

Sixty years old in 1978, the Finnish Air Force originated in the civil war between 'white' and 'red' factions which broke out when post-Revolution Russian forces remained in the newly-independent state. Its first aircraft were a motley assortment of imports from France, Germany and elsewhere, and were often flown by Swedish volunteer pilots fighting on the side of the Finns. First efforts to re-organise it on a firm basis were made by a French

OSTERREICHISCHE HEERESFLIEGERKRAFTE – AUSTRIAN ARMY AVIATION, 1979

Fliegerregiment I		
I Hubschraubergeschwader		
I Staffel	Agusta-Bell 204B	Linz-Horsching
II Staffel	Bell OH-58B	Tulln-Langenlebarn
III Staffel	Agusta-Bell 206A	Tulln-Langenlebarn
		(detachment at Klagenfurt)
Flachflugelstaffel	Skyvan, Turbo Porter, Safir,	Linz-Horsching
	O-1E Bird Dog	
Fliegerregiment II		
II Hubschraubergeschwader		
I Staffel	Alouette III	Aigen-in-Ennstal
		(detachment at Schwaz)
II Staffel	Alouette III	Aigen-in-Ennstal
Uberwachungsgeschwader		
I Staffel	Saab 105 Ö	Linz-Horsching
III Staffel	Saab 105 O	Linz-Horsching
Fliegerregiment III (HQ: Graz-Thalerhof)		
III Hubschraubergeschwader		
I Staffel	Agusta-Bell 204B	Graz-Thalerhof
II Staffel	Sikorsky S-65 Ö	Graz-Thalerhof
Jagdbombergeschwader		
II Staffel	Saab 105 Ö	Graz-Thalerhof
IV Staffel	Saab 105 Ö	Graz-Thalerhof
(Duzenflugstaffel – OCU)		
Schulgeschwader	Safir	Zeltweg

Above: II Hubschraubergeschwader operates 21 Aérospatiale Alouette IIIs on SAR, casevac, observation and liaison duties.
Left: one of the two heavily utilised Short Skyvans of the Flachflugelstaffel which also operates 16 Saab-91Ds, 12 Pilatus PC-6s and 7 Cessna O-1Es.
Below left: the two Sikorsky S-65Ös of III Hubschraubergeschwader assist in Alpine civil engineering projects

military mission in 1919, followed in 1924 by one from Britain, and an aircraft manufacturing capability began to make the Finnish Air Force less dependent upon direct imports of aircraft from these countries. Licences were obtained to produce selected foreign types, but home-designed aircraft also began to appear from the mid-1920s onward. Like most other European nations, Finland saw the need to modernise and expand its armed forces in the latter 1930s, and this brought into the Air Force inventory imported modern British and American types such as the Bristol Blenheim and Brewster F2A, plus licence-built aircraft such as the Dutch Fokker D XXI fighter. A victim alternately of the USSR and Germany during World War II, the Finnish Air Force fought gallantly during both the early 'Winter War' of 1939–40 and the later 'Continuation War', but suffered heavy losses. After the end of hostilities in Europe it was obliged to reduce its already-depleted strength to no more than 60 first-line aircraft, chief among which was the German Messerschmitt Bf 109G fighter.

That figure is still not exceeded today, despite the

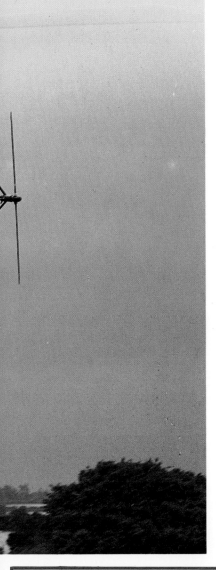

fact that the Ilmavoimat (Finnish Air Force) is the second largest of Finland's three primary services, with a manpower force of about 3,000 personnel. Some 80 per cent of all Finland's servicemen are conscripts, national service varying from eight months for other ranks to 11 months for officers and NCOs. An independent arm of the defence forces, the Ilmavoimat is administered from Tikkakoski, with Major General Rauno Meriö as Commander-in-Chief and Colonel K. Korttila as Chief of Air Staff. Principal air bases are at Kauhava, Luonet-jarvi, Pori, Rissala, Rovaniemi, and Utti. Training establishments include the Ilmasotakoulu (air force academy) at Kauhava, the Ilmavoimien Viestikoulu (Air Force Signals School) at Tik-kakoski, and the Ilmavoimien Teknillinen Koulu (Air Force Technical School) at Halli.

Rovaniemi is the home of the Lapin Lennosto, or Lapland Wing, responsible for the air defence of the northern half of the country. South-eastern Finland is covered by the Karelian Wing (Karjalan Len-nosto), based at Rissala, while the southern and western areas are the responsibility of a third air defence wing, the Satakunnan Lennosto, based at Pori. Each of these wings has one hard-core combat unit. At Rovaniemi, the fighter squadron is HävLv 11, which has been flying Saab 35 Drakens since the summer of 19 2. Initially these were aircraft leased from Sweden, but were replaced from the spring of 1974 by the locally-built Saab 35S version assembled by the state-owned Valmet aircraft factory.

At Pori, HävLv 21 operates a few Drakens, though a substantial number remains of the original 80 Potez Magister armed trainers imported or licence-built by Valmet in the late 1950s/early 1960s, and many of these are available for the light strike role. It is to replace the Magisters in the joint roles of weapon training and attack that Finland is now awaiting delivery of 50 British Aerospace Hawks; most of these, too, will be assembled by Valmet, deliveries being scheduled to begin in 1980.

The Rissala-based third air defence wing has as its combat unit HävLv 31, which was equipped from the mid-1960s with the Soviet MiG-21F Fishbed-C. This sector was perhaps the weak spot in the Finnish air defences, since this early model of the MiG-21 is suitable for clear-weather interception duties only, lacking an all-weather capability. However, de-livery of Mig-21bis fighters during 1979 has remedied this deficiency.

Operational conversion training for the three combat squadrons is carried out on tandem two-seat versions of the now-ancient MiG-15, the MiG-21 and the Draken. Including these, the total combat aircraft strength of the Finnish Air Force is fewer than 50 machines.

Utti is the base for the Kuljetuslentolaivue, a unit which embraces a multitude of assorted aircraft and duties. Its primary purpose is to provide the trans-port requirements of the Finnish Air Force, a task which it accomplishes mainly with about eight aged Douglas C-47s, backed up by two more recently acquired Cessna 402s and five Piper Cherokee Arrows for communications and liaison work, an assortment of helicopters and a couple of Ilyushin Il-28 jet bombers now employed for target towing. The helicopter flight is about a dozen strong, com-prising about six Russian Mil Mi-8s, three smaller

ILMAVOIMAT – FINNISH AIR FORCE, 1979		
HävLv 11	Saab Draken	Rovaniemi
HävLv 31	MiG-21F, MiG 21bis	Rissala
HävLv 21	Potez Magister	Pori
Ilmasotakoulu	Saab Safir	Kauhava
(Central Flying School)	Potez Magister	
Kuljetuslentolaivue	Douglas C-47	Utti
(Transport Squadron)	Mil Mi-4 and Mi-8	

Below: the Potez Magister armed trainer is flown by the Finnish air force's HävLv 21 in the weapons training role and is also available for light attack.
Bottom: one of the six Saab 35BS Drakens flown by HävLv 11, the fighter element of the Finnish air force's Lapland wing. They were leased from Sweden in 1972 for training and later bought outright following deliveries of six J 35Fs, three S 35C trainers and 12 Valmet-assembled S 35XS

Mi-4s, and one example each of the Aérospatiale Alouette II, Agusta-Bell Jet Ranger and Hughes 500M.

Primary flying training is given, at the air force academy at Kauhava, on about two dozen Saab 91D Safir piston-engined trainers. Student pilots then progress to advanced training and weapon training on the Magister before operational conversion and assignment to a front-line squadron. The Safirs are about to be replaced by the first of 30 examples of the Valmet Leko-70 Vinka, an indigenously-designed and manufactured type ordered in the spring of 1973 and first flown in prototype form in July 1975. As already indicated, the Magisters will eventually give way to the BAe Hawk in the advanced training role.

The size of Finland's air force may be modest, and the modernity of some of its equipment somewhat less than it would wish, due mainly to considerable political and financial limitations. Its training, however, is of a high standard, and its operational efficiency and readiness are worthy of respect.

Having created an observation unit using tethered balloons in 1896 (the Servicio Militar de Aerostación), Spain can claim to be one of the earliest countries to recognise the potential uses of the air for military purposes. This force was first employed

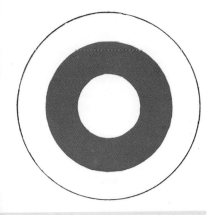

Left: the helicopter flight of the Ilmavoimat's Transport Squadron flies single examples of the AB 206A Jet Ranger (illustrated), Alouette III and Hughes 500, alongside Mi-8s and 4s. Below left: HävLv 31, the Karelain wing's fighter squadron, operated the MiG-21F, 22 of which were supplied by the Soviet Union in 1963–65, introducing the Ilmavoimat to supersonic flight. All-weather MiG-21bis are currently re-equipping HävLv 31. Bottom: a Saab 35BS of HävLv 11, assigned defence of northern Finland

in an actual conflict in Morocco during 1909–11, and it was during the latter year that Spain acquired, from France, its first four fixed-wing powered aircraft. In March 1911 the government created an embryo air force, the Aeronáutica Militar Española, increasing its aircraft strength to 12 by the end of that year. During World War I an Aeronáutica Naval was also formed, although little development of either service was possible during the war years. After the Armistice, however, the victorious combatant nations had large quantities of war-surplus combat aircraft for disposal, enabling the Spanish government virtually to take its pick of several excellent French, British and Italian aircraft. Many of these were employed operationally against the Riff tribesmen in Morocco, until the latter's capitulation in 1926.

By this time, Spain had a healthy home aircraft industry which was providing an increasing proportion of the AME's equipment, but the 600 aircraft or so which it boasted in 1927 were run down during the four years before the monarchy was deposed and the country became a republic. When a revolution broke out in Morocco in the summer of 1936, the AME's aircraft totalled less than half of its 1927 strength, and in the ensuing three-year civil war in Spain both of the opposing air arms–those of the Republican government and the Nationalist forces under General Franco–attracted a tremendous influx of newer foreign aircraft, depending upon the political sympathies of the supplier. After his accession to power as the new President, Franco set up a Spanish Air Ministry in August 1939 and a new, autonomous air force, the Ejército del Aire, three months later. That title is retained by the Spanish Air Force of today.

The wide assortment of types that made up the nearly 1,000 aircraft of the EdA was hardly conducive to giving the new service a cohesive start in life, but with most of the rest of Europe at war it had little choice other than to try to rationalise its equipment at least partially by the addition of domestically-built aircraft. There were predominantly of German and Italian design, but by 1945 the service was still operating a hopelessly large mixture of mainly obsolete aircraft, and output of nationally-designed types was depressingly low. Salvation appeared, however, with the signing of a defence treaty with the USA in 1953, whereby in return for allowing Strategic Air Command the use of air bases in Spain, the US Government provided large-scale financial and technical assistance enabling the Spanish Air Force to expand, modernise and re-equip with jet aircraft.

This began with the establishment of a jet school (Escuela de Reacción) at Radajoz. From here the first Spanish jet pilots graduated in mid-1955, going on to fly North American F-86F Sabre jet fighters from the following autumn. Over the next few years the EdA formed nine Sabre squadrons and also took delivery of appreciable numbers of North American T-6 Texan piston-engined basic trainers, Douglas C-47 transport aircraft, Grumman Albatross amphibians for air/sea rescue, and Sikorsky and Bell helicopters. The Texans were regarded purely as a stopgap, for Spain's Hispano Aviación company flew a domestic replacement for the T-6 in the shape of the HA 100 Triana in late 1954;

within a year it had also flown a basic jet trainer, the HA 200 Saeta. Spain's old Heinkel He 111 (CASA 2.111) bombers were turned into trainers for the aircrew of multi-engined aircraft. Its other major aircraft manufacturer, CASA, began producing the Halcón twin-engined light transport for the EdA. In 1956 a new air defence command was established at Torrejón de Ardoz, assuming all responsibility for the interceptor fighter force, aircraft control and warning stations and anti-aircraft defences.

Since 1953 the US-Spanish base agreements have been renewed about every five years, the most recent renewal being in 1975. This guaranteed the continuing use by SAC of the air bases at Torrejón (Madrid), and Zaragoza, as well as the naval base at Rota (Cadiz). However, the US Air Force has had to forgo its former flight refuelling tanker base in Spain and can no longer base nuclear submarines or nuclear weapons in the country.

Now 40 years old in its current guise, the EdA is a strong and effective air force with 200-plus combat aircraft, most of them modern types, and a considerable back-up of support, general-purpose and training aircraft. Manpower totals about 36,000, of whom some 9,000 are conscripts on a 15-month term of national service. Chief of the Air Staff is Lieutenant General Don Ignacio Alfaro Arregui, under whom the air force is organised into main commands: Air Defence Command (Mando de la Defensa Aéreo), Tactical Air Command (Mando de la Aviación Táctica), Air Transport Command (Mando de la Aviación de Transporte), Training Aviation (Aviación de Entrenamiento), and the Air Search and Rescue Service (Servicio Aéreo de Busqueda y Salvamento). Geographically, aircraft operate in one or other of the four Regiones Aéreas (air regions) into which the country is divided; these have their headquarters at Madrid, Seville, Zaragoza, and Las Palmas in the Canary Islands.

The Mando de la Defensa Aéreo, or MDA, has two squadrons each of Dassault Mirage III-E and McDonnell Douglas F-4C Phantom fighters, and four squadrons have formed on the swept-wing Mirage F1C, of which Spain ordered 72 in the mid-1970s. The Mirage III squadrons (*escuadrones*) are Nos 111 and 112, forming the 11th Fighter Wing (Ala de Caza) at Manises, the Phantom equips Escuadrones 121 and 122 (Ala de Caza 12) at Torrejón and the Mirage F1.C entered service with Escuadron 141 of Ala de Caza 14, based at Los Llanos-Albacete. Co-ordinating the air defence network of interceptors, ground radar warning and control stations, surface-to-air missile batteries (Nike-Hercules, Hawk and Improved Hawk), and anti-aircraft artillery, is the recently-introduced Combat Grande semi-automatic computer-based air defence system, funded largely by the United States and compatible with both the French Strida and NATO Nadge systems employed for similar purposes. Additional elements of the MCA provide jet conversion training at the Escuela de Reactores at Badajoz, while Esc 744 and 745 equip the Escuela de Polimotores at Matacan to provide multi-engine experience on the Beech Baron and King Air, CASA Aviocar and Douglas C-47. Helicopter instruction, and training in aerial photography, are

EJERCITO DEL AIRE–SPANISH AIR FORCE, 1979

Mando de la Defensa Aérea–Air Defence Command

Ala de Caza 11	Dassault Mirage III	Manises
Ala de Caza 12	McDonnell Douglas F-4C	Torrejón
Ala de Caza 14	Dassault Mirage F1C	Los Llanos
Grupo 41	Lockheed T-33	Valenzuela

Mando de la Aviación Táctica–Tactical Air Command

Ala 21	Northrop SF-5A	Morón
	Hispano HA220	
Ala 46	Douglas C-47	Canary Islands
	CASA Aviocar	
	Hispano HA200	
	North American T-6	
	Northrop RF-5	
Escuadrón 221	Grumman HU-16	La Parra
	Lockheed P-3	
Escuadrilla de Enlace 407	Cessna O-1	Tablada

Mando de la Aviación de Transporte–Air Transport Command

Ala de Transporte 35	CASA Azore	Getafe
	Douglas C-54	
Ala de Transporte 37	DHC Caribou	Villanubla
Escuadrón de Transporte 301	Lockheed C-130	Valenzuela

*Left: ex-USAF McDonnell Douglas F-4Cs equip
121 and 122 Squadrons, Ala de Caza 12, which
also flies four RF-4C reconnaissance aircraft.
Below: 60 CASA C.101 Aviojet basic/advanced jet
trainers are on order for the Ejército del Aire. The
first prototype is pictured*

Above right: the Mirage F1CE interceptor equips Ala da Caza 14 at Los Llanos, an aircraft of this unit's No 141 Squadron being pictured.
Above: the KC-130H Hercules tankers of No 301 Squadron are used to refuel the McDonnell Douglas F-4C Phantoms of Ala de Caza 12.
Left: operational conversion onto the Spanish air force's advanced jet aircraft is by way of the Lockheed T-33A trainers of Grupo 41.
Below left: the Grumman HU-16 Albatross anti-submarine amphibians of No 221 Squadron were superseded by the Lockheed P-3 Orion in 1977, but the type continues in service for air/sea rescue work

dispensed at Cuatro Vientos, at the Escuela de Helicopteros and the Servicio Cartografico y Foto respectively, while paratroops train at the Escuela de Paracaidistas at Alcantarilla.

Naval aviation in Spain originated during World War I with the creation of a small Aeronáutica Naval, equipped mainly with Curtiss Type F flying boats. The force was expanded postwar with the acquisition of Dornier and Macchi flying boats, Supermarine Scarab amphibians and Blackburn Velos torpedo-bombing floatplanes. Responsibility for naval aviation was assumed by the Ejército del Aire upon its formation in November 1939, and did not revert to a separate service until June 1954 when, as a result of the Spanish-American defence treaty of the previous year, it was re-established as the Arma Aérea de la Armada with US-supplied Bell 47G helicopters.

Today the Spanish naval air arm, although small, has a fair amount of muscle, including in its inventory the British Aerospace Harrier V/STOL strike aircraft and the Bell HueyCobra gunship helicopter. There are six active squadrons, two others – Escuadrilla 002 and Essla 004 – having been disbanded or deactivated within recent years.

The only fixed-wing unit is Eslla 008, which was formed at Rota (Cadiz) at the end of 1976 to equip with the first of an initial order for six AV-8A and two TAV-8A Harriers, to which the Spanish Navy has given the name Matador. An additional five Matadors were ordered in 1977. All other naval aviation squadrons are helicopter units, two of the

more important being Essla 003, with the Agusta-Bell 212AS for anti-submarine warfare, and Eslla 007, operating Bell AH-1G HueyCobras in the anti-shipping strike role. These three units form a mixed-mission group operating either from shore base at Rota or from the Spanish Navy's only current aircraft carrier, the PH-01 *Dédalo* (formerly the US Navy escort carrier *Cabot*). Essla 003 also retains some of its earlier anti-submarine Agusta-Bell 204AS helicopters, now stripped of their ASW gear and employed instead for transport and light strike duties.

Also operating alternatively from the *Dédalo* or from a shore base is Escuadrilla 005, which flies the Spanish Navy's largest helicopters, Sikorsky SH-3D and SH-3G Sea Kings. Eslla 005 was formed in 1966, initially with six of these excellent ASW helicopters, later increasing this establishment to 24. Budget limitations, influenced also by the political changes in Spain following the death of President Franco, have postponed indefinately plans for a second aircraft carrier, or sea control ship, the *Almirante Carrero*. This was to have entered service in the late 19′0s, but in its absence the *Dédalo* is now scheduled to continue its active life until 1982 or even later.

The two remaining Spanish Navy helicopter squadrons each fly smaller aircraft, the Bell 47Gs of Eslla 001 being used for shore-based communications and liaison flights from the airfield at Marin. Eslla 006, flying the Hughes 500M, is yet another anti-submarine unit, formed in the late 1960s and

SVENSKA FLYGVAPNET – SWEDISH AIR FORCE, 1979

F1	Saab J35F Draken	Västerås
F4	Saab J35D Draken	Ostersund
F5	Saab Sk60	Ljungbyhed
	Scottish Aviation Sk 61	
F6	Saab AJ37 Viggen	Karlsborg
F7	Saab AJ37 Viggen Lockheed C-130 Douglas C-47	Såtenäs
F10	Saab J35F Draken	Angelholm
F11	Saab S32C Lansen	Nyköping
	Saab S35E Draken	
F12	Saab J35F Draken	Kalmar
F13	Saab J35F Draken Saab SH37 Viggen	Norrköping
F13M	Saab J32 Lansen	Malmslätt
	Saab Sk60	
F15	Saab AJ37 and SH37 Viggen	Söderhamn
F16	Saab J35 and Sk35 Draken	Uppsala
F17	Saab J35F Draken	Ronneby
	Saab SH37 Viggen	
F18	Saab Sk60	Stockholm
F20	Saab Sk60	Uppsala
F21	Saab S35E and J35D Draken	Luleå
	Saab Sk60	
	Saab SH37 Viggen	

Left: the AJ 37 version of the Swedish air force's Saab Viggen is intended for all-weather ground attack missions. Two such aircraft from F7 at Såtenäs fly in echelon over typical Swedish terrain.
Inset far left: an AJ 37 from F 15 is pictured with a selection of the stores carried by the Viggen.
Inset left: one of F6's Viggens climbs away from its base at Karlsborg with afterburner lit

operating from the Navy's D-40 and D-60 class destroyers (one helicopter each) with a shore base at Ferrol.

Youngest of all Spain's three separate air arms, the Fuerzas Aeromoviles del Ejército de Tierra (Army Mobile Air Forces) came into existence in July 1965 as a small force equipped with light observation and tactical transport helicopters. Still an all-helicopter air arm, although much larger and undertaking a far wider range of duties, it assumed its present title in March 1973.

Now possessing somewhere in the region of 150 helicopters, it divides these into five main operating units, each attached in support of one military region. Nearly half of the helicopters operated are Bell UH-1 series Iroquois general-purpose types. Heavy-lift capability is provided by 10 Boeing-Vertol Chinooks at Los Remedios, while the principal light observation helicopter is the Bell OH-58A Kiowa, of which about 30 are in service. Several of the smaller helicopters are equipped for combat duties: some UH-1s, for example, have cabin-mounted machine guns for gunship operation, while some of the Alouette IIIs are equipped as anti-tank aircraft with rocket launchers and wire-guided missiles.

Fostered by such air-minded pioneers as Dr Enoch Thulin, Baron Carl Cederström and Oscar Ask, the Swedish armed forces took an interest in aviation from as early as 1910. Both the Army and the Navy maintained an air service throughout World War I and afterwards, but on 1 July 1926 Sweden became one of the first nations to establish a fully-autonomous air arm, the Kungl Svenska Flygvapnet or Royal Swedish Air Force. Equipment of this small but independent service comprised a mixture of domestic and foreign designs, but its planned expansion programme never took place. As a result, by the early 1930s some hasty re-equipment with de Havilland Moth trainers became necessary to maintain the training of new pilots, while Hawker Hart and Osprey combat aircraft had also to be

imported from Britain. Swedish production of Harts and the indigenous Jaktfalk fighter helped to tide the service over for a few more years until, in 1936, the European political and military situation at last persuaded the Government to inaugurate a more realistic programme of expansion and modernisation.

This was achieved by a mixture of direct imports of some foreign aircraft and licence manufacture of others, the main sources being Germany, Great Britain and the United States. However, by the outbreak of World War II the Flygvapnet still possessed fewer than 150 aircraft instead of the nearly 260 planned. Supplementary orders for US fighters and dive bombers were curtailed when the American government stopped all military aircraft exports, and Sweden turned to Italy for replacements. Once Italy, too, became involved in the war in Europe, Sweden was obliged to look to its domestic industry to fill the gaps. It responded admirably to this task, producing such types as the FFVS J 22 fighter and the Saab B 17 medium bomber. By 1944 the Flygvapnet had seven fighter and seven bomber wings in its combat inventory, a large proportion of these aircraft having been produced by the home aircraft industry. The indigenous Saab J 21 piston-engined fighter was converted postwar into the jet-powered J 21R to give the Swedish industry its first design and construction experience with jet aircraft. The first J 21Rs were delivered in 1949, joining the British Vampire jet fighters which had been imported two to three years earlier.

From necessity, Sweden had evolved during the war years an effective system for observing and reporting foreign aircraft trespassing in its neutral airspace. This system was placed entirely under Air Force control in 1947–48, and development began

Opposite: this Saab J 35F Draken serves with F13, the Draken being the main Swedish interceptor.
Above right: the Aérospatiale Alouette II is known as the HKP-2 in Swedish air force service, a float-equipped example being shown.
Centre right: the Sk 60 jet trainer is operated by F5, the Flight Training School at Ljungbyhed.
Below right: the Swedish navy uses Kawasaki-built Boeing-Vertol 107 helicopters for anti-submarine warfare, minesweeping and rescue work.
Below: the Scottish Aviation Bulldog trainer complements the Sk 60 with F5

of a new radar surveillance network to assume the brunt of this responsibility. Newer and more complex radar defence networks were evolved during later years, culminating with the current, all-Swedish, highly-automated STRIL 60 system introduced in 1960. Tailored specifically to Swedish conditions, this provides the fighter force controllers and their staff with inputs from numerous coastal and inland radar stations, AA sites, radio stations and other sources, from which they can calculate any given threat and deploy their defence forces to the best advantage. All modern Swedish military aircraft are designed for complete compatibility with the STRIL 60 system.

The other major factor affecting the present status of the Swedish Air Force is the fact that, since World War II, its growth has been linked intimately with that of the domestic aircraft industry. The Saab J 21R has already been mentioned as an ingenious conversion of a piston-engined design to jet power, and this provided Saab with valuable experience for its later jet aircraft, of which the next into Flygvapnet service was the J 29 fighter. First flown in 1948, the type had a de Havilland Ghost turbojet engine; it was the first production fighter in western Europe to feature a swept-back wing and was a near-contemporary of the American Sabre and Soviet MiG-15. More than 660 J 29s were built by Saab, serving with the Flygvapnet from 1951 onwards in fighter, attack and reconnaissance versions. Shortly afterwards came the Avon-powered Saab 32 Lansen, an attractive transonic attack, reconnaissance and all-weather fighter aircraft of which 450 were built; a number were still in service in the late 1970s.

Sweden's unrelenting neutrality for so many years has tended to build up an image of a nation whose forces are untried under general operational conditions. It is perhaps forgotten that, particularly in the late 1950s and early 1960s, its air force played a prominent part in assisting the United Nations in various trouble zones. In the spring of 1958, for example, it provided air reconnaissance and surveillance over totally alien terrain during frontier clashes between Syria and the Lebanon. A more exacting task lay ahead in 1961–63, when air-to-air combat with hostile forces was only one ingredient of its contribution to UN attempts to restore peace and order following the insurrection of the breakaway province of Katanga against the former Congolese republic (now Zaïre).

Meanwhile, the biggest single leap forward in postwar Swedish design accomplishment had been represented by the Saab 35 Draken, which first flew in October 1955. Design of this unique 'double delta' fighter, for speeds of Mach 2 and above, began as early as 1949, when no other aircraft in the world – except the air-launched, rocket-powered Bell X-1 – had flown at Mach 1, much less at twice the speed of sound. Saab eventually produced nearly 700 Drakens during the 1950s and 1960s, including, for the first time since the war, aircraft for export to Denmark and Finland. Those for the Swedish Air Force were in five versions for interceptor, reconnaissance and combat training roles.

Equally radical in its design is Saab's latest product for the Flygvapnet, the Saab 37 Viggen (Thunderbolt). In essence, this also presents a 'double delta' configuration, but this time in the form of a single large-delta main wing, with smaller, delta-shaped 'canard' foreplanes at the front, designed to reduce the lengths of runway required for take-off and landing and improve manoeuvrability in the air. Like its predecessors, the

*Above: the Dassault Mirage IIIS interceptor forms
the backbone of the Uberwachtungsgeschwader, or
surveillance wing, and equips two flieger staffeln.
Left: the latest warplane in Swiss service is the
Northrop F-5E, which will fly with four staffeln.
Below left: this DH Venom carries a reconnaissance
pod underwing and serves with Flieger Staffel 10.
Opposite: the Patrouille Suisse aerobatic team flies
the Hawker Hunter, as does nine front-line units*

Viggen is an adaptable basic platform for all kinds
of weapons or other equipment, and is being built
in versions optimised for interception, (JA 37),
attack, (AJ 37), reconnaissance (SF 37 and SH 37),
and operational training.

The ability of the Draken and Viggen to take off
and land within about 500 metres reflects yet another
Swedish defence tactic–that of designating any
straight stretch of highway or motorway of sufficient
length as a provisional runway in the event of an
emergency. A system of underground hangars,
blasted out of the solid rock of the mountains
adjoining the highways, simultaneously provides the
Flygvapnet with virtually impregnable accommodation for its aircraft, personnel and supplies,
plus a high degree of mobility for its squadrons in
the event of hostilities.

In 1965 the Flygvapnet had some 10 fighter
groups, six attack groups and three reconnaissance/
observation groups forming its operational component, each group having up to three squadrons.
Total front-line aircraft numbered well over a
thousand. Since then, however, changes in the
political climate have brought considerable reduc-

tions in successive defence budgets. Symptomatic of this was the cancellation in 1979 of the promising new B3LA light strike aircraft project, intended to replace the Saab 105 (Sk 60).

As a result, in 1979 Sweden's air force numbered about 950 aircraft in all, of which 450 were first-line operational types. Other aircraft are in store, including about 100 Lansens. In quantity and quality, this is still sufficient to place Lt General Dick Stenberg's Flygvapnet among the top ten air forces in the world in terms of size and effectiveness, though reductions are scheduled to continue towards a target of 20 operational squadrons in 1985, compared with about 30 in early 1978.

These squadrons are divided between seven geographical air defence sectors, each administered by one of six regional commands under tri-service direction. Four of the regional commands Boden, Kristianstad, Ostersund and Stockholm–have Draken or Viggen interceptor squadrons assigned to them. Six squadrons serve with Wings F10, 12 and 17 in southern Sweden at Angelholm, Kalmar and Kallinge/Ronneby respectively, five with F1, 13 and 16 at Västeras, Norrköping and Uppsala in the east, three with F4 at Ostersund/Frösön in South Norland and one with F21 at Luleå in the extreme north.

The seven attack and four reconnaissance squadrons, under the separate control of Eskader 1, Attack Command, are deployed mainly by F6 Wing at Karlsborg, F7 at Satenäs, F11 at Nyköping and F15 at Söderhamn. In addition, a fifth Wing–F21 at Luleå–operates a mixture of reconnaissance Viggens and Saab Sk 60B and C light strike aircraft; and one squadron of F17 flies the SH 37.

Eskader 1 also incorporates the Flygvapnet's principal transport squadron, a part of F7 Wing, which has a small number of Lockheed C-130 Hercules and a slightly greater number of ageing Douglas C-47s. Wing F13 at Malmslätt also has a few C-47s, plus a couple of Sud Aviation Caravelles for ECM and experimental flying; it also operates about 35 Lansens for ECM and target towing.

The Navy was the first Swedish service to own an aeroplane, a Blériot monoplane presented to it in 1911; by the end of World War I it had some two dozen assorted aircraft. All aviation became the responsibility of the Flygvapnet when this autonomous service was created in 1926 and the Navy did not again operate aircraft of its own until 1957, when it received four Vertol Model 44 helicopters ordered from the United States.

Currently, the Kungl Svenska Marinen (Royal Swedish Navy) operates between 35–40 helicopters, most of them shore-based. The 1st Helikopter Division (squadron), which moved to Berga in 1961 from its former base at Bromma, was the Navy's only helicopter unit until the formation of the 2nd Helikopter Division at Torslanda, near Gothenburg, in 1969. During the 1960s the Vertol 44 continued to be the main type operated, along with about a dozen smaller Aérospatiale Alouette IIs.

In 1969 the 2nd Helikopter Division was transferred to a new base at Save, and the Vertol 44s began to be replaced during the 1970s by the larger Boeing-Vertol 107, also a tandem-rotor type but with shaft-turbines instead of piston engines. When Boeing manufacture of this aircraft ended in the early 1970s, production and development were

SWISS KOMMANDO DER FLIEGER UND FLIEGABWEHRTRUPPEN

Flugwaffe Brigade 31:

Flieger Regiment 3 (Uberwachungsgeschwader)

Flieger Staffel 1	F-5E	Dubendorf
Flieger Staffel 10	Mirage IIIRS	Dubendorf
Flieger Staffel 11	F-5E	Dubendorf
Flieger Staffel 16	Mirage IIIS	Payerne
Flieger Staffel 17	Mirage IIIS	Payerne

Flieger Regimenten 1 & 2 (Units exist on a cadre basis only, allocated to 12 dispersed airfields in the Alps)

Flieger Staffel 2	Venom FB54
Flieger Staffel 3	Venom FB54
Flieger Staffel 4	Hunter F58/F58A
Flieger Staffel 5	Hunter F58/F58A
Flieger Staffel 6	Venom FB54
Flieger Staffel 7	Hunter F58/F58A
Flieger Staffel 8	Hunter F58/F58A
Flieger Staffel 9	Venom FB54
Flieger Staffel 13	Venom FB54
Flieger Staffel 15	Venom FB54
Flieger Staffel 18	Hunter F58/F58A
Flieger Staffel 19	Hunter F58/F58A
Flieger Staffel 20	Venom FB54
Flieger Staffel 21	Hunter F58/F58A/T68

Fliegerschule I	Pilatus P-2/P-3	Magadino
Fliegerschule II	Vampire T55/Venom FB54	Emmen
Zeilfliegerkorps	FFA C-3605	Sion
Transportfliegerkorps	Twin Bonanza and Ju 52/3mG4e	Dubendorf

ARMY AVIATION

Leichte Fliegerstaffel 1	Alouette II/III
Leichte Fliegerstaffel 2	Alouette II/III/Porter
Leichte Fliegerstaffel 3	Alouette III
Leichte Fliegerstaffel 4	Alouette II/Super Cub
Leichte Fliegerstaffel 5	Alouette III/Dornier Do 27
Leichte Fliegerstaffel 6	Porter
Leichte Fliegerstaffel 7	Porter

*Right: Alouette III helicopters serve with the
leichte fliegerstaffeln and their work includes
alpine search and rescue and civil aid projects.
Below right: three Junkers Ju 52 trimotors of
World War II vintage serve as transports and
paratroop trainers, based at Dubendorf.
Bottom right: the Pilatus P-3 flies as a primary
trainer with Fliegerschule I at Magadino*

continued by Kawasaki in Japan as the KV-107/II,
and an additional batch from Japanese production
was acquired by the Swedish Navy under the
designation Hkp 7. To maintain commonality with
Air Force and Army 107s, the Navy's examples are
also powered by Rolls-Royce Gnome turboshaft
engines, and equipped with Decca navigation.
Currently employed chiefly for anti-submarine
patrol, minelaying, minesweeping, and search and
rescue, they were joined in 1976 by a further 10
BV 107s transferred from the Air Force.

Only about half of the original Alouette IIs
remain, and these are kept in reserve. The light
helicopter force is based now upon the Agusta-Bell
206A JetRanger, of which 10 were acquired in the
late 1960s. These cover a range of miscellaneous
duties which include reconnaissance, fire control,
and communications/liaison. Operations are carried
out both from shore bases and from deck platforms
on two of the Navy's coastal defence destroyers.

Army aviation in Sweden dates from 1911, its
first aeroplane being a donated Nieuport IVG
monoplane. From four aircraft at the outbreak of
World War I it rose to a strength of about 50 at the
end, most of them produced domestically either by
the Thulin factory or by the Army Aviation Workshop. The latter was expanded after the war,
supplying part of the Army's needs until the creation
of a fully-independent air force in 1926. In later
years an Air Observations Corps formed part of the
Army's anti-aircraft defence force, but this too was
transferred to Flygvapnet control in 1948. However,
only a few years later a small Army Aviation
Department was created, equipped with light
aircraft for communications, liaison flying and
artillery spotting.

Initially, these were all fixed-wing types, comprising the Piper Cub, Super Cub, Dornier Do 27A,
and, for evaluation, a couple of Swedish-designed
Malmö MFI-10s. Of these, only a handful of the
Do 27s remain, the principal fixed-wing type for
liaison and artillery spotting now being the Scottish
Aviation Bulldog, of which 20 were acquired in the
early 1970s. The Bulldogs also fulfill a valuable
additional role as trainers, and all fixed-wing
aircraft are based at the Artillery Flying School
(Artilleriflygskolan) at Nyge, near Nyköping.

Expansion of the Army Aviation Department to
include helicopters for battlefield support began in
the 1960s, and 30 or more light rotorcraft are now
employed: a dozen piston-engined Agusta-Bell
204Bs and about 20 turbine-engined Agusta-Bell
JetRangers. The latter is the Army's standard light
observation type and is distributed to provide
battlefield support for individual Army formations.
Army rotorcraft pilots receive their basic training
from the Air Force, after which specialised training
on helicopters is given at Boden, at the Armen
Helikopterskolan (Army Helicopter School).

Top: this Mirage IIIS carries the badge of No 17 Squadron on the port side of the fin, with No 16 Squadron's on the starboard, the two units' aircraft being pooled at their Payerne base.
Above: the Soko Jastreb is Yugoslavia's main ground attack aircraft

Switzerland was among the first nations in Europe to encourage early aviators, and formed its own air force a few days before the outbreak of World War I. The Fliegertruppe, as its name suggests, had support of the army as its primary function, a role especially valuable in a country consisting of high mountains over some three-quarters of its total area. Switzerland's own designer, D. Haefeli, gave the country some of its first indigenous aircraft types; another Swiss, Marc Birkigt, was responsible for initiating the famous line of aero-engines produced in France by Hispano-Suiza.

Links with France resulted in the adoption of several types of Dewoitine fighters during the inter-war years, the balance of the Fliegertruppe's first-line aircraft being drawn from Great Britain, the Netherlands and, in the late 1930s, from Germany. By the end of that decade Switzerland had a small but capable aircraft industry, based chiefly upon the Federal Aircraft Factory at Emmen, the Dornier factory at Altenrhein, and the newly-formed Pilatus

Flugzeugwerke at Stans, near Lucerne. The Flieger-truppe was reorganised in 1936, becoming a fully-autonomous service with the new title of Schweizer-ische Flugwaffe.

By the end of World War II the Flugwaffe had more than 500 first-line aircraft, divided among 24 squadrons grouped into four air regiments. Man-power amounted to nearly 1,000 full-time person-nel, with a further 2,700 in reserve. Combat aircraft were chiefly German Messerschmitt Bf 109 and Swiss-built D-3801 (Morane-Saulnier MS 406) fighters. Switzerland's first brush with the jet age was the purchase in 1946 of a handful of de Havil-land Vampires for evaluation, but it was to be a few years more before jet aircraft became a regular part of the Swiss Air Force inventory. Meanwhile, the acquisition of substantial numbers of North Ameri-can P-51 Mustang fighters and T-6 Texan trainers was reinforced by a national product in the form of the Pilatus P-2 advanced flying and weapon training aircraft. In the late 1940s the FFA–Flugund Fahrzeugwerke Altenrhein, successor to the former Dornier factory–began designing its own jet fighter, as did the Federal Aircraft Factory; both projects were, however, shelved after a few years through lack of funds and Vampire fighter-bombers were purchased instead.

In early 1951 the Flugwaffe's first-line aircraft strength was streamlined into 21 squadrons (400 aircraft). The number of combat squadrons in 1979 remains at about the same level, although the num-ber of aircraft has been slimmed down still further to a figure of about 350, with an additional 375 or so for second-line duties. Switzerland's armed forces exist on a very small cadre of regular officers and men, the vast majority of their manpower being drawn from reservists who, after initial training, perform a few weeks' refresher training each year and are on a 48-hour call-up standby. From this reserve Switzerland can mobilise more than half a million troops if the occasion demands, of whom some 45,000 are Flugwaffe personnel. All mainten-ance of Swiss military aircraft is carried out by civilian engineers.

Air defence bases in the country are situated at Alpnach, Altenrhein, Belp, Dübendorf, Emmen, Grenchen, Interlaken, Locarno, Meiringen, Pay-erne, Sion, Stans and Turtman. Administratively, the air force is under the control of the Kommando der Flieger und Fliegerabwehrtruppen (Swiss Air Force and Anti-Aircraft Command) at Berne, head-ed by Oberstkorpskommandant Kurt Bolliger. The KFFAT is further sub-divided into three brigades, responsible respectively for all flying units (Flug-waffe Brigade 31), logistic, airfield and personnel support (Flugplatz Brigade 32) and anti-aircraft defences (Fliegerabwehr Brigade 33).

Brigade 33 is equipped with two battalions of Bloodhound surface-to-air missiles and 22 bat-talions of 35mm twin-Oerlikon anti-aircraft guns. The main airborne defence task falls to Flugwaffe Brigade 31, whose 19 combat squadrons are divided between three air regiments (*fliegerregi-menten*) and perform the primary roles of inter-ception, ground attack/close support and tactical reconnaissance. Interception and tactical recce are undertaken by Fliegerstaffeln 10, 16 and 17, flying from Dübendorf and Payerne with Mirage IIIs and a

small number of aged but specially-modified recon-naissance Venoms.

Numerically, however, the main emphasis is on tactical aircraft, and nearly 300 of the Flugwaffe's 350 first-line aircraft, in almost equal quantities, are fighter/ground attack Hunter 58s and de Havilland Venom fighter-bombers of various marks. Two of the Hunter squadrons (Nos 1 and 11), based at Dübendorf, are equipped with Sidewinder air-to-air missiles and share the air defence interception task with the Mirage squadrons. These five squadrons are manned by full-time regular personnel and are at instant readiness at all times. The remaining Hunter squadrons, also with a Sidewinder capability, are divided about equally between defence/inter-ception and close support duties.

The Venoms, although maintained faithfully and refurbished since their acquisition in the 1950s, are now being retired or replaced, but a substantial number still remains in service. To take their place the Swiss Government, after a long and agonising period of indecision, decided finally in favour of the Northrop F-5E Tiger II, placing an initial order (since increased) for 72. These began to enter service in late 1978, and are being assembled in Switzerland by the Federal Aircraft Factory.

In 1966 Switzerland invested a considerable sum of money in a computer-based radar detection and control system, the Hughes 'Florida' for its air defence network. This modular system, supple-mented by additional British units from Ferranti and Plessey, cost the Swiss Government more than $100 million. It became operational in 1970, and now co-ordinates all air defence operations by the anti-aircraft SAM and gun batteries as well as the interceptor squadrons.

The air transport unit of the Swiss Air Force, based at Dübendorf, is the Transportfliegerkorps, virtually the last military operator of the celebrated Junkers Ju 52/3m tri-motor. Three of these seem-ingly indestructible veterans survive, strange bed-fellows for the TFK's other aircraft, a pair of Beechcraft Twin Bonanzas which are a mere 20 or so years old. Light transport and communications/ liaison flying is done by seven LFS (leichtflieger-staffeln or light aviation squadrons), four of which are attached to four Army Corps for general support duties. Equipment of these units is a mixture of fixed-wing lightplanes and light helicopters.

Training of Flugwaffe pilots starts at Magadino on the piston-engined Pilatus P-3, successor to the earlier P-2, although these will probably be replaced

in the early 1980s by Pilatus' latest design, the turbo-prop-powered PC-7 Turbo Trainer. Advanced flying training and weapon training then follows on 1950s-vintage Vampire T Mark 55s, other marks of the Vampire/Venom series, or (for helicopter pilots) the Alouette II and III. Advanced flying training takes place mostly at Emmen, while weapon training involves also the Sion-based Zeilfliegerkorps, a target-towing unit equipped with some two dozen examples of the C-3605, a pre-war two-seater modernised in the early 1970s by conversion to turboprop power. Operational conversion training takes place on two-seat Hunters, Mirages and Northrop F-5Fs.

A feature of Switzerland's air defence network are the kavernen, in which, like Sweden, the country has exploited the natural mountainous nature of its terrain to provide additional security for its defence forces. In addition to the air defence bases already listed, the Flugwaffe has about a dozen tactical air bases tucked away in the Alps in central Switzerland for use as dispersed operating sites in the event of an emergency. Only taxiways, valley runways, or straight stretches of motorway designated as emergency runways, are above ground. Command/communications centres (linked to the 'Florida' system), combat aircraft, weapons and ammunition, maintenance, food and medical facilities are all accommodated in subterranean caverns hewn from the solid rock of the Alps and capable of withstanding nuclear attack. A 100-ton steel and concrete door seals off each kaverna, inside which, in several tiers, can be housed the men and equipment of two complete combat squadrons, with each squadron capable, it is claimed, of being airborne within one-and-a-half minutes.

The first steps towards military aviation in Yugo-slavia were taken in 1912 when a handful of officers of the Serbian Army – the state of Yugoslavia was not created until 1918 – were sent to France to learn to fly. A unit formed with French aircraft was involved in the second Balkan War in 1913, and formation of a national Yugoslav air arm followed shortly after the end of World War I. The 1920s saw the establishment of a small domestic aircraft industry – the Ikarus and Rogozarski factories – and the creation of a small naval air service. Licence-built foreign aircraft, mostly of French origin, kept the Yugoslav Air Force supplied until the mid-1930s, when a programme of modernisation and expansion introduced aircraft imported from Brit-ain, Germany and Italy. By 1941 the YAF strength

Top left and upper centre: the indigenously designed and built Soko Galeb is the Yugoslav air force's basic jet trainer, some 60 serving in this role. The type also operates with tactical reconnaissance units alongside its single-seat counterpart the Jastreb. Yugoslavia is currently co-operating with Romania to produce a new light attack and training aircraft, the Orao, which is likely to enter service in the early 1980s

Top right: more than 100 MiG-21 interceptors contribute to Yugoslavia's air defences, a MiG-21PF variant being illustrated. The older North American F-86D and F-86K Sabre all-weather fighters also serve in this role. Yugoslavia's acquisition of warplanes from both Soviet and Western sources emphasises the country's neutral position, although a Communist state. Above: a single Ilyushin Il-18 is flown as a government transport

was some 400 first-line aircraft, plus about 50 more with the naval air service; a high proportion of these, however, were by then obsolete. During the Axis occupation the country was partitioned, Italy setting up a separate, but short-lived, Croatian Air Force. After liberation, many YAF aircraft and personnel became part of the RAF Balkan Air Force until the end of World War II. After the country became a socialist republic in late 1945, the postwar air force was remodelled with Soviet advisers and equipment.

Immediate steps were taken to create a strong air force, using as a nucleus the experienced and not inconsiderable elements of the Yugoslav Air Force (Jugoslovensko Ratno Vazduhoplovstvo) which had survived the war. Much on the Soviet pattern, the country was divided into six regions; a national aeronautical union was formed, dedicated to fostering sport flying, the interchange of aeronautical knowledge and the creation of a huge reserve of flying personnel. The national aircraft factories, hard hit by war damage, were refurbished and in action again by late 1945 or early 1946, their first production task being the partial manufacture of Il-2 Shturmovik ground attack aircraft for the USSR. By the end of 1947 the JRV was by far the strongest air force in the Balkan states of Europe, with about 40 front-line squadrons and some 10,000 personnel.

These totals were estimated in the late 1970s to be in the order of 24 combat squadrons (not counting Navy units) and a manpower of some 40,000, including 7,000 conscripts. In 1948, however, a major political change occurred in the country when Yugoslavia severed diplomatic relations with the USSR. Although it remains a Communist state, it has stayed outside the Soviet Bloc during the ensuing three decades, and this is reflected in the selection of aircraft which now equip its squadrons; many are still drawn from the Soviet Union, but a substantial number are of domestic design and manufacture, while others are US or French.

The interceptor force is totally Russian-supplied, comprising two divisions (eight squadrons) forming one air corps with its headquarters at Zagreb. These are equipped with Mikoyan MiG-21F, PF and M Fishbed-C, D and J fighters, which are armed with Atoll air-to-air missiles and are employed together with eight Soviet Guideline surface-to-air missile batteries at various bases for air defence. Yugoslavia has about 20 military airfields.

Biggest contingent of the air force is the ground

attack air corps estimated at 15 squadrons in 1978, which has its headquarters at Zemun. One squadron of ageing Republic F-84G Thunderjets and one of the domestically-designed Soko P-2 Kraguj may now have been disbanded; the remainder are equipped with more than 100 of Soko's larger product, the J-1 Jastreb and its two-seat counterpart, the G2-A Galeb. These are expected to be replaced in the early 1980s by the new Orao strike fighter, currently being developed jointly with the Romanian aircraft industry.

The transport element of the JRV is an extremely mixed bag, about half of its aircraft being ancient Douglas C-47s and Ilyushin Il-14s. A few jet transports of assorted sizes are used for presidential and VIP transportation, but the most modern military transports are nine Antonov An-26 freighters supplied in recent years from the USSR. Also in service for light transport, liaison and aeromedical duties is the Yugoslav-designed UTVA-66, a utility aircraft with a secondary capability for light attack.

About 100 helicopters are in service for a variety of support and utility roles, the most numerous type in 1979 being the Mil Mi-8; about 50 are in service, including some fitted with Sagger anti-tank missiles. About 20 Mi-4s are also flown, but deliveries are well under way of more than 100 Aérospatiale Gazelles, these being built under licence by Soko at Mostar. A second user of helicopters is the Yugoslav Navy, whose principal types are the Mi-8, Kamov Ka-25 Hormone and Gazelle. The Ka-25s undertake the specialist anti-submarine warfare role, while the larger Mi-8s are assigned to assault, support, and offshore patrol duties.

Flying training, at Mostar and various other subordinate units, is carried out primarily on the Galeb and the residue of a large number of Lockheed T-33As which Yugoslavia received from the USA in the early 1950s. About 30 T-33As are used, plus about twice that number of the more modern Galeb.

Western–and Soviet–eyes will no doubt be directed with increasing interest during the next few years at the progress of the Orao strike fighter. Already the Yugoslavian aircraft industry provides approximately one-third of its air force's needs in the form of nationally-developed aircraft, as well as manufacturing others, such as Gazelle helicopters, under licence. This degree of self-sufficiency is likely to increase still further if, as is widely believed, up to 200 examples of the Orao are to be produced for the air force of each of the two nations involved in the programme.

CHAPTER TEN

Warsaw Pact

Soviet Union / Poland / East Germany
Czechoslovakia / Hungary / Romania / Bulgaria

On 6 September 1976, Lieutenant Viktor Belenko of PVO Strany, the Soviet Air Defence Command, overshot the runway at Hokodate civil airport in Northern Japan. In his first abortive attempt to land his Mikoyan MiG-25 fighter he had streamed and lost his brake parachute; on his second he came to rest only feet away from a radar installation and Western Intelligence had received its biggest aircraft prize direct from the Soviet Union.

In Britain, Western Europe, North America and many other parts of the world it is possible to pick up specialist magazines and trade journals which explain and illustrate in great detail all but the most secret aspects of new aircraft and equipment. Governments publish details of defence expenditure, the organisation of their Armed Services and the regular promotions and postings of officers. Not so in the countries of the Warsaw Pact. There, the smallest military detail is classified. For example,

on 30 August 1977 Colonel of Justice Tokarev wrote in the Russian military newspaper Red Star: 'On another occasion, I happened to hear a Lieutenant who was talking on the phone, name the full designation of the unit he served in. Meanwhile, the Company Commander, the Lieutenant's immediate supervisor, present during this conversation did not reprimand him.' Because the Soviet press is completely controlled and publishes only what the government wishes its people to read, description and assessment of Soviet military strength demands a painstaking process of adding thousands of small pieces of information together. Hence the enormous value of the receipt, literally out of the blue, of Lieutenant Belenko's MiG-25, NATO code-named Foxbat.

In fact, from close examination of this and other Warsaw Pact aircraft, from eye witness reports, from a study of many aspects of Soviet military

and industrial developments, from Soviet exercises and operations beyond her frontiers and from shrewd assessments by civilian aviation professional writers, it is possible to construct a very comprehensive picture of the 13,000 Soviet and some 3,000 other Warsaw Pact military aircraft, their organisation, training, strengths and weaknesses.

As these figures suggest, Warsaw Pact airpower is dominated by the Soviet Union. The national air forces of Poland, East Germany, Czechoslavakia, Hungary, Romania and Bulgaria fly aircraft which are almost entirely Russian in origin, they use Russian as a common military language, they conform to Russian operating and training procedures and, despite the presence at various levels of national commanders, they are under the overall command and control of the Soviet Union exercised from headquarters in Moscow. Therefore, despite the modernisation of the national air forces in the last decade, and recent modifications to the command structure to give greater prominence to national commanders, Warsaw Pact airpower remains primarily Soviet Airpower.

In the Soviet Union there are three quite separate air forces with a fourth readily available to support military operations. Air defence of the Soviet Union is the responsibility of the Protivo-vozdushnaya oborona strany (PVO Strany), The National Air Defence Force. Manned by over 500,000 troops, it comprises manned interceptors, an extensive network of surveillance radars and surface-to-air missile batteries. The organisation usually referred to as The Soviet Air Forces has three operational components. The first is Fronto-vaya Aviatsiya (Frontal Aviation), organised and equipped to support the Soviet Army but becoming increasingly capable of longer-range missions. The second is Dal'naya Aviatsiya (Long-Range Aviation), which was originally designed as a long-range strategic nuclear force but is now capable of carrying out conventional and nuclear bombing attacks in addition to photo reconnaissance, electronic intelligence and air-to-air refuelling missions. The third element in the Soviet Air Forces is Voenno-Transportnaya Aviatsiya (The Military Transport Command).

The third separate air force is the Soviet Naval Air Force (SNAF), controlled by the Soviet Navy and responsible for all maritime operations including reconnaissance, conventional and nuclear weapon attack, carrier operations, mining and anti-submarine warfare. The state-owned airline, Aeroflot, is the fourth Air Arm. Commanded by an air marshal, the airline is regarded as a full-time reserve transport force which may be called upon at any time to support military operations.

In World War II, the Soviet Union was taken completely by surprise by the German attack of Sunday, 22 June 1941. Seventy per cent of Soviet fighter aircraft were concentrated on some 60 airfields well within range of the Luftwaffe. They were neither dispersed nor concealed, nor was there any radar network to give a last-minute warning of

Below left: although the Tupolev Tu-95 Bear first entered service in 1956, it remains a front-line strategic bomber with Long Range Aviation in the late 1970s and also undertakes maritime reconnaissance duties.

Below: the most controversial Soviet aircraft of recent years (even its Soviet designation being in doubt) is the Tupolev Backfire. The second phase of the Soviet-US strategic arms limitation negotiations has been bedevilled by arguments over the bomber's range capabilities, the Americans insisting that Backfire has an intercontinental range, while the Russians deny this.

Bottom: Tupolev's Blinder bomber does not serve in large numbers, despite its supersonic performance

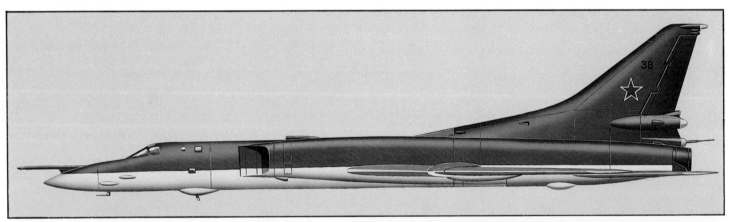

the impending *blitzkrieg*. Actual dimensions of the disaster are still difficult to assess but the Germans claimed 1,200 Soviet aircraft destroyed by lunchtime, and a further 600 by nightfall of the first day. By the end of the week the claims had risen to 4,000. Whatever the actual numbers of losses, the Soviet air defences were smashed, leaving ground forces and countryside open to the free-ranging Luftwaffe. The lessons were bitter and were well-remembered after 1945.

With the onset of the Cold War, Soviet armies in Europe were challenged only by the ability of the Royal Air Force and United States Air Forces to take any war back to the Soviet heartland. The USAF Boeing B-29s, of course, could carry atomic weapons. In the face of this threat, Stalin placed strong emphasis on the construction of a comprehensive air defence network. At first, this depended heavily on manned interceptors of World War II vintage, but was soon strengthened by the large-scale production of MiG-15s using an engine copied from the Rolls-Royce Nene 2, exported to the Soviet Union in September 1946. Over the next decade, PVO Strany grew in strength until by 1960 it included some 6,000 manned interceptors, mainly comprising the all-weather MiG-19 Farmer and Sukhoi 9 Fishpot as well as the earlier marks of the most widely-used day-fighter in the world, the MiG-21 Fishbed.

However, on 1 May 1960 a new era of air defence began. While Premier Khrushchev was watching a military parade in Red Square Moscow, Marshal Biryuzov, Commander-in-Chief of Soviet anti-aircraft defences, mounted the rostrum and spoke quietly to him. Diplomats present noted that the Marshal was wearing ordinary uniform and speculated that something important must have happened. However, none could have foreseen the impact of the news. That morning, Gary Powers had left Peshawar airfield in Pakistan to overfly Central and European Russia on a photographic reconnaissance mission in his Lockheed U-2. His flight plan was intended to take him over Chelyabinsk and Sverdlovsk northwards to Bodö airfield in Norway. Such flights had been made many times before at heights well beyond the capability of the PVO Strany manned interceptors, but that day he only reached Sverdlovsk, where he was shot down from some 75,000 feet by a radar-guided surface-to-air missile: an SA-2 Guideline of the kind subsequently deployed in large numbers in North Vietnam. Four years previously the first surface-to-air missile sites had been built round Moscow, but the SA-2 batteries were, according to Khrushchev's evidence, deliberately placed below the regular track of the U-2.

In the last twenty years, Soviet air defence has come to rely increasingly on surface-to-air missiles (SAMs) rather than the previous single concentration on manned interceptors. Now, nine separate systems protect large areas of Warsaw Pact territory, control being shared between PVO Strany and Frontal Aviation. The most numerous is still the medium- and high-altitude SA-2. At lower altitudes it is complemented by the SA-3 Goa which can engage targets flying at 150m (500ft). SA-3 sites naturally tend to be located where the Soviet Union would expect low-level intruders, such as

the Baltic and Black Sea coastal areas.

The SA-4 Ganef was the first practically mobile missile in that its supporting radars are also mounted on vehicles. Like the earlier missiles, Ganef is guided to its targets by radar which could be engaged at heights roughly comparable to those of the SA-2. Ganef is deployed in large numbers with Soviet army formations. SA-5 Gammon is a long-range, high-altitude weapon designed to operate against high-speed aircraft. It has an effective ceiling of more than 26,000m (85,000ft) and can carry a nuclear warhead for use against missiles or massed aircraft. It carries its own target-seeking radar and is designed to provide area defence over several regions of the USSR.

SA-6 Gainful had been in Soviet service for two years before it took the Israeli Air Force by painful surprise in the October war of 1973. Its careful siting and well co-ordinated response in the Suez area forced General Peled temporarily to withdraw his aircraft from beyond its range. Gainful's continuous-wave radar guidance system had not been detected by the Israeli aircraft whose electronic warfare equipment was designed to identify the more standard pulsed radar emissions. SA-6 is primarily a low- and medium-level fully mobile system which can engage targets down to 90m (300ft) or, if its target acquisition radar is supplemented by optical aids, possibly down to 30m (100ft) under conditions of good visibility. It has replaced anti-aircraft guns in the Soviet Union and is widely-deployed with Soviet and national Warsaw Pact armies.

Another aspect of surface-to-air missile systems, and of SA-6 in particular, emerged in the October war. The Egyptians had to commit 150,000 highly-trained men to their SAM batteries: the equivalent of 15 divisions which could not be deployed to other battle areas. Moreover, when the SA-6s did move forward, losing both their prepared sites and their close co-ordination, their effectiveness was considerably decreased. Nevertheless, the presence of SA-6 clearly complicates the task of hostile offensive support aircraft.

SA-7 Grail is a shoulder-fired, infra-red heat-seeking missile which has earned international notoriety as 'the terrorist weapon', almost certainly being responsible for the shooting down of the Rhodesian Viscount airliner in 1978. It has been in service with Warsaw Pact armies since 1969 and is designed to give the infantryman some means of low-level defence against air attack. It has been widely-exported and frequently accompanied by very exaggerated claims for its capabilities. Its kill-rate in the October war was astonishingly low when compared to the several hundred which were fired. The Strella, as it is known in the Soviet Union, is certainly effective against slow-flying targets such as helicopters or civilian aircraft, but far less so against high-speed jet aircraft whose major heat emission point is naturally at their rear at the jet efflux.

This limitation produces an uncomfortable situation for the soldier using the Strella in that it is much less effective against an aircraft flying towards him than it is against one at ninety degrees or leaving him. Moreover, it remains vulnerable to flare decoys and is impaired both by bad weather

*Above left: the Tu-16 Badger serves in significant numbers with Naval Aviation units, some aircraft being armed with anti-shipping missiles.
Above: a Mya-4 Bison flies on ocean patrol. As with other Soviet long-range types, it carries a nose-mounted flight refuelling probe.
Below: radar-equipped Bears provide guidance information for over-the-horizon missile attacks*

Top: these MiG-23S variable-geometry fighters belong to a regiment based at Kubinka, near Moscow. Code-named Flogger by NATO, this fighter has been supplied both to PVO Strany and to Frontal Aviation.
Above: the national insignia of Romania, whose air arm is modest in comparison with those of other Warsaw Pact satellites

and by sunlight behind a target, although recent modifications have slightly improved its performance under such circumstances, and against high-speed crossing targets.

The SA-8 Gecko as observed in the November Red Square parade of 1975 is a fully-autonomous mobile system; unlike SA-4 or 6, it incorporates its missiles, acquisition radar, fire control radar as well as a television camera for optical tracking all on the same wheeled amphibious vehicle. It carries four missiles on firing positions and is designed to provide low-level coverage down to 30m (100ft). As with all radar-guided SAMs, range and target acquisition is influenced by the vehicle's position relative to local terrain and the radar cross-section of the target, but it could be expected to acquire low-flying aircraft from 11–13km (seven to ten miles) out to under 1·6km (one mile) from the launcher. It is now being deployed among the Warsaw Pact armies.

The ninth system is the vehicle-mounted SA-9 Gaskin, which resembles the SA-7 Grail in its infra-red guidance and general performance characteristics. It is a low-level weapon which could be expected to defend army formations, headquarters or 'chokepoints' which would otherwise be attractive targets for aircraft. Like Grail, it presents a far greater threat to receding aircraft than to one approaching and is therefore particularly vulnerable to cluster or precision air-launched weapons.

The introduction of so many different SAM systems in such great numbers since 1960 has clearly strengthened Warsaw Pact air defences considerably, although at the cost of a high proportion of military manpower and resources. Moreover, as events in both Vietnam and the Middle East demonstrated, SAM systems are vulnerable to electronic counter-measures, to decoys and to straightforward destruction by specialist weapons and tactics. Although SAMs are deployed in areas where most needed, the

Soviet Union would be a vast area to defend against any enemy who made full use of the flexibility of airpower. Consequently it is not surprising that, although the numbers of aircraft in PVO Strany declined during the 1960s as the SAMs were introduced, the USSR continued to design and deploy increasingly high quality manned interceptor aircraft.

Among the first interceptors was the all-weather Yak-28P Firebar, which carries Anab air-to-air missiles with either infra-red or semi-active radar warheads; Firebar is now beginning to be phased out of front-line service. Arriving in the same period was the Tu-28P Fiddler, a long-range high-altitude interceptor which operates mainly in Northern Russia, equipped with four large air-to-air Ash missiles, again with either semi-active radar or infra-red warheads. Fiddler, however, has not been deployed in very large numbers.

In the late 1960s, two other all-weather fighters made their appearance: the Sukhoi Su-11 Fishpot-C–which, as its designation indicates, was an improved version of an earlier model–and the Sukhoi Su-15 Flagon. Both aircraft carried improved radars and the Anab long-range air-to-air missile. Flagon continues in service in large numbers with PVO Strany but Fishpots are also now being withdrawn along with the last of the MiG-17s and MiG-19s.

The 'third generation' aircraft which is replacing the earlier types has already achieved an international reputation. The prototype MiG-23 Flogger was first demonstrated publicly at the 'Day of the Soviet Air Fleet' at Domodedovo airfield near Moscow on 9 July 1967. From that prototype two similar, but operationally quite distinct aircraft have been developed: one is primarily an interceptor, the MiG-23B, the other primarily for offensive support, the MiG-27.

The single-seat Flogger-B is an all-weather

fighter with a variable-geometry wing which permits it to operate with greater effectiveness either as a high-level interceptor or in the air superiority role at lower altitudes. Its High Lark radar gives it a limited ability to engage targets below it and, unlike any of its PVO Strany predecessors, it can also attack from abeam as well as head-on or from the rear. Its radar is supplemented by infrared target acquisition equipment which aids low-level detection in good weather. It normally carries four air-to-air missiles: two long-range Apex with either infra-red or radar guidance, and two short-range Aphids with infra-red warheads. Both these missiles have greater manoeuvrability and are better-suited to low-level operations than their predecessors. In addition, Flogger carries a fuselage-mounted double-barrelled 23mm cannon. In common with its contemporaries it is fitted with a receiver to give warning of hostile radar emissions. Flogger B is being introduced into national Warsaw Pact air forces and, in a modified form, has been exported to several other countries. Despite its low-level limitations it has considerably strengthened Warsaw Pact air defences.

It is, however, MiG-25 Foxbat which has caught the imagination of aviation enthusiasts. Also flown at Domodedovo, it later claimed a world altitude record of over 36,600m (120,000ft) and a speed record over a closed circuit of 2,900km/h (1,800mph). The reconnaissance version frequently flew with impunity over the Middle East well beyond the reach of any potential interceptor. Inevitably, like the SA-7 missile, it began to acquire a fearsome reputation to such an extent that supersonic shockwaves heard over the South coast of England in 1976 prompted nervous questions to the Ministry of Defence. Needless to say, the sonic bangs had not been produced by Foxbat.

In September 1976, Lieutenant Belenko's unsolicited gift allowed the first real assessment of the aircraft to be made in the West. It was not, in fact, built of heat-resistant alloys, but almost entirely of steel, with titanium only at high-temperature points. Theoretically, it could exceed Mach 3, but the red warning on the machmeter above Mach 2·8 discouraged the pilot from sustaining it. It had two extremely powerful Tumansky engines which not only allowed it to climb very rapidly to high altitude, but also generated the power for its Foxfire radar which had a range of some 80km (50 miles) and comparatively good anti-jamming qualities. With its four long-range Acrid missiles it was clearly a high-altitude interceptor of considerable effectiveness – but only against high-level intruders who would present no problems of identification. Foxbat had been designed in the early 1960s in anticipation of a USAF B-70 bomber programme and its ability to carry out interceptions at very high speed might also give it a limited capability against high-level air-to-surface missiles. In short, Foxbat is an effective weapon system but with a very restricted role, compared, say, to the wide-ranging activities of Flogger-B.

This complex array of missiles and aircraft is almost entirely co-ordinated and controlled from ground control interception units (GCIs) drawing information from some 5,000 ground radars plus a handful of Tupolev 126 Moss airborne early warning aircraft. Moss can in no way be compared either to the USAF Boeing E-3A AWACS or to the RAF's imminent AEW Nimrod. Moss cannot detect low-flying targets over land and is not much more efficient over the sea. It probably operates most effectively in areas such as Northern Russia with aircraft such as Fiddler and is primarily concerned with extending ground-based radar cover against high-level threats in the Polar regions.

Although in battle conditions interception could be expected to devolve on independent individual aircraft, in peacetime training they work closely

Top: an early MiG-23S pictured just before touchdown, with the wing swept fully-forward and the ventral fin hinged to starboard to obtain the necessary ground clearance.
Above: Bulgaria's air force comprises some 250 combat aircraft and a personnel strength of 22,000 men

with the ground-control units. In view of the number of surface-to-air missiles now deployed in Warsaw Pact territory, it must be reassuring for pilots to be under such close control. That reassurance would, no doubt, have been welcomed by the large number of aircrew in the Yom Kippur war on both sides who were shot down by 'friendly' missiles.

The interceptor force is still heavily dependent upon voice control, but increasingly automated, duplicated communications are being introduced. The basic principles remain the same: ground control to place a fighter in a position where it can acquire its own target within missile range, thereby giving the victim minimum warning time. Now, however, Foxbat and probably the later fighters can to a certain extent be manoeuvred from the ground into the attack position. This broad principle of control is a direct descendant of that used by the Royal Air Force in the Battle of Britain and in theory will permit the swift concentration of air defence at any threatened point. In practice, however, the system is clearly vulnerable to electronic countermeasures and to specialist anti-radar weapons of the kind already in service with NATO air forces. Moreover, until PVO Strany aircraft are equipped with efficient 'look down, shoot down' systems, their capability will continue to be restricted.

Nevertheless, the introduction of the increasingly effective SAMs and the modernisation of the manned interceptor force has greatly enhanced the strength of PVO Strany in the last decade and continued improvements can be expected. Assuming that the USSR seeks to remedy obvious weaknesses, it will continue to improve the quality and security of communications, develop a real low-

level target acquisition airborne radar and may well introduce a new low-altitude air superiority fighter. An improved version of Foxbat is also a possibility, equipped with digital as opposed to vacuum tube avionics, a loiter capability and improved downward-looking radar. As in the West, improvements will also continue to be sought in weapon effectiveness, particularly to improve weapon agility at lower altitudes.

There is already evidence to suggest that these improvements are in hand. In December 1978 Dr W. J. Perry, United States Under Secretary of Defence for Research and Engineering, announced that a modified Foxbat had already shot down a low-flying drone the size of a small fighter aircraft in a series of tests monitored by the United States. He also said that a Soviet counter-part to the E-3A AWACS was in a 'very early development phase'.

United States press sources have asserted that two new interceptor prototypes are under evaluation at Ramenskoye Test Centre twenty miles south-west of Moscow. One, from the Mikoyan design team, may well be the Foxbat derivative with an air-to-air weapon system capable of tracking four targets simultaneously at a range in excess

Top and right: the MiG-23S all-weather fighter is known to NATO as the Flogger-B. A significant variant of this design is the MiG-27 (Flogger-D), which is a specialised ground-attack fighter.
Above: this variable-geometry Sukhoi Su-20 strike aircraft is pictured in Polish air force markings.
Top right: an example of the MiG-25 Foxbat-A was flown to Japan by a defecting pilot in 1976.
Above right: the Hind-D 'gunship' version of the Mi-24 assault helicopter entered service in 1975

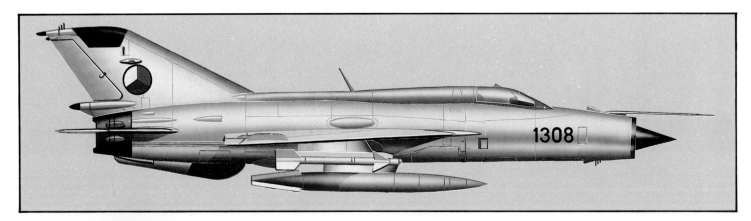

Top: Czechoslovakia flies an estimated total of 170 MiG-21 interceptors in six regiments. A MiG-21PF Fishbed-D is illustrated.
Below: late production MiG-21s, such as this MiG-21MF, have a limited all-weather and ground-attack capability, unlike the initial models.
Below right: although superseded by later types in PVO Strany, many MiG-21s serve on in Frontal Aviation

of 40 km (25 miles). The other, from the Sukhoi bureau could be a smaller high-performance fighter designed to carry a new generation of smaller and more agile air-to-air weapons.

These reports are not unreasonable. In the last few years the Soviet aviation industry has been regularly producing 1,000 current generation fighters annually and has maintained a strong research and development sector. It would, therefore, be most surprising if the development programme did not now include a 'fourth generation' of aircraft twelve years after the arrival of Flogger and Foxbat. However, the presence of prototypes is not automatically followed by production and entry into squadron service, so Western observers will continue to watch developments with great interest.

Clearly, the security of the home base is a primary responsibility of any government and, therefore, an increase in the effectiveness of air defence is likely to have an influence on defence and foreign policy. Yet, while PVO Strany retains its precedence in the Soviet military hierarchy, the capability of Frontal Aviation has also been considerably enhanced in the last decade.

The element of the Soviet Air Forces known as Frontal Aviation can best be compared to a Western tactical air force in that while organised to support ground operations it contains its own air defence, ground-attack, light-bomber, reconnaissance and light-transport aircraft. It is the direct descendant of the Air Armies of World War II which finally wrested control of the Eastern Front skies away from the Luftwaffe. With ever-increasing numerical superiority, aircraft such as the Il-2 Shturmovik ranged ever more deeply behind German lines returning the medicine endured in June 1941: harassing troop movements, interdicting supplies, adding firepower to the advancing Russian ground forces and generally wreaking havoc.

As we have seen, Soviet emphasis for several years after World War II was laid primarily on air defence. Consequently, although the 16 Air Armies of Frontal Aviation maintained some 6,000 aircraft, the ground-attack elements such as the MiG-17 Fresco or the Su-7 Fitter-A had short range and limited load-carrying capacity. Then, Frontal Aviation effectiveness was reduced still further by the cuts in conventional air strength made by Mr Khrushchev as he increased reliance on short and medium-range surface-to-surface missiles. In the last decade, however, this picture has changed

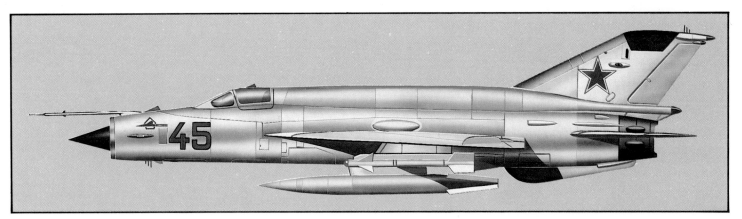

with considerable implications for the balance of military power in Europe. First, and most obviously, has been the introduction of a new generation of aircraft designed for longer-range offensive operations.

Yet another prototype displayed at Domodedovo in 1967 was modified to receive the designation Su-17 Fitter-C. Fitter-C retained a large part of the Fitter-A airframe, but the presence of a much more powerful engine, a variable-geometry outboard wing section and greatly improved avionics produced virtually a new aeroplane with approximately double the range and double the payload of its predecessor. A later mark, Fitter-D, which entered service in 1976 is distinguished by a blister under the nose which carries enhanced avionic equipment including a laser range-finder. Consequently, Fitter-D has a greater navigational and weapons accuracy than its predecessors. It is armed with two 30mm cannon and can carry a variety of free-fall bombs, surface-to-air missiles and nuclear weapons.

The second aircraft is the offensive support variant of the Flogger: Flogger-D or MiG-27 which began to enter Warsaw Pact squadron service in 1975. It differs so considerably from the interceptor Flogger B that it may be said to be a genuine specialised ground-attack aircraft. It is equipped with laser range-finder and computerised navigation/attack system. It has a rapid-fire, six-barrel 23mm Gatling gun and can carry a large variety of conventional and nuclear ordnance. Its cockpit has additional armour plate protection and the undercarriage is fitted with large, low-pressure tyres to facilitate operations away from conventional runways.

Deployed deeper in Warsaw Pact territory, on Russian bases, is a more recent acquisition to Frontal Aviation's offensive strength: the Sukhoi Su-19 Fencer. Although in squadron service in 1974, comparatively little information has been released about this twin-engined, two-seat light bomber. It resembles the USAF General Dynamics F-111 in shape and size, and like the General Dynamics aircraft carries a specialist weapons system operator whose presence suggests a complex avionics fit conferring all-weather navigational and attack accuracy greater than hitherto possessed by aircraft of Frontal Aviation. From bases in Western Russia it can intervene in and beyond a land battle in Central Europe. Deployed forward at short notice into Poland or East Germany it

Top: the main innovation introduced on the MiG-21SMT (Fishbed-K) is the fitting of a tail-warning radar.
Above: Poland has the second-largest air arm in the Warsaw Pact, with a strength of almost 900 aircraft.
Below left: a formation of Sukhoi Su-7 ground-attack aircraft fires salvoes of unguided rockets.
Below: the Mil Mi-24 assault helicopter carries eight fully-equipped troops

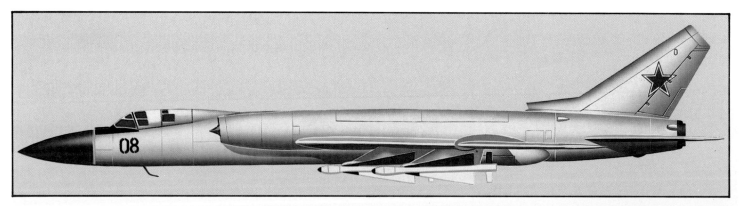

The Tupolev Tu-28P, code-named Fiddler by NATO, is a two-seat all-weather interceptor which provides a platform for four Ash air-to-air missiles (top). When employed in the reconnaissance role (above), a ventral pack may be fitted containing ground-mapping radar. Few examples of the Tu-28P saw service

could attack NATO reinforcements and bases throughout Western Europe, including those in the United Kingdom. In the near future, the light bomber force of Frontal Aviation will probably be completely re-equipped with Fencer.

There are two other elements in the increase of Frontal Aviation's offensive strength. First is the fact that the introduction of surface-to-air missiles has freed interceptor aircraft to place more emphasis on ground attack. For example, the later marks of the MiG-21 Fishbed have been modified by improved avionics and the carriage of bombs, rockets or even a tactical nuclear weapon. Second, whereas ten years ago there were no helicopters in Frontal Aviation, there are now over 3,000 and their numbers are increasing. Not only do they provide the traditional battlefield mobility, but two in particular, the Mil Mi-8 Hip and the Mil Mi-24 Hind-D, have been modified to operate as short-range attack helicopters in close support of ground operations. They deliver concentrated firepower from a combination of machine guns, anti-tank guided weapons, rockets or bombs and, by their increasing presence, they also free conventional fixed-wing aircraft for more wide-ranging operations.

To assess the real significance of this dramatic

increase in the range and hitting power of Soviet Frontal Aviation in the last decade it is necessary to recall the basic principles of NATO and Warsaw Pact strategies. Since 1967, NATO strategy has been based on the concept of 'flexible and appropriate response', which postulates that any Warsaw Pact encroachment on NATO territory will be met by an appropriate level of force sufficient to check it and to force the invaders either to withdraw or to risk an unacceptable level of retaliation which may be nuclear. In practical terms, this requires NATO forces to be able to fight a conventional war for an unspecified but limited period. But, as recent UK Defence White Papers have illustrated, NATO is heavily outnumbered in the Central Region of Europe in tanks, armoured personnel carriers, artillery and aircraft.

It is generally assumed that should the Warsaw Pact ever decide to use military force in Europe, the West would be able to identify some key activities which would provide a warning. Depending on the Pact's own preparation time, which would inevitably be accompanied by increased signal traffic, unusual aircraft or ground force deployments and other indicators, NATO could expect to receive a minimum warning of forty-eight hours of impending attack, but more likely several days.

Therefore, for the strategy to work, NATO must be able to reinforce a threatened or attacked area very quickly. In the short term this could only be done by air, either by moving troops and supplies rapidly across or into a region, or by direct air attack on advancing enemy ground formations.

Against this background, the significance of the powerful and still improving Frontal Aviation can be clearly seen. Its aircraft can now threaten NATO reinforcement bases, supply routes, airfields, headquarters and ground forces to a greater extent than ever before. And if Warsaw Pact aircraft should ever be able to achieve air superiority in the region of their choice, then the threat to NATO strategy would become very serious indeed.

Numbers, however, are only one factor when assessing the capability of an air force, and Frontal Aviation is no exception. On the one hand, of the 9,000 aircraft of all types deployed in twelve military districts inside Russia and four air armies in Eastern Europe, a high proportion are multi-role, usable either as fighters or for ground attack. Moreover, they could be supported by many of the 3,000 aircraft of the national Warsaw Pact air forces which are steadily being re-equipped with 'third generation' aircraft such as Flogger-B and Fitter-C. They also benefit from the enforced

standardisation of equipment and operating procedures in the Pact. Aircraft hangars and other installations are being 'hardened' on many airfields and frequent deployment exercises to secondary airfields occur. In war, there is little doubt that operations would take place within a dense electronic warfare environment. And, as a last consideration, Soviet aircraft could deliver chemical weapons which would obviously still further complicate the defensive task of NATO.

In the face of such formidable military power, it is dangerously easy to exaggerate Soviet strengths and under-estimate their weaknesses. The rigid command and control system employed by Frontal Aviation facilitates the rapid reinforcement of one army, division or regiment by another, but as in the case of PVO Strany, makes the organization very vulnerable to countermeasures. Yet, in an environment so thickly populated by both aircraft and SAMs and probably associated with a fast-moving land campaign, some kind of control must be maintained. Recent articles in Soviet journals suggest that, while the problem is recognized, the answers are not proving very easy; not least because of the reluctance of junior officers to display initiative by disregarding or modifying pre-planned air operations.

Top: since the mid-1970s, Sukhoi's Su-15 has been the most numerous Soviet interceptor fighter. Some 1,000 examples of the Flagon are estimated in service in 1979.
Above: the Tupolev Tu-126 Moss provides limited early-warning cover to supplement ground radar. Its efficiency when compared with the Boeing E-3A is considered doubtful

Moreover, while frequent 'off-base' operations are practised, it is not easy to prepare sufficient airfields to support detached Floggers or Fitters for any length of time. Consequently, and for the foreseeable future, the disruption of flying from a relatively small number of Warsaw Pact airfields would seriously degrade Warsaw Pact offensive fixed-wing air operations. Nor are there signs of any extension of V/STOL or STO/VL operations on land.

The Hip and Hind helicopters would undoubtedly add a powerful punch to any armoured threat if they were allowed to operate unopposed, just as the Junkers Ju 87 Stuka contributed to the success of *blitzkrieg* in World War II. But the careers of Stuka pilots were considerably shortened by the Hurricanes and Spitfires of RAF Fighter Command. In a European conflict of the 1980s, Soviet helicopters operating without the benefit of local superiority would face the same fate as the Stukas at the hands of NATO fighters and ground defences. However it is likely that the Soviet forces would be prepared to accept very high attrition rates to press home an offensive.

Nor should the attitude of Soviet aircrew be under-estimated. Thoroughly politically indoctrinated, imbued with a high degree of professional pride and well versed in the exploits of their predecessors in World War II, Soviet fighter and ground attack crews are unlikely to be lacking in determination. This, of course, may not be the case in all the national air forces. In Poland, Czechoslovakia and Hungary, for example, the armed services have not always shown unqualified enthusiasm for their Soviet 'colleagues'.

In any clash with NATO air forces, the Warsaw Pact aircrew would require all their determination because for the foreseeable future, despite their new equipment, they are likely to be out-ranged, or out-gunned, or out-manoeuvred, or all three by NATO aircraft such as the McDonnell Douglas F-15, Panavia Tornado ADV, General Dynamics F-16, Grumman F-14 and Northrop F-18. This is quite apart from NATO's increasing surface-to-air missile defences. More comprehensive and rapid application of micro-processors has allowed the West to maintain the technical superiority of its equipment and, provided that research and development efforts are not allowed to deteriorate, an edge in this critical area should be maintained. Moreover, in terms of aircrew qualities, NATO pilots would not only match their adversaries' determination, but would be able to draw on the confidence of superior equipment and training without any political restraints on individual initiative.

Nevertheless, it is quite clear from such analysis that the strategy of the Warsaw Pact ground forces, based on surprise, rapid movement and heavy concentration of force, has been considerably enhanced both by the improvements in quality of the equipment of Frontal Aviation during the last decade, and by its likely numerical superiority. Nor is the improvement coming to an end. A third prototype, recently reported at Ramenskoye may well be a specialist ground support aircraft which has been compared to the Northrop A-9, an unsuccessful competitor with the Fairchild A-10 for the recent USAF close air support contract. American reports indicate that this prototype will be equipped with 30mm cannon, laser guided missiles and anti-radar emission weapons. If so, it will be a further formidable close support weapon but, like the helicopter, it will be uneasy in a hostile air environment.

While this tense and extremely important competition for potential tactical air superiority in Europe continues largely unnoticed outside the specialist press, one aircraft has captured many headlines during the recent negotiations between the United States and the Soviet Union to limit strategic arms. The Tupolev Tu-26 Backfire entered

Top: the Antonov An-24 superseded the Ilyushin Il-14 in both civil and military service with several Warsaw Pact countries. The type is pictured in Bulgarian air force markings.
Centre: the Russian airline Aeroflot plays a major role in supporting Military Transport Aviation, operating types such as the Antonov An-30.
Above: the helicopter has assumed an ever-increasing importance in Soviet battlefield strategy during the past decade. The Mil Mi-6 is representative of troop transport rotorcraft employed

Top: as with its Western counterpart the Bristol Britannia, the Ilyushin Il-18 turboprop has found military application as a long-range transport. The Polish air force are among its operators.
Centre: although largely superseded by more modern helicopters, the Mil Mi-4 Hound continues in service with the Soviet Union and its satellites. Ten are flown by the Romanian air force.
Above: the Ilyushin Il-76 is expected to provide a large proportion of the Soviet Union's long-range transport capability in the 1980s

regimental service with Long Range Aviation (LRA) in 1975. Since then it is estimated that about three have been built each month and joined the Long Range Aviation and Soviet Naval Aviation in equal proportions. The Soviet Union recently confused its category slightly by asserting that it carried the bureau designation Tu-22M, the number hitherto ascribed to the Blinder, to which it bears absolutely no resemblance. Whatever its actual designation, Backfire is the most modern aircraft in an area of Soviet military aviation which has not seen many distinguished predecessors.

Only two bombers, the four-jet Myasishchev M-4 Bison and the Tupolev Tu-95 Bear ever had

even a limited unrefuelled high level inter-continental radius of action. Both entered service in 1956 but the combined total probably never exceeded 250. Far more numerous was the medium range Tu-16 Badger which, entering service in 1954, could reach targets at low level throughout Western Europe from its bases in the USSR. Then, in 1961, the twin-jet Tu-22 Blinder was expected to replace Badger in the ratio of one to three. However, despite a supersonic dash capability, it also was not produced in large numbers and production ceased in 1969.

In 1975, reports were first seen in the Western Press of a new variable geometry, supersonic inter-continental bomber from the Tupolev bureau. Although only twin-engined, Backfire has an un-refuelled range (carrying some 9,000 kg [20,000 lb] of weapons), which permits attacks on all Western Europe, including the United Kingdom, and targets at sea in Eastern Atlantic. Whether it could reach the United States on a one-way unrefuelled mission with possible recovery in Cuba or elsewhere is a subject which has been hotly debated in Washington.

When first deployed, Backfire carried an in-flight refuelling probe, but in 1979 several aircraft have been observed without it. Whether this removal indicates a fundamental change of policy or is only a temporary gesture to the US-USSR strategic arms limitation negotiations remains to be seen. Like all the other bombers except Bison, Backfire carries air-to-surface missiles in addition to free-fall bombs. Soviet air-to-surface missiles have been deployed since the early 1960s and have used a variety of guidance systems. Some are accurate enough to be used with conventional warheads. Kangaroo AS-3, Kitchen AS-4, Kelt AS-5 and Kingfish AS-6 may all be launched from 160 km (100 miles) or more from their target. Backfire has been observed carrying AS-4s which probably embody improved guidance systems.

In addition to their long range nuclear strike and attack responsibilities, LRA aircraft also, carry out photo-reconnaissance (Badger-E, Bear-E), electronic intelligence (Badger-F), electronic counter-measures, and air-to-air refuelling (modified Bisons) sorties.

Therefore, in the event of hostilities between the Warsaw Pact and NATO, it can be assumed that long range aviation aircraft would probably attack targets on sea and land beyond the range of Frontal Aviation aircraft, using both nuclear and conventional weapons, as well as providing reconnaissance and electronic warfare support. No national Warsaw Pact air force possesses medium or long range bombers nor is there any evidence to suggest that the Soviet Union has ever encouraged their development or purchase, although some Blinders and Badgers have been sold to other, non-European, countries.

In the light of previous Soviet bombing experience, and the controversy which has surrounded the future of the long range manned bomber because of its expense and possible vulnerability, further press reports of developments at Ramenskoye are intriguing. Some suggest that a bomber variant of the delta-wing, supersonic Tu-144 'Concordski' airliner is being flight tested. If so, its

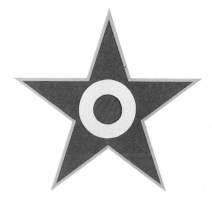

Above: the Hungarian air force insignia. Above right: crew members of a Soviet Navy Beriev Be-12 return from a mission. Opposite top: the VTOL Yakovlev Yak-36 Forger was deployed aboard the Soviet navy aircraft carrier Kiev in 1976. Regarded by Western observers as an interim development, it may have interceptor, attack and anti-submarine potential, but it is considered to be inferior to the British VTOL BAe Sea Harrier in terms of performance and combat effectiveness.
Opposite top right: the insignia of the East German air force.
Opposite centre: the Soviet navy's Ilyushin Il-38 long-range anti-submarine warfare aircraft is a conversion of the Il-18 airliner.
Opposite below: the Beriev Be-12 Tchaicha, code-named Mail by NATO, is a turboprop version of the Be-6 operated in the coastal reconnaissance and anti-submarine warfare roles. Now obsolescent, the Be-12 is among the last amphibians in military service in the world

aerodynamic design can fit it only for high speed, or high altitude penetration, or cruise missile delivery. Others report a more realistic four turbo-fan, variable-geometry longer-range and faster successor to the Backfire. Yet another describes a transonic successor to the Tu-95 Bear, which would primarily be a cruise missile launcher for inter-continental missions. Also, in typical Soviet manner, engine and airframe modifications seem to be under test for Backfire.

Again, the presence of such development work should be expected. Design decisions must have been taken about a decade ago and, bearing in mind the problems obviously met by the Tupolev bureau with the Tu-144, it should not be surprising if prototype development is proceeding only very slowly. There has never been any doubt that air-power in general, and the manned aircraft in particular, offers the statesman far more flexible support than a ballistic missile and it appears that the Soviet Union has recently come to a very full appreciation of that fact. What is doubtful is whether the Soviet Union can afford, or can choose to afford, a weapons system of a kind which in the B-70 and B-1 the United States has twice rejected. Consequently, and particularly in view of the strategic arms limitations negotiations, develop-ments of a new generation Soviet manned bomber will repay the most careful observation.

The third component of the Soviet Air Forces commanded by Chief Marshal of Aviation Pavel Kutakhov is Military Transport Aviation, which is comparable in role to the USAF's Military Airlift Command. There is, however, one major difference in that Military Transport Aviation is the largest military air transport fleet in the world, comprising some 700 main transport aircraft and it is frequently supported by ostensibly civil aircraft of the state airline, Aeroflot.

From the earliest days of the Soviet state, Transport Aviation has been seen as an important element in promoting the unity and economic development of the vast areas of the Socialist Republics, as well as providing air support to the other branches of the Armed Services. Con-sequently, military and civil aviation have main-tained a far closer association than in either Britain or the United States.

Although a strong airborne force had been constructed before World War II, its operations

met with little success on the Eastern Front and on several occasions paratroops re-converted to in-fantry. On the other hand, conventional rein-forcement by air was widely practised, as for example in the support of Stalingrad in October 1942. After World War II, Stalin emphasised the construction of fighters and, to a lesser extent bombers, rather than transports. Only in the later 1950s, with the introduction of new turbo-prop aircraft, did Military Transport Aviation begin to become a modern air support force and have increased opportunities to re-develop airborne forces.

The first such aircraft was the twin-engine Antonov An-8 'Camp' which is now withdrawn from front-line service with Military Transport Aviation but is scattered about in small numbers in support of PVO Strany and Frontal Aviation units. The four-turbo-prop Antonov An-12 Cub however still forms the backbone of Military Transport Aviation with over six hundred still in active service. Cub may be compared in range and payload with the Lockheed C-130 Hercules – some 1,370 km (2,200 miles) carrying 20,000 kg (44,000 lb) of freight or one hundred troops or a combination of each. Like the Hercules it performs a variety of support tasks: air-drop of men and supplies, troop and equipment carrier, all with an ability to operate from short, secondary runways. However, unlike the Hercules, only its crew compartment is pressurised, so transport of troops is normally restricted to altitudes below 4,500 m (15,000 ft) with consequent restrictions on range. Nevertheless, Cub demonstrated its abilities during the swift invasion of Afghanistan in 1979. Such an operation would have been much more difficult for the Soviet Union to mount without such an aircraft.

In the last decades, however, Military Transport Aviation's reach has been considerably extended by the introduction of some fifty massive four turbo-prop An-22s (code-named Cock). This type was seen at Prague in 1968, but it only entered regular service between 1970 and 1974, when production ceased. Cock can lift close to 45,000 kg (100,000 lb) of heavy equipment in excess of 6,400 km (4,000 miles). It has been observed to carry armoured vehicles, surface-to-surface missiles and surface-to-air missiles with their supporting equipment. It has suffered from a variety of technical defects which may have contributed to the relatively short

production run. For example, two were lost in 1970 while on long range relief missions – one off Iceland en route to Peru and one near Calcutta while returning from East Pakistan. But the Cock enabled the Soviet Union to move large quantities of supplies to Egypt and Syria in 1973 and played a prominent part in the military intervention in Ethiopia, Angola, Mozambique and Afghanistan between 1977 and 1980.

The most recent and perhaps the most important acquisition by Military Transport Aviation, however, is the Ilyushin Il-76 Candid, a four-jet, long-range transport which resembles the USAF's Lockheed C-141. Candid appeared at the Paris Air Show in 1971 and since 1974 some 50 have entered service with Military Transport Aviation and the numbers are steadily increasing. It has a payload twice that of Cub and a considerably superior range. During recent Warsaw Pact exercises it has been observed dropping both armed paratroops and armoured personnel carriers.

Together, Cock and Candid comprise a significant addition to Soviet airpower. Ten years ago, Soviet military intervention, either directly or by proxy, in East or Central Africa would have been virtually impossible. However, the USSR may well have learned from Western example about the ability of long range flexible airpower to intervene swiftly and precisely in a disturbed local situation to create more favourable circumstances for the diplomat to exploit. Bearing Aeroflot markings, Soviet civilian and military transport aircraft now range the world. Aeroflot aircraft rotate naval crews and deliver 'students' to and from Moscow and troubled African areas. Aeroflot aircraft carry out the twice-yearly rotation of troops from Eastern Europe to the motherland. An-22s retain their tailgun turrets while wearing Aeroflot livery.

Overall, the co-ordinated use of Military Transport Aviation and Aeroflot by the Soviet Union illustrates most vividly the value of closely integrated national military and civil airpower and the

capability continues to grow. Current speculation focuses on the wide-bodied 350-seat, four turbo-fan Ilyushin Il-86 Camber. At present, Camber is only entering Aeroflot service but as has been explained, that is only a technicality as far as its use as a troop carrier is concerned. It would, however, also seem to have a potential for development either as an air-to-air refuelling tanker or as the airframe for a Soviet equivalent to the Boeing E-3A AWACS. Camber, therefore, will also join the group of aircraft whose future development will be keenly studied by Western observers.

Near the coasts of the United Kingdom, a different kind of observation regularly takes place. Maritime reconnaissance Bear-D aircraft of the Soviet Naval Air Force are frequently intercepted by Phantoms of No 11 Group of Royal Air Force Strike Command. The Bears no doubt observe the reaction time of the Phantoms, while the Phantom crews note any new addition to the Bear's intelligence-gathering equipment. But Bear-D is not primarily interested in the speed of RAF fighter response, as it is the major provider of maritime reconnaissance to the Soviet Navy.

Czechoslovakia and Poland are among the Warsaw Pact countries which have designed and built their own training aircraft. The Czech L-39 (top) is the intended successor to the L-29 Delfin, which served with nine air forces from mid 1963. The TS-11 of Poland (above) was flown solely by the air force of that country and was named Iskra (Spark)

But the change of Western carrier task forces from ally to potential enemy forced the Soviet Union to make provision for their destruction beyond the launch range of their carrier-borne aircraft. This task remains a major responsibility of the bombers of Soviet Naval Air Force.

SNAF bombers are variants of aircraft in service with Long Range Aviation. Badgers and Backfires are equipped with long range air-to-surface missiles, some of which may have been modified to attack radar emitting targets, and operations would be supported by various kinds of electronic counter-measures. While the bulk of the force is composed of Tu-16 Badgers, the advent of Tu-22/26 Backfire, as in Long Range Aviation, obviously adds considerable strength by its increased range, speed and ability to deliver later models of the family of air-to-surface missiles. When the importance to NATO strategy of timely reinforcement is remembered, the threat of this maritime offensive airpower to NATO shipping is clearly very important.

Intelligence for both air attacks and 'over the horizon' attacks on shipping by submarines or surface vessels is provided by the specialist Bear-D, or by modified Badgers. It is possible that a reconnaissance variant of Backfire may enter service, just as reconnaissance variants of earlier bombers have appeared. Precedent may not however on this occasion necessarily be followed, because of apparently increasing reliance by the Soviet Union on satellites for maritime intelligence gathering and communication.

Shorter-range reconnaissance is provided by the ship-borne Kamov Ka-25 helicopter, known as Hormone, which is deployed singly on *Kara*-class and *Kresta*-class cruisers as well as in squadrons on the *Kiev*-class aircraft carriers and *Moskva*-class helicopter carriers. In coastal waters the land-based Mil Mi-14 Haze undertakes this role. Both helicopters and aeroplanes would be expected to detect hostile shipping 'over the horizon', relay its exact position to Naval Attack Forces and then to provide post-attack information.

As in Western Navies, the helicopter would also make a major contribution to anti-submarine warfare working sonobuoy or magnetic anomaly detection patterns for depth charge attack by surface vessels, or by the helicopters themselves. Longer-range anti-submarine operations are carried out by the Bear-F, or by the specialist anti-submarine Ilyushin Il-38 May. The obsolescent amphibian Beriev Be-12 Mail is limited to coastal waters, where it still serves a useful purpose.

Partly because the Soviet Union lacked the World War II anti-submarine experience of the Western allies, and partly because of the emphasis laid elsewhere by Stalin, Soviet anti-submarine warfare operations appear to be considerably less effective than those of NATO, whether measured by quality of equipment or tactical exercises. Despite Admiral Gorshkov's affirmation, it may well be that the allocation of Soviet naval funds to aviation has been of comparatively low priority. It is generally assumed that in the event of hostilities, Soviet naval aviation could be supported by aircraft of Long Range Aviation.

In two other areas, however, the Soviet naval aviation has been considerably strengthened. The

Admiral Gorshkov, Commander-in-Chief of the Soviet Navy, is well known not only for presiding over the construction of the modern Soviet 'blue water' fleets but for his writing on seapower. 'No naval operation,' he has stated, 'is conceivable without air forces.' Unlike the British arrangement, where support to the fleet from bases on land is provided by the Royal Air Force, in the Soviet Union all maritime air operations except air defence are the responsibility of SNAF.

The 1,200 or so aeroplanes and helicopters of SNAF are equipped to provide reconnaissance, to attack surface shipping, to co-operate in anti-submarine warfare and to support amphibious operations. Naturally, some of the weapons carried for use against surface shipping could, if necessary, be turned against coastal targets such as radars or installations associated with maritime operations.

The growth in importance in Soviet naval aviation has taken place at the same time as that of the surface fleets, but not always for the same reasons. Until the end of World War II, Soviet naval aviation was largely concerned with in-shore operations, either defending land force flanks, or working with the 'white water' (coastal) naval units.

Top: Soviet pilots are pictured with their Sukhoi Su-9 Fishpot all-weather fighters. The Su-9 entered service in 1961, with PVO Strany, the Soviet Air Defence Force and in smaller numbers with Frontal Aviation, the Soviet tactical air force. Large numbers remain in service with PVO Strany, despite the appearance of more recent all-weather interceptors.

Above: designed as a supersonic medium-range strike and reconnaissance aircraft for the European theatre, the Tupolev Tu-22 Blinder entered service in 1963. A new version appeared in 1967 with an in-flight refuelling probe, and armed with an AS-4 Kitchen stand-off missile, greatly improving strike range and potential

most obvious is in the commissioning of the *Kiev* class of aircraft carriers. Two such ships, *Kiev* and *Minsk*, are now operational, while a third, *Kursk*, is nearing completion. *Kiev* attracted a lot of attention in 1976 when she appeared carrying a dozen Yakovlev Yak-36 Forger vertical take-off and landing fighters. However, the development of these aircraft still gives rise to speculation. They do not appear to have been deployed on *Minsk* and one wonders whether Forger is still in a slowly moving development stage, or whether Soviet deployment policy has been modified.

Theoretically, Forger considerably strengthens Soviet naval power. It appears capable of delivering air-to-air, infra-red homing missiles, air-to-surface rockets, cannon fire, air-to-surface guided missiles, depth charges and torpedoes. It could provide limited air cover to Hormones in anti-submarine operations and it could destroy unarmed NATO surveillance aircraft such as BAe Nimrod, Lockheed P-3 Orion, or even the E-3A AWACS. In fact, in a conflict with NATO naval or air forces, Forger's limited radius of action, lack of both defensive radar and long range air-to-air missiles could reduce her to little more than nuisance value, particularly as the exact position of the two or three *Kiev*-class ships deployed at any one time would be known to Western Intelligence.

What the Forger-*Kiev* combination does offer the Soviet Union is a further method of exerting military influence and pressure a long way from Soviet frontiers. It is another element in Admiral Gorshkov's 'balanced navy' used in peace and war to underpin the foreign policies of the Soviet government. They form an impressive weapons combination in the eyes of a developing, relatively-small country. In a trouble spot, the simple presence of *Minsk* or *Kiev* would inhibit competitive military activities. Would, one wonders, United States Marines have swept so blithely ashore in the Lebanon in 1958 if *Kiev* and her Forgers had been cruising off Beirut? World-wide interests have traditionally demanded world-wide power and the Soviet Union has been quick to learn the lesson.

Closer to home, 'white water' power has been enhanced by the deployment in Soviet naval aviation of the Sukhoi Su-17 Fitter-C, which could not only attack hostile shipping at short range, for example closing amphibious forces, but could also give strong support to the Warsaw Pact's own amphibious operations. Finally, one other 'white water' experience could well be put to good use in view of the ingredients of NATO strategy. According to a recent statement in the Soviet journal *Air Defence Herald* by Major General Vishensky, Deputy Chief of Staff of Soviet Naval Aviation, in World War II the Naval Air Arm laid 2,428 mines. Most modern Soviet naval aviation aeroplanes could lay mines and, depending on other priorities, a proportion of the Blinders could be expected to carry out that role.

Overall, therefore, although in some respects Soviet naval aviation remains the Cinderella branch of Soviet Military Aviation, there are clear signs that it too is being considerably strengthened. Three new aircraft have, for the first time, gone straight into Naval service–Hormone, Backfire, and Forger. While United States Navy submarines have withdrawn closer to their own coasts, as their ballistic missile ranges have increased, so Soviet anti-submarine warfare capability has had to develop a longer reach and, using the ship-borne Hormone, it has done so. By attacking NATO shipping from the air, the warning time, which the deployment of additional submarines or surface vessels for that purpose would inevitably give NATO, would be considerably reduced. In *Kiev*, they have a long range military instrument which can roam the world and encourage the provision of base facilities in countries such as Guinea, Aden or Angola which, in turn, permits their maritime aircraft to range ever more widely over the oceans.

The impact of all these developments in the last decade on Warsaw Pact military aviation has been enormous. The airspace of the Pact is now more densely defended by better equipment than ever before by the addition of the nine surface-to-air missile systems and the introduction of new manned interceptors. The tactical air forces can fly further carrying more ordnance and both air defence and close support aircraft in the national forces are being modernised to the standards of their Soviet ally. Despite the Pact's heavy reliance on new generations of surface-to-surface missiles, the manned bomber continues to be developed and deployed. The new transport aircraft are carrying Soviet equipment and allies in peacetime thousands of miles from the Soviet frontier and in war would provide large-scale mobility to the Warsaw Pact armies. And from above the oceans, Western naval superiority is increasingly challenged by the land and sea-based aircraft of Soviet Naval Aviation.

A real assessment of military strength cannot be made by allusion simply to a few aircraft and weapons systems. Political training, which runs from the Chiefs of the Political Administration down to the *zampolit* (political officer) in all units in the Soviet Air Forces may be strongly motivating. On the other hand, it may be inhibiting and counter-productive. Why, one might ask, should a Warsaw Pact airman need constant political education, unless his motivation is thought to be suspect? Perhaps, because of the large numbers of conscripts, or perhaps because the Soviet Union regards continual political education as essential for the production of a good Communist citizen and ally.

When assessing operational proficiency, what will be the results of aircrew tending to remain with one aircraft type in one role for very long periods, perhaps even for all their Service career? On the one hand, it should produce a high degree of tactical ability, but what will the impact on senior command be of limited type and role experience? Or, if several flight authorisations are required for each sortie and if most sorties (including long-range bomber and maritime) are very carefully pre-planned, how well would aircrew respond to the uncertainty of war, degraded communications and minimum briefing periods? Perhaps a clue may lie not so much in the laborious preparations which finally brought down Gary Powers' Lockheed U-2, but in the oddly delayed and uncertain PVO Strany response to the Korean airliner which by accident penetrated a most sensitive region of Soviet airspace for several hours in 1978.

How much further can the Warsaw Pact afford

Above: the Kamov Ka-25 Hormone anti-submarine warfare and short-range maritime reconnaissance helicopter carries a variety of sonar and radar and two torpedoes. Some 250 Ka-25s are in Soviet navy service, deployed aboard the anti-submarine helicopter carriers Leningrad and Moskva, the carriers Kiev and Minsk and singly aboard Kresta-II and Kara class cruisers. Right: the Tu-22 Blinder is powered by two 12,250 kg (27,000 lb) thrust turbojets which give it supersonic dash capability. Its defensive armament consists of one radar-controlled cannon in the tail, replacing earlier Soviet bombers' tail gunner position

to develop its airpower? A MiG-23 costs a lot more than a MiG-19 and the Soviet Union does not give equipment to her allies. Indeed, serious debates must take place in the Soviet Central Committee about how far the country can continue to spend twelve to fifteen per cent of its national resources on defence. On the other hand, existing equipment could be further improved at comparatively small expense. For example, USAF electronic counter-measures equipment almost certainly found its way into Soviet hands from North Vietnam and it may be assumed that such expertise will be translated into Warsaw Pact missiles and aircraft. It is highly probable that details of F-14 Tomcat and Phoenix air-to-air missiles have leaked across the Caspian Sea after the upheavals in Iran. And, generally, improvements to avionics and weapons induced by the micro-processor/computer revolution can be made more cheaply than hitherto. Again, however, full exploitation in the USSR of that revolution may be inhibited by lack of a broad, highly skilled industrial base and by bureaucratic restriction on design ingenuity and development.

But, in whatever way these questions are answered, there is no doubt that the airpower of the Warsaw Pact, largely represented by the airpower of the Soviet Union, is a formidable military instrument second only to that of the United States. There is equally no doubt that the Soviet Union has a very thorough grasp of the wide-ranging potential of that airpower both as a peace-time instrument of foreign policy for the statesman, and as an instrument of war for the commander. Should ever Western policies of deterrence, defence and detente fail, airpower would almost certainly decide the outcome of any conflict. This is the underlying significance of recent developments in Warsaw Pact military aviation.

CHAPTER ELEVEN

Middle East
Israel/Egypt/Iraq/Morocco Lebanon/Libya/Algeria/Jordan

Scarcely a day passes without some reference to the Middle East in the news headlines. From Morocco's anti-guerilla war in the Western Sahara, through Colonel Ghadaffi's Libya, to Syria's conflict with Israel, the Middle East abounds with potential flash-points. Ironically, it is Israel which provides the one unifying force for the Arab nations, as the ideological and territorial differences between them would otherwise almost certainly have led to war long ago. It is the West's dependence on Arab oil which, to a great extent, determines allegiances and

shapes loyalties in this volatile region.

Seeking to extend its sphere of influence, the Soviet Union has been liberal in aid to the Middle East, particularly to left-wing military dictatorships, but as a counterbalance, Jordan and Morocco remain monarchies and Saudi Arabia prefers financing Western arms, to hinder the spread of Soviet influence. The United States is also deeply involved, particularly with Israel, and possibly in the future with Egypt unless something can be salvaged from the abandoned Arab Organisation

Right: following the French embargo on the sale of Dassault Mirage 5s to Israel, the IAI Kfir was designed to provide Israel with an indigenous, low-cost, high-performance multi-role jet in the event of similar embargoes. Based upon the Mirage and powered by the General Electric J79 engine, production problems were minimised. Some 150 Kfirs are in service, mainly deployed in the air superiority role, but the type can be rapidly reconfigured to operate as a long-range fighter, interceptor or attack fighter. The improved Kfir-C2 (illustrated) was revealed in 1976, all earlier Kfirs being updated to C2 standard.
Below: the McDonnell Douglas F-4E equips some seven Israeli strike/reconnaissance squadrons

Above: the IDF/AF operates 12
Lockheed C-130Es, 12 C-130Hs and
two KC-130H tankers. They form the
backbone of the transport and support
wings. Israel's C-130s became famous
when they carried commandos to
Entebbe airport, Uganda in 1976 to
release and evacuate hostages held by air
hijackers.

Right: some 80 IAI-built Magisters
serve as trainers at Hatzerim. They are
available for light attack duties. The
Israeli decision to replace them has been
postponed and they are scheduled to
serve into the 1980s.

Below: the IDF/AF operates 30
Sikorsky S-65C-3s for heavy transport
and assault duties. Together with 12
Aérospatiale SA 321K Super Frelons,
they constitute the heavy lift force

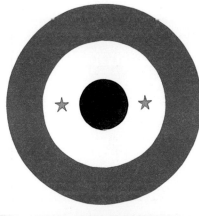

for Industrialisation, upon which so much depended following the break with the USSR. For the moment, however, Arab air power continues to grow.

Throughout its short, 31-year history Israel has been in a state of war with its Arab neighbours. On three occasions, in 1956, 1967 and 1973, the world held its breath while Israeli ground and air forces battled with numerically superior opponents. These were only milestones, however, on the long road of guerilla raids, counter-strikes, terrorist bombings and indiscriminate shootings. Honed to perfection through constant alertness, and backed by the mighty arsenal of the United States, the unified Israeli Defence Force (IDF) is a formidable deterrent.

The recent Camp David peace agreement between Israel and its former principal enemy, Egypt, has led to a reduction in defence spending for the current year, but the 1979–80 total of £1,877 million continues to represent a crushing financial burden on a population of less than four million. In the light of its *rapprochement* with Egypt, the United States is carefully balancing its arms deliveries to the two countries. However, Israel has learnt its lesson from previous arms embargoes and established its own defence equipment industry, extending to advanced aircraft and avionics plants.

Conclusive proof of Israeli accomplishment in weapons technology is provided by the IAI Kfir, a potent conjunction of the well-proven Dassault Mirage and the General Electric J79 engine, now serving the IDF/AF in significant numbers. More sinister is the development of a nuclear capability, admitted by President Katzir in 1974, and estimated to total some 10–20 warheads each of 20 kilotons. The IDF has received 109 LTV MGM-52C Lance launchers from the United States, fitted with a 1,000lb high-explosive cluster warhead, although this may easily be replaced with a more potent weapon.

Israeli requests for the longer-range Pershing missile were vetoed in 1975 and instead the IDF received further defensive equipment in the form of 25 McDonnell Douglas F-15 Eagles from December 1976 onwards, and four Grumman E-2C Hawkeye airborne early warning and control aircraft between December 1977 and August 1978. Eagles participated in their first offensive mission in a raid on Palestinian camps in Lebanon during March 1978, and scored their first aerial victories in the course of an air battle in June 1979, during which the destruction of six Syrian MiG-21s was shared with a squadron of Kfirs.

The potency of the Hawkeye's General Dynamics AN/APS-125 overland radar was clearly demonstrated when the force of MiGs was picked up on take-off – according to some reports, as they were still gathering speed on the runway. Vectored by the watching Hawkeye, six Eagles and four Kfirs intercepted the eight MiGs, destroying five with air-to-air missiles released from the Eagles and one with 30mm cannon fire from a Kfir. A seventh was damaged, but all Israeli aircraft were claimed to have returned unscathed.

A further 15 Eagles have been authorised for delivery in 1981–83 and this number may be increased to 35 to give parity with Saudi Arabia's allocation of 60. Israeli strike and defensive capa-

bility will be further reinforced from 1980 with the delivery of 75 General Dynamics F-16s, a year earlier than originally promised due to the cancellation of the Iranian order. Despite Israeli hope of obtaining 250 F-16s, most of which were to have been built under licence, US approval has been given for only 75 to date. Although Israel maintains that the benefits of peace with Egypt are counterbalanced by increasing Soviet aid to Iraq and Syria, American aid is given with circumspection. Associated armament and support facilities for the F-16s in the form of 600 AIM-9L Sidewinder and 600 Maverick missiles, together with a small number of KC-135 flight-refuelling tankers, has been indefinately witheld.

Further changes brought about by the peace agreement concern Israeli bases in the Negev desert. The forward airfield at Refidim will be vacated by the end of 1979, with Etam, Etzion and Ophir following by 1982. In return, the United States is financing the considerable cost of two new bases at Matred and Ovda.

Although the bases facing Egypt will now see little operational activity, the remainder will continue to function as before. Following the French military equipment veto after the 1967 war, the IDF/AF was converted to largely American equipment. Pending the transition, Israel supplemented the original spearhead of 72 Mirage IIICJ fighters and five Mirage IIIBJ trainers with some 40 locally-built versions designated Nesher. The Nesher took a prominent part in the 1973 fighting but Israeli Aircraft Industries was already working on a more ambitious development known as the Kfir.

From its earliest days, Israel gained experience in obtaining through unorthodox methods what was not freely available through more normal channels. The challenge of rapidly producing a complex strike-fighter without the necessary drawings and data was beyond even the most determined IAI designer, but several crates of paper were eventually 'obtained' via Switzerland, and the Nesher and Kfir quickly appeared in the IDF/AF front line. The Kfir, however, was no straight copy of the Mirage. A complex re-engining programme involved major internal changes to accommodate the GE J79 with its bigger dimensions, higher thrust and greater jet-pipe temperature, and to these were added Israeli instrumentation and avionics.

Some 150 Kfirs have been produced for home use, air defence variants having the indigenous Shafrir infra-red, air-to-air missile in place of the Sidewinder. The aircraft has additionally been offered for export, at half the price of the Mirage III and with 40 per cent better performance, although US restrictions on the J79 have hampered and frustrated potential orders, and none has been sold outside Israel to date.

With the Kfir firmly established in the air superiority role, the Phantom is being increasingly used for strike duties. Israel is believed to have received a total of 242 F-4E Phantoms and a further 12 RF-4Es for tactical reconnaissance. Most have been delivered new from the manufacturer, but losses during the 1973 war were replaced by aircraft hastily transferred from USAF squadrons in the front line. Phantom armament comprises

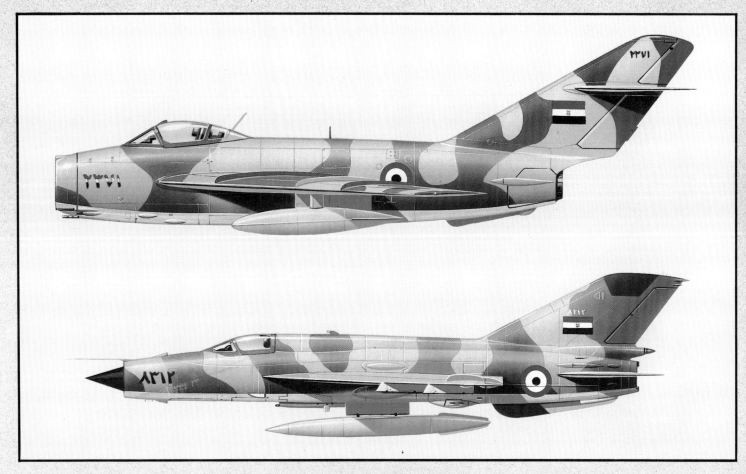

Top: the subsonic MiG-17 has been latterly employed as a ground attack aircraft by the Egyptian air force. Above: representing the later marks of the ubiquitous MiG-21 interceptor, the MiG-21MF can carry K-5 Alkali radar-guided missiles. The break in relations with Russia has hampered the MiGs' continued operation. Below: an alternative source of funds and assistance utilised by Egypt was Saudi Arabia, which funded the purchase of a number of Dassault Mirage fighters

locally-built 30mm DEFA cannon, AGM/RGM-84A Harpoon anti-ship missiles, Hobos TV-guided 'Smart' bombs, AGM-45 Shrikes, AGM-62 Walleyes and AGM-65A Mavericks.

Primary strike weapons of the IDF/AF is, however, the redoubtable Douglas A-4 Skyhawk, which was employed in considerable numbers in 1973, but suffered the highest casualties (53, as opposed to 33 Phantoms and 11 Mirages/Neshers). Vulnerability to SAMs has been decreased by the installation of an extended jet-pipe, protruding several feet beyond the rear fuselage. In theory, a heat-seeking missile scoring a direct will remove part of the superflous ironmongery and not damage the more important fin and tailplane.

Some 287 A-4E, -H, -M and -N variants of the Skyhawk have been delivered to Israel, together with 25 TA-4H trainers, and all are now fitted with

the Pratt & Whitney J52 turbojet and 30mm cannon. Allowing for combat and training losses, the IDF/AF therefore has a current strength of some 630 operational types, backed by 125 transport and liaison aircraft, 120 trainers and 150 helicopters. Personnel totals 21,000, including 2,000 conscripts, plus 25,000 available on mobilisation.

Additional air defence capability is provided by 15 SAM batteries comprising 90 MIM-23 Hawks, aided by army-operated MGM-72A Chaparral tracked SAM launchers, GD FIM-42A Redeye infantry SAMs and radar-directed, self-propelled cannon. A small number of former French combat

Above: the Israelis' effective use of troop-carrying assault helicopters was undoubtedly a factor which prompted Egypt's purchase of the Westland Commando in the mid and late 1970s. These have become increasingly important as Soviet helicopters have been hit by lack of spares.
Right: this Soviet-built Antonov An-12 heavy transport is operated in quasi-civil markings, despite the presence of a rear gunner's tail turret.
Below right: serving alongside the An-12 is the familiar shape of the Lockheed C-130 Hercules, the camouflage of which reflects its tactical role. Six C-130H Hercules were delivered in December 1976

types, Vautours, Super Mystères (retro-fitted with J52 engines) and Mystère IVAs, are retained for operational training.

Mainstay of the transport and support forces is the Lockheed C-130 Hercules, of which 24, including 12 ex-USAF C-130Es and two KC-130H tankers, have been received. Further air-refuelling capability is provided by ten Boeing 707s, modified to tanker/freighter configuration by Bedek Aircraft. Lighter transport and communications types include 14 IAI Aravas, two IAI 1123 Westwind liaison jets, four Islanders, nine Dornier Skyservants, 25 Cessna U206s and 16 Queen Airs.

IDF/AF also uses a large fleet of helicopters centred on original deliveries of 12 Super Frelons and 30 Sikorsky S-65Cs for heavy transport. Lighter types include some 60 AB.205/UH-1D Iroquois, 20 Jet Rangers, and a dozen Bell 212s. The Iroquois have provision for Hughes TOW anti-tank missiles, but are soon to be augmented by 30 TOW-equipped Hughes 500M helicopters, due for delivery from mid-1979, joining a dozen Bell

are to IAI 1123 standard with General Electric turbojets, whilst the third is a 1124 with AiResearch turbofans.

In addition to aircraft, Israel has devoted considerable attention to development of an indigenous missile armoury to complement, and replace, weapons supplied by the United States. For anti-SAM site strike Phantoms and Kfirs are equipped with the Rafael Luz 1, a TV-guided, 80 km (50 mile) range missile, whilst Rafael Shafrirs are used for air defence. Shafrir is the Israeli equivalent of the Sidewinder, the Mark 2 being a parallel to the AIM-9D/G and the Mark 3, an AIM-9L. Despite repeated reports of a long-range surface-to-surface missile (SSM) capable of carrying a nuclear warhead, the only locally-produced SSM so far identified is the Ze'ev short range – 4.8 km (three miles) – battlefield bombardment rocket used by the IDF/Army.

During the past four years, Egypt has made the welcome change from war to peace, yet both Soviet and Arab-backed plans for military re-equipment have disintegrated into resentment and mistrust. Although numerically strong on paper, the Egyptian Air Force and Air Defence Command (Al Quwwat Aljawwiya Ilmisriya) administers a force of Soviet aircraft virtually grounded through lack of spares and vainly looks to the shattered remnants of the Arab Organisation for Industrialisation for its next generation of aircraft.

Following the expulsion of 20,000 technicians and advisers in 1972, Soviet-Egyptian relations gradually deteriorated, culminating in the abrogation of the mutual Friendship Treaty in 1976, shortly after delivery of 48 MiG-23 and MiG-27 Flogger B/D air superiority and strike aircraft. Despite claims that the Soviet Union had not made good losses incurred in the 1973 war with Israel, the EAF at that time possessed some 550 first-line

AH-1G Hueycobra helicopter gunships. Battlefield surveillance is the task of two Grumman OV-1 Mohawks, received in 1976.

Pilot training is undertaken at Hatzerim on IAI-built Magisters, after about $7\frac{1}{2}$ hours initial grading on some 25 remaining Piper Super Cubs. Beech Queen Airs are used for twin conversion.

In co-operation with the IDF/Navy, three IAI Westwind Sea Scans have been in service since 1977 for off-shore surveillance. Unlike the proposed export model of the Westwind, with its nose radome, the IDF/AF Sea Scans have their Litton 360 degree search radar in a retractable ventral dome. Two

aircraft including 200 MiG-21F/PFM/MF and M Fishbeds for air defence, 120 MiG-17 Fresco C fighter-bombers, 50 Su-20 Fitter C variable-geometry strike fighters, 120 Su-7 Fitter As, and 25 Tu-16 Badger D/G missile-armed medium bombers.

To these were added 32 Mirage 5SDE strike interceptors and six Mirage 5SDD trainers paid for by Saudi Arabia, to which Egypt contributed funds for a further 14 and is soon to take delivery of a second, similar batch. Egyptian combat elements are organised on the Soviet pattern with regiments, or wings, each with up to three squadrons with a nominal strength of 20 aircraft.

The main strike element, spearheaded by Tu-16s carrying two AS-5 Kelt cruise missiles, has now been disbanded through spares problems, leaving a single squadron of 24 MiG-27 Flogger Ds with terrain-following radar and laser range-finding to shoulder the burden. Tactical strike and reconnaissance forces are completed by Sukhoi Su-7s and Su-20s, together with a few obsolescent MiG-17s, as attempts to find a successor have been frustrated by the ramifications of the peace agreement with Israel.

The chosen replacement was to have been the Northrop F-5E and F-5F trainer, for which Saudi Arabia was providing finance. With imminent delivery of the first ten examples (from an embargoed Ethiopian order) and negotiations proceeding at Camp David, the Saudis found it appropriate to query the purchase price of the F-5s. By the time pricing had been justified, Israel and Egypt were at peace, and Saudi Arabia withdrew its backing. After suggestions by President Sadat that the necessary funding could be provided by the novel method of public subscription, the F-5 issue was allowed to lapse following the promise of 35 Phantoms from the USAF. Pilots and ground technicians began training in the United States in June 1979 in

preparation for the first deliveries in September.

Ground-attack roles are also undertaken by the EAF's MiG-21s, carrying two 500kg or four 250kg bombs on underwing pylons, or four pods each with 12 rockets as an alternative to air-superiority missile armament. Only 120 of the 210 remaining Fishbeds are in an airworthy condition. Older MiG-21F/PFM variants have been replaced by the MiG-21M/MF, these differing mainly in armament.

Both British and Chinese aid has been sought in

Above: the Antonov An-12 partially equips two transport squadrons of the Iraqi air arm, in which Soviet types provide the backbone of the operational force. The Transall is among the types being considered to augment existing equipment.
Right: the markings of Syria (upper) and Morocco.
Opposite top: Mirage F1 operators in the Middle East include Iraq, Libya, Morocco and Jordan.
Below: the Lebanese air force flies Mirage IIIELs

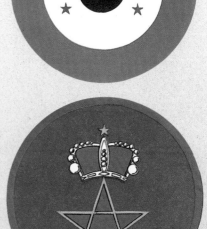

keeping the MiG-21 flying, although after much urging, the Soviet Union has returned some of the 170 engines on overhaul at the time of the break in relations. Rolls-Royce and British Aerospace are servicing engines and airframes, whilst revised avionics are provided by the United States. Spares have been obtained from Yugoslavia, but a new avenue of co-operation has been opened with another disaffected Soviet ally, China.

With experience of Soviet aircraft, the Chinese initially provided replacement engines for the MiG-17 force, and later assisted in the overhaul of other types and equipment. Desperately seeking examples of modern technology on which to base their flagging aviation industry, the Chinese were presented with a MiG-23 Flogger, Sagger anti-tank missile and Soviet tanks, and reciprocated by shipping a gift of 40 Shenyang F-6 (MiG-19) fighter-bombers to Egypt early in 1979.

A further 40 F-6s will be made available at advantageous prices, and assistance has been promised in returning the grounded fleet of 21

Above and left: the Dassault Mirage 5 equips two squadrons and an operational training unit of the Libyan Arab Republic Air Force, with which they have served since 1971. Newer equipment in in the shape of the Soviet MiG-23 has since been acquired to augment the delta-wing Mirages

MiG-23s to airworthy condition. Although dated in design, the F-6 combines trans-sonic performance with good manoeuvrability and heavy firepower from three 30mm cannon, and is still an effective fighter-bomber. Egypt plans to retain its MiGs, including 90 MiG-17s and MiG-15UTIs, for a further 10–15 years, and continued liaison with China is almost certainly assured.

In a further parallel of Soviet organisation, the Egyptian Air Defence Command was established as a fourth service (after the army, navy and air force) in 1968, with nine squadrons of MiG-21s and 75,000 personnel, representing a quarter of the total defence forces. EADC comprises about 100 missile and anti-aircraft battalions, each with 200–250 men, plus supporting units, organised on a regional basis into brigades or regiments. Brigades comprise between four and eight battalions, and are deployed within one or two divisions defending Cairo and the Suez Canal Zone.

Current strength of the EADC is believed to comprise about 80 SA-2 Guideline SAM sites along the West Bank of the Suez Canal, in conjunction with P-35 Bar Lock, Thin Skin, P-12 Spoon Rest and Fan Song E early warning radars. There are additionally some 65 sites for SA-3 Goa TV-guided SAMs and associated P-15 Flat Face and Squint Eye acquisition radars, together with several dozen tracked triple launchers for SA-6 Gainful SAMs with Straight Flush radar guidance and Low Blow fire control radars. This intensive SAM belt is backed by several hundred anti-aircraft guns up to 100mm with Fire Can and Whiff radar direction and a similar number of tracked ZSU-23-4 multiple cannon with Gun Dish radar. Egyptian infantry also has SA-7 Grail shoulder-launched or jeep-mounted SAMs.

During the 1973 war, the effectiveness of the Egyptian air defences clearly came as an unwelcome surprise to the Israeli air force. Manned interceptors claimed only a small proportion of the aircraft shot down, those IDF/AF aircraft attempting to fly below the effective height of the SAMs falling victim to the deadly ZSU-23-4 cannon. First replacements for the 600 SA-2, SA-3 and SA-6s remaining in the Egyptian inventory arrived early in 1979 in the form of an initial batch of 20 French-built Shahine systems, basically the Crotale mounted on an AMX-30 tracked vehicle. MIM-23B Improved Hawks have recently been offered by the United States, but plans for the licence production of Shahines have been held in abeyance in the aftermath of the Arab boycott. A similar fate has befallen the British Swingfire anti-tank missile.

Although Saudi Arabia is the spiritual leader of the Arab world, Egypt was, until recently, the military leader. Backed by $1,043 million of oil revenue, the Arab Military Industrial Organisation (later Arab Organisation for Industrialisation) was established in Egypt in May 1975 to provide equipment, principally for the continued fight against Israel. Acting as a holding company, normally with a 70–75 per cent stake, the AOI formed subsidiaries to handle production of a wide range of licence-built military equipment with a view to gaining basic experience before embarking on advanced, and eventually indigenous, programmes.

These ambitious plans were abandoned when Saudi Arabia, principal source of finance, withdrew its backing in retaliation for what it viewed as the defection of Egypt from the Arab cause after the peace agreement with Israel. The AOI was dissolved on 1 July 1979, although Egypt hopes to salvage something from the ruins by freezing assets held in local banks and convincing Western participants that continued co-operation is viable.

In the meantime, it is to the United States that Egypt will look for military equipment. To complement the annual defence budget, which totalled £1,400 million for 1978–79, America is providing £638 million worth of armaments on highly favourable credit terms over the next three years as part of the Middle East 'peace package'. Requests for up to 300 Northrop F-16s were denied, but Phantoms and more Hercules will be some of the earliest deliveries.

To supplement the original transport force of about 30 Antonov An-12 freighters, some 40 Ilyushin Il-14s and three An-24s, the United States made its first deliveries to Egypt in December 1976 in the form of six C-130H Hercules (including two equipped for ECM), together with a dozen Firebee reconnaissance drones. By early 1979, the EAF Hercules fleet had risen to 20, and a further 11 on order will virtually replace the earlier Soviet equipment. Sharing the main transport airfield at Cairo West is the Presidential and VIP Flight with single examples of the Boeing 707, 737 and Dassault Falcon 20 and two Commando 2Bs.

Commandos have taken over a large proportion of helicopter transport tasks from the remaining 20 Mil Mi-4s and 80 Mi-8s, a further ten Mi-6s having been placed in storage through spares shortages. Five Commando Mark 1s delivered in 1974 have been augmented by 23 Mark 2s and six Sea King Mark 47s, the latter equipped for ASW work, France contributing 64 Gazelles comprising four SA-342Ks and the balance in SA-342Ls. Normal Gazelle equipment is the HOT anti-tank missile, but the Egyptian aircraft are apparently used mostly for liaison, only 12 having HOTs, and a further dozen, the AS 12 missile.

Prior to the rift with the Soviet Union, pilot training for the EAF was undertaken in the USSR, this now being carried out in an expanded Egyptian operation. Students begin training at the age of 18 on the Helwan Goumhuriah at Bilbeis, later transferring to the Yak-18 for the basic stage. Jet conversion takes place after 80 hours, when prospective pilots then fly a further 170 hours on the Aero L-29 Delfin.

During the 1973 war, Delfins were used with notable success in their alternative role of light strike aircraft, resulting in the issue of a requirement for a follow-up type for which the Alpha Jet was chosen. Bilbeis also houses the navigation school, equipped with a small number of Il-14s. Technical training for ground crew is the responsibility of the engineering institute at Helwan, the airfield also accommodating the Goumhuriah production line which has assembled some 300 aircraft, latterly equipped with the Rolls-Royce/Continental O-300 engine.

In addition to the Sea King, the Egyptian Navy possesses a small missile capability centred on ten

Left: the Aérospatiale SA 321 Super Frelon has borne the brunt of the Libyan heavy lift helicopter requirement since the early 1970s. The arrival of Italian-built examples of the Boeing Vertol CH-47C twin-rotor type will ease the load on the SA 321, which also undertakes SAR duties.

Below: heavy logistic support is the task of the Lockheed C-130 Hercules turboprop transport. Two batches of eight Hercules were ordered, but an embargo prevented delivery of the second batch to Libya

patrol boats with Styx missiles obtained from the Soviet Union. The Navy also uses three SRN-6 hovercraft, and has a further three on order.

Paradoxically, the future of the Egyptian Air Force is less certain in this time of peace than it was during hostilities with Israel. Shunned by the rest of the Arab world, it must now decide whether to persevere with a reduced self-sufficiency programme, which in the long-term would benefit the country through increased industrialisation, or take the softer option of relying on the United States for its aircraft and missiles. A compromise may be made, whereby overseas aid is accepted for the immediate future whilst the aircraft industry moves towards more advanced projects as it gains experience.

Iraq's love-hate relationship with the USSR and Syria (its nominal ally in the fight against Israel), oil-for-arms deals with France and small-scale purchases in Britain, Germany and Switzerland, have seen the Iraqi Air Force (Al Quwwat Aljawwiya Aliraqiya) equipped with a multi-national mixture of aircraft.

After fighting a civil war against Kurdish tribesmen, Iraq indulged in brief border disputes with Iran and Kuwait and later backed both Eritreans and Somalis in their attacks on Ethiopia, incurring Soviet displeasure in the process. This issue resolved, Iraq has in the past year taken delivery of

£1 billion worth of Soviet aircraft and missiles to boost its present defence spending of approximately £3,500 million per year.

In 1975, Iraq began negotiations with France for a large-scale arms package centred on the Mirage F1, taking delivery of 32 single-seaters and four trainers and ordering a further 24 early in 1979. French helicopters feature prominently in the IAF inventory, and after obtaining 12 Westland Wessex Mk 52s in 1965, later complemented by Soviet Mil Mi-4s, Mi-6s and Mi-8s, Iraq turned to Aérospatiale for its requirements.

Following the delivery of 16 Alouette IIIs equipped with AS 11 or AS 12 missiles a further 31 were obtained, some fitted-out as gunships. Additional anti-tank capability is provided by 40 HOT-firing SA-342K Gazelles, deliveries of which were completed in December 1977. Larger types comprise 12 SA-321 Super Frelons, the last received in October 1977, and three VIP transport SA-330 Pumas delivered in August 1976, one of the latter having been lost in an accident shortly afterwards. Super Frelons are equipped to carry the Aérospatiale AM 39 long-range, anti-shipping missile, an initial quantity being ordered in late 1978.

Spearhead of the IAF strike force, however, is a squadron of 12 Tupolev Tu-22 Blinder supersonic bombers backed by some four squadrons of MiG Floggers. Forty MiG-23s were joined by a similar number of MiG-27s during 1978, complementing five squadrons of Sukhoi Su-7 and Su-20 Fitters and three squadrons of Hunters. From 46 Hunters delivered directly from the RAF or refurbished by Hawker Siddeley, attrition and combat losses against Israel have reduced the number in service to approximately 30.

For air defence, Iraq has four squadrons, each with 20 MiG-21PF/PFM or PFMA day fighters. An extensive SAM system based on V750K Guidelines, SA-3 Goas and SA-6 Gainfuls is backed by Soviet radars and communications equipment, but the Mirage F1 is rapidly becoming the principal all-weather interceptor, fitted with Matra R.550 Magic AAMs.

Transport forces are equipped with predominantly Soviet types–although the IAF has been offered, but has not taken up, the Hercules–and has additionally expressed interest in the Transall. Two transport squadrons fly six Antonov An-12s, ten An-2s, eight An-24s, two An-26s, two Tupolev Tu-124s, a Heron and two Islanders. Eight Iraqi helicopter squadrons concentrate on assault and anti-tank roles, whilst three recently delivered Mil Mi-10 Harkes operate as 'flying cranes'.

Pilot training schedules are currently being revised, following the delivery of new basic and advanced training aircraft from Czechoslovakia and Switzerland. The IAF College at Rashid is replacing the Yak-11 with the FFA AS202/18A Bravo, the first of 40 having been received in May 1979. IAF Bravos feature a new wing with strongpoints for varying combinations of bomb-load up to 660lb for alternative light-strike operations. During 1978, 24 Aero L-39 Albatros' replaced the earlier L-29 Delfins and Hunting Jet Provosts, although a small number of MiG-15UTIs are retained for advanced instruction before students undertake operational conversion on the MiG-

· 21UTI, Su-7U, MiG-23U, Hunter T Mk 66/69 or Mirage F1B.

Faced with the defection of Egypt from the anti-Israeli cause, Syria has embarked on a massive build-up programme for its armed forces with a 1979 defence budget of £1,000 million, 66 per cent higher than the previous year. With the resolution of earlier differences culminating in a mutual defence treaty in October 1978, Iraq is financing much of the increased expenditure, with the USSR as principal supplier. In an attempt to weaken the Soviet monopoly, Saudi aid is directed more towards European equipment.

Despite heavy and continued combat losses from the 1967 and 1973 Israeli wars, plus many clashes in between, the Syrian Air Force (Al Quwwat al-Jawwiya Arabia as-Suriya) is maintained at full establishment by rapid replacement with modern Soviet equipment. In the 1973 war, Israel claimed the destruction of 149 Syrian aircraft and six helicopters, but within a year, 300 new combat aircraft, including 52 MiG-23 Flogger strike-interceptors, had increased the SAF strength by 25 per cent.

Manned by some 25,000 personnel, the SAF has in excess of 400 combat aircraft, plus a further 250 transports, trainers and helicopters. The armed forces are augmented by several thousand Russian, Cuban, Vietnamese and North Korean advisers and have assumed the additional responsibility of providing a peace-keeping force to separate the rival factions in the Lebanese civil war.

Syria was the first non-Communist country to receive MiG-23 Floggers, these equipping two squadrons shortly after the 1973 war. A further dozen MiG-27s have been added, and Syria is believed to be negotiating with the Soviet Union for a further 50 Floggers paid for by Iraq. One squadron of 16 MiG-23 Foxbats has been based in Syria for high-altitude interception and reconnaissance, but these, together with their Soviet crews, have been withdrawn.

Twelve interceptor squadrons fly some 220 MiG-21PFs and MFs complementing an extensive SAM system, estimated by Israel as comprising 70 SA-2, SA-3, SA-6 and SA-9 sites around the Damascus area and along the Israeli border on the Golan Heights. Other Soviet missile equipment includes Atoll air-to-air missiles for MiG-21s, SA-7 infantry SAMs and Styx ship-to-ship rounds deployed on 12 naval patrol boats. Air defence Command is a separate organisation, under army control, but incorporating elements of ground and air forces.

Six squadrons of strike aircraft operate some 50 MiG-17s and 60 Su-7BMs for ground attack, backed by army-controlled Frog and Scud surface-to-surface missiles, first deliveries of the latter taking place in the final stages of the 1973 conflict. To combat Israeli tanks, Snapper, Sagger and Swatter ATMs were used in considerable numbers in 1973, and further capability has since been provided in the form of 50 SA-342L Gazelles fitted with HOTs, together with 2,000 MILAN anti-tank missiles from France. Nine Kamov Ka-25 Hormone ASW helicopters (the first to be exported by the USSR) are operated by the SAF on behalf of the navy. The transport force presently comprises

An-12s, An-24/26s, Il-14s, Il-18s, Yak-40s and Piper Navajos.

SAF pilot training begins on the Yak-11/18 and SIAT Flamingo, 32 of the latter being delivered, followed by an additional order for 16 to complete replacement of the Chipmunk. Jet experience is gained via the L-29 Delfin, and it would appear logical that L-39s will be received in due course. After advanced instruction on the MiG-15UTI, pilots convert to operational types through the MiG-21U and Su-70.

Whilst Arab states at the Eastern end of the Mediterranean continue their hot and cold wars against Israel, Morocco has become increasingly involved in operations against Polisario guerillas in the former Spanish colony of Western Sahara. The position has become more acute following the abandonment of Mauritania's claim to the disputed territory in August 1979, and the Polisario will henceforth be able to increase the scale of their operations against Moroccan forces, indicating a further escalation of the conflict.

Founded in November 1956, after the attainment of Moroccan independence, the Aviation Royale Cherifienne–Al Quwwat Aljawwiya Almalakiya Marakishiya, or Royal Moroccan Air Force received its first combat aircraft from the Soviet Union in 1961, in the form of 12 MiG-17s and two MiG-15UTIs. These now remain in storage, but the flirtation with Communism was brief. In 1966, the RMAF took delivery of 23 Northrop F-5As, two RF-5As and three F-5B trainers from the United States. They are operated by two squadrons from Kenitra, a further six following from Iran.

Morocco had earlier received eight Fouga Magisters from the French Air Force, and to these were subsequently added a further 24 ex-Luftwaffe aircraft, refurbished by Aérospatiale and capable of operating with attached armament. The United States further supplied 18 Fairchild C-119 Boxcar transports, whilst 48 Agusta-Bell 205s were obtained from Italy from 1970. SAR facilities are catered for by six Kaman HH-43B Huskies transferred from

Right: the national marking of Algeria. Top: the Algerian air force currently operates six Fokker F-27 Friendships, comprising four Series 400 with strengthened freight floors, and one Series 600 (illustrated) on government duties. A further six are on order.
Above: although the United States government vetoed the sale of Beech T-34C Turbo-Mentors to the Algerian air force, six were delivered to the civilian flying school in early 1979

the USAF, these being followed by 12 C-130H Hercules delivered in two equal batches between 1974 and 1977 as C-119 replacements.

Whilst this was sufficient for a peace-time air force, the Sahara dispute which erupted in 1976 prompted a considerable increase in armaments expenditure, this now averaging in the region of £350 million per year. Expansion had, however, begun late in 1975 with the placing of an order for 25 Mirage F-1CH fighter-bombers, subsequently increased to 50, all of which have now been delivered, equipped with MATRA R.550 Magic air-to-air missiles. RMAF strength thus comprises some 85 operational types, 50 transports and trainers and over 100 helicopters, excluding those of the Gendarmerie, manned by 6,000 personnel.

Despite a doubling of US military credits in the three years before 1977, further hardware has been embargoed through American dissatisfaction with the use of F-5s against the Polisario, in contravention of their terms of supply for internal use only. The Algerian-supported guerillas appear capable of looking after themselves and have claimed the destruction of several aircraft, the latest being an F-5 in February 1979.

Previous United States offers of 24 F-5E/Fs were passed over in favour of the second batch of Mirages, and in place of 20 T-2D Buckeye jet trainers, the RMAF ordered 24 Alpha Jets, the first of which was flown in May 1979, with deliveries scheduled to begin in August. Additional purchases from France have included Crotale low-level SAM systems.

Approval for the sale of more specialist equipment for anti-guerilla use – 24 OV-10 Broncos and a similar number of AH-1 helicopter gunships – has been withheld by the US Congress, although Saudi Arabia has financed 40 SA-330 Pumas (including some for the Gendarmerie) and six SA-342 Gazelles to augment Gendarmerie Alouettes. Nevertheless, the United States has agreed to the delivery of six Meridionali-built CH-47C Chinooks from Italy from February 1979 onwards, and to options on

two further batches of six, and is to begin work on a defensive net of 3D radar sites and associated communications stations late in 1979.

Training operations have recently received a much-needed boost with the arrival of ten FFA AS202/18 Bravos and a dozen T-34C Turbo-Mentors in 1977–78 as replacements for the North American T-6 Harvards and T-28 Fennecs used hitherto. The arrival of Alpha Jets will complete the modernisation of the pilot training syllabus, although their use in the strike role is not excluded.

For the future, the RMAF would appear committed to continued operations in Western Sahara, where half the 80,000 strong army is already engaged against the Polisario. King Hassan II has recently announced the formation of a National Defence Council to assist in the formulation of future policies, but with costs now in the region of £500,000 per day and little help forthcoming from the United States, Morocco will become increasingly dependant on Saudi money and European arms.

Despite its central position in the Middle East, Lebanon has maintained neutral policies to avoid direct involvement in Arab-Israeli conflicts, only to be convulsed for 19 months in 1975–76 by civil war between right-wing Maronite Christians and left-wing Palestinians and Muslims. Although the regular forces, including the Force Aérienne Libanaise – Al Quwwat Aljawwiya Allubnamiya, or Lebanese Air Force, were little involved in the civil conflict, the army effectively disintegrated as troops joined the opposing sides. Lebanese forces are now being re-organised with US assistance to take over from Arab League peace-keeping units, although heavy artillery and combat aircraft are not to be supplied, and Lebanon is presently seeking tanks, missile boats and Puma and Gazelle helicopters from France.

Defence expenditure is now increasing despite the shattered state of a once thriving exonomy, this totalling £80 million in the past year, with particular emphasis being placed on increasing the army to 15,500 from a present strength of less than half that number. Military assistance from the United States has been limited to some $65 million in credits per year.

Even before the civil war, the nominal status of the FAL was illustrated by the fact that of ten Mirage IIIEL strike-interceptors and two Mirage IIIBL trainers ordered from France in 1965, together with 15 Matra R.530 air-to-air missiles, only half were put into service and the remainder stored. The Mirage fleet has been periodically offered for sale without finding a buyer, and thus continues to provide limited air defence capability.

Principal combat type of the FAL is effectively the Hawker Hunter, which equips one fighter/ ground attack squadron. Lebanon originally received six Hunter F Mark 6s from the RAF as part of a UK military aid programme and supplemented these in 1965 with four F Mark 70s and three two-seat T Mark 66C trainers refurbished by Hawker Siddeley. After the loss of five aircraft in training accidents, contracts for a further six FGA Mark 70s were placed in 1975 to bring total Hunter procurement to 19. The first three were delivered in the spring of 1976, but the remainder were held by the British government whilst the internal situation was stabilised after the civil war, and did not arrive until December 1977. It now appears the Hunter will continue in service well into the 1980s.

A modest five-year upgrading programme initiated in 1972 has so far only resulted in further equipment from France. As a follow-up to previous procurement of three SE313B Alouette IIs, seven SE3160 Alouette IIIs and four Magisters, Lebanon obtained seven more Alouette IIIs and four Magisters together with four Agusta-Bell AB212 helicopters from Italy. Britain's contribution involved six SAL Bulldog Mk 126 trainers delivered in 1975 to replace the FAL's veteran Chipmunks.

Lebanon's air force is thus one of the smallest in the Middle East, with 25 combat aircraft, including reserves, and 35 other types. Personnel would appear to be under the normal authorised strength of 1,000 and transport capability is restricted to one de Havilland Dove and one Turbo Commander for communications duties. It is unlikely that any large-scale re-equipment plans will be attempted in the immediate future.

Dominated by the volatile Colonel Ghadaffi, the

were eventually delivered in 1971, by which time France had become the principal arms supplier following an order for 110 Mirages – 32 5DE strike-fighters, 58 5D ground-attack fighters, ten 5DD trainers and ten 5DR tactical reconnaissance aircraft. Helicopter contracts concerned nine Super Frelons for SAR and heavy-lift duties, and ten Alouette IIIs, of which four and one respectively have been transferred to the Maltese Armed Forces' Helicopter Flight. Twelve refurbished Magisters were received for basic training, but when the last Mirage was delivered in May 1974, only 25 pilots were reportedly able to fly the aircraft despite plans to train 200, plus 600 technicians.

Soon after the 1973 war, Libya signed an arms and assistance agreement with the Soviet Union, its provisions including the establishment of an integrated air defence network with early-warning and communications links with SAMs, training of 1,500 Libyan personnel and secondment of Soviet technicians. In return, the USSR was granted facilities at Okba bin Nafi (formerly Wheelus) for the basing of reconnaissance aircraft to shadow NATO forces in the Mediterranean, and use of

Above right: the Royal Jordanian air force's transport unit, No 3 Squadron, operates two Lockheed C-130Bs and one C-130H. The unit also operates three CASA C.212A utility transports and one C.212C executive transport for government duties. No 3 Squadron is based at King Abdullah Air Base, Amman, the air force's headquarters and the base for its transport, liaison and rescue elements. No 3 shares the base with No 7 Squadron, flying Aérospatiale Alouette IIIs.
Left: King Hussein's personal Boeing 727 pictured while visiting West Beach, California, USA, in November 1977. A Riley Dove is also at his disposal.

Libyan Arab Republic spends more per capita on armaments than any other state in the world, whilst maintaining close links with the Soviet Union and providing bases for Russian aircraft and ships operating in the Mediterranean. Spurred by a fierce hatred of Israel and financed with vast oil revenues, Libyan arms purchases are apparently outstripped only by its ability to provide trained crews for the operation of complex hardware. Considerable numbers of Soviet and Pakistani specialists have therefore been called-in to help operate the air force (Al Quwwat Aljawwiya Al Libiyya), as relations with neighbouring Arab states are far from good.

After limited operations with United States assistance from its establishment in 1959, the LARAF underwent a considerable expansion following the overthrow of the monarchy in September 1969. The Revolutionary Arab Socialist Government cancelled a large defence package from Britain and evicted the USAF from its base at Wheelus, but sought to take delivery of a further eight Northrop F-5s to complement its original deliveries of eight F-5As and two F-5B trainers, together with eight C-130 Hercules. The latter

Libyan ports for the Soviet Navy.

By mid-1974, between eight and ten batteries of SA-2, SA-3 and SA-6 SAMs were installed for airfield defence and the first Tu-22 Blinders were operating over the Mediterranean in Libyan markings, but with Soviet crews. In May 1975, an initial batch of 13 MiG-23 Floggers (including five two-seat Flogger B trainers) was shipped to the LARAF, followed by a further 16 shortly afterwards. These formed the equipment of two air defence squadrons, again with Soviet pilots. Supplies from the USSR also included a dozen Mi-8 helicopters, and gradually the aircraft were accepted into the Libyan inventory as pilots were trained to fly them.

Mirage 5Ds now equip two ground-attack fighter squadrons, plus an OCU with 5DD trainers, whilst a further two strike squadrons with 5DEs each have an attached tactical reconnaissance flight with Mirage 5DRs. Squadrons are believed to have over 50 per cent Pakistani personnel, but despite this national deficiency, a further order for Mirage F1s was placed in 1974 covering 16 each of the F1AD and F1ED variants plus six F1BD trainers, with options on a further 50. Mirage F1s have been issued to two squadrons and are equipped

with Matra Super R.530 and R.550 Magic air-to-air missiles.

France has also supplied Crotale low-level SAMs, with Italy emerging as a second large-scale supplier of Libyan arms, including OTOMAT long-range, ship-to-ship missiles for ten fast patrol boats of the navy. To satisfy an urgent transport aircraft requirement (after a second batch of eight Hercules was embargoed by the United States and support facilities for the existing fleet withdrawn), Aeritalia undertook to supply 20 G.222s. Plans were thwarted by US restrictions on the General Electric T64 turboprop, but a version with licence-built Rolls-Royce Tynes has recently flown and will be immune from export restrictions.

Similar problems attended the sale of SIAI-Marchetti SF-260W Warriors through their Collins avionics, and at one time some 80 aircraft were stored in Italy pending arrival of alternative systems. A first batch of 100 has now been delivered, and a further 130 will be assembled in a newly-built plant established in Libya with Italian assistance. No restrictions have been placed on 12 Meridionali-built CH-47C Chinook medium-lift helicopters delivered from 1976 onwards, or a single Agusta-Sikorsky S-61A-4 for VIP use to join the Boeing 707, two Falcon 20s and two Jet Stars reserved for presidential and government communications.

Chinooks now equip a fourth helicopter squadron as a follow-on to Super Frelons, Alouette IIIs and Mi-8s, plus independent flights of AB.47Gs, providing logistic support for border posts and army-manned missile sites and radar installations. Army Aviation has six AB.47Gs, a similar number of AB.206 Jet Rangers and a few Cessna O-1 Bird Dogs transferred from Italy. Recent deliveries from the Soviet Union have included a squadron of Mi-24 Hind helicopter gunships, apparently flown by Russian personnel, and 12 more Blinders, plus a small number of Blinder D trainers. Some SA-342 Gazelles have been obtained early in 1979 for police and military use, and Libya is understood to have requested 50 more Floggers.

Most LARAF primary and helicopter flying training is currently undertaken at foreign civil schools, from where, after about 100 hours instruction, students begin a two-year course at the Zawia air force academy. Staffed mainly by Yugoslav and Pakistani instructors, the Academy operates a dozen Magisters and 38 Soko Galebs. Operational tactical and weapons training is completed at the Mirage OCU, manned mainly by PAF personnel and the MiG-23 squadron with exclusively Soviet instructors. OCU courses of eight-to-ten months include combat radar and navigational training on the Mirage nav/attack system with two modified Falcon ST trainers. Multi-engine conversion is provided by C-47 Dakotas in one of the two transport squadrons, both predominantly manned by Pakistanis.

Algeria is yet another example of a North African country which has turned to the Soviet Union after a military take-over and is now being supplied with the most modern forms of combat aircraft and equipment. After a long and bitter struggle with France, Algeria gained independence in 1962 and established Al Quwwat Aljawwiya Aljaza'eriya or l'Armée de l'Air Algérienne with Egyptian assistance, receiving five MiG-15s, 12 Czech-built Yak-11 advanced trainers and 12 Goumhuriah primary trainers from Egypt in November 1962.

A large base-extension programme followed, although the AAA was well-provided with ex-French airfields including the massive complex at Mers-el-Kebir, nuclear test facility at Reggane and missile and armaments trials station at Colomb-Bechar.

A major build-up of the AAA occurred during 1965 after the overthrow of the government by a military junta, resulting in first deliveries of six MiG-21s interceptors, 15 Il-14 and An-12 transports, 24 Il-28 jet bombers, 40 Mi-4 helicopters and over 30 V750VK Guideline SAMs. Soviet instructors joined other East European personnel at the MiG base at Ouargla, increasing the Russian presence in Algeria to around 1,500. Algeria now has some 200 combat aircraft and 130 other types and 5,000 personnel. Defence expenditure for 1978 was estimated at about £225 million, or 30 per cent up on that of two years previously as a result of concern over the guerilla war in neighbouring West Sahara.

Present interceptor forces comprise two squadrons of MiG-23 Flogger fighters with 20 aircraft each, backed by three squadrons of MiG-21Ps totalling 80 aircraft for day interception. Recent reports indicate the AAA has also received six or more MiG-25 Foxbats which may be operated on reconnaissance duties, presumably in Soviet hands. A single light bomber squadron with 20 Il-28 Beagles is complemented by a fighter-bomber regiment comprising five squadrons: two sharing some 30 MiG-17Fs, two with ten Su-7BM Fitters each and one with about 15 MiG-15bis and MiG-15UTIs as armed advanced trainers. Limited procurement from France in 1970 provided 28 refurbished Magister armed trainers for two squadrons and five SA-330 Pumas as assault transport helicopters.

The Transport Wing provides logistic support, with its mainly Soviet equipment now being replaced by aircraft from Western sources. A dozen Il-14s are giving way to a similar number of Fokker F-27 Friendships, backed by eight An-12 Cub heavy freighters, four Il-18 Coots and six Beech twins. Three Fokker F-28 Fellowships are on order, apparently to form a naval transport squadron. Supporting units include the Helicopter Wing with over 20 Mi-4s, four Mi-6 heavy-lift types, 12 Mi-8s, Pumas and six Hughes 269As for training.

Fixed-wing instruction is undertaken on Yak-11s, Magisters, MiG-15UTIs and MiG-21Us with three Queen Airs for twin-conversion. Many Algerian pilots are trained in the USSR and elsewhere in the Middle East, but despite a US refusal to supply Beech T-34C Turbo-Mentors to the AAA, six examples were delivered to the civilian flying school early in 1979.

Following its formation after the Arab-Israel war of 1948, Al Quwwat Aljawwiya Almalakiya Alurduniya, the Royal Jordanian Air Force, was initially administered by British personnel under the direction of King Hussein, who qualified as a pilot with the RAF. In recent years, Jordan has been more dependent on United States aid, which began shortly before the 1967 war with Israel in which the RJAF lost its entire strength of Hunters, together with six transports and two helicopters.

Re-building of the air force with British, US and Arab assistance began with deliveries of 18 Lockheed F-104A Starfighters and two F-104B trainers and some 40 Hawker Hunters. First to arrive were Hunter F Mark 6s, FGA Mark 9s and FR Mark 10s from the RAF to which F Mark 73As and F Mark 73Bs were later added from contracts placed with Hawker Siddeley for refurbished aircraft. Over 500 Short Tigercat SAMs were ordered in 1969 for airfield protection.

Main re-equipment, however, began in 1972 and included the supply of 44 new Northrop F-5E Tiger II fighter-bombers plus two F-5F trainers, 30 surplus F-5As and four F-5Bs from Iran, three Fairchild C-119Ks and two Lockheed C-130B Hercules, supplemented in 1972 by a single Dassault Falcon 20 for royal communications.

Jordan then sold 555 Tigercats, 162 practice rounds and associated equipment to South Africa in 1974, but Egyptian objections to the sale of 31 Hunters, plus spares, to Rhodesia resulted in the Hunter fleet being presented to Oman.

After the initial delivery of 20 F-5As and two F-5Bs from Iran in late 1974, replacing the Hunters of No 1 Squadron and establishing a new No 2 Squadron at King Hussein Air Base, Mafraq, F-5Es and F-5Fs began arriving in 1975 for No 17 Squadron at Prince Hassan Air Base, H-5, for air defence. A second F-5E unit has subsequently formed, joining the 20 survivors of 39 Starfighters supplied to No 9 Squadron at H-5, all three interceptor squadrons being earmarked for transfer to three new bases under construction well behind the Israeli border for adequate warning of attack.

Little has been heard of the new air superiority fighter promised for 1979–80 as a Starfighter replacement and to permit the F-5Es to be used for ground attack. Despite King Hussein's prediction of a future combat strength of 176 aircraft, the RJAF presently fields only half that number, together with some 50 transport, training and helicopter types.

Air defence missile capability has, however, been improved with the recent delivery of 532 MIM-23B Improved Hawks to equip 14 batteries in the Amman-Zerka area and at airfields and radar sites east and south of Amman. Jordan had previously considered a Soviet alternative, but the Hawks were funded by Saudi Arabia in pursuance of its aim to minimise Soviet influence in the Middle East. The Jordanian army has additionally obtained 270 FIM-43A Redeye infantry SAMs.

Main transport base is King Abdullah Air Base, Amman International Airport, which also houses the RJAF headquarters. Fairchild C-119s were replaced from 1976 by two more C-130Bs, although these were transferred to Singapore the following year prior to the arrival of a single C-130H in 1978. For logistic support of desert outposts and installations, liaison and communications were undertaken by three CASA C.212A Aviocars from late 1975, together with one personnel-carrying C.212C, replacing a similar number of Douglas Dakotas.

Fifteen Alouette IIIs based at Amman, with various detachments, a Boeing 727 and two de Havilland Doves complete the transport force. Jordan has requirements for additional helicopters including four Sikorsky S-76s, ten TOW-armed Bell AH-1 Cobras and a small number of medium-lift types, but none have yet been received.

Primary and basic pilot training begins on 12 SAL Bulldogs, comprising 70 hours dual and 30 hours solo flying in eight months. Students then proceed to No 6 Squadron at Mafraq with eight Cessna T-37s for a further seven months and then convert to the F-5B/A with No 2 (OCU) Squadron at the same base.

Personnel of the Royal Jordanian Air Force total 6,500 and they serve in eight departments, comprising Operations, Training, Technical, Procurement, Civil Mechanics, Electronics and Communications, Administration and Air Police. Defence expenditure for 1978 was some £150 million and the aircraft inventory comprised about 140 machines.

CHAPTER TWELVE

The Gulf & India
Saudi Arabia/Iran/Pakistan/Gulf States/ Afghanistan/India

AL QUWWAT ALJAWWIYA ASSA'UDIYA–ROYAL SAUDI AIR FORCE		
Unit	Aircraft	Base
No 1 Sqn	Various	Riyadh
No 2 Sqn	Lightning	Tabuk
No 3 Sqn	F-5E	Taif
No 4 Sqn	C-130	Jeddah
No 7 Sqn	F-5F	Dharan
No 8 Sqn	Cessna 172	Riyadh
No 9 Sqn	Strikemaster	Riyadh
No 10 Sqn	F-5E	Khamis
No 11 Sqn	Strikemaster	Riyadh
No 12 Sqn	AB 205, AB 206	Taif
No 14 Sqn	AB 205, AB 206	Taif
No 15 Sqn	F-5B	Dharan
No 16 Sqn	C-130	Jeddah

Commanding a central position in Arab affairs as one of the principal Islamic states and financial backer of several of the less affluent Middle Eastern nations, Saudi Arabia nevertheless looks to the West to provide equipment for its air force (Al Quwwat Aljawwiya Assa'udiya), shunning the Soviet influence which has pervaded many of its more radical neighbours.

With defence expenditure approaching £5,000 million (a quarter of the national budget) in the last year, the Saudis rank seventh in terms of world military procurement, and following the recent changes in Iranian policies, second only to Israel in dependence on United States' weaponry. Future deliveries of the McDonnell Douglas F-15 Eagle as part of the Middle East 'peace package' and the recent Saudi pledge to increase oil supplies to the United States in compensation for the Iranian shortfall, are likely to see increased interdependency in forthcoming years.

In its present form, the Royal Saudi Air Force dates from 1950, when a British military mission began reorganising the national forces with the

Left: Dassault Mirages purchased by Saudi Arabia for Egypt pictured in Saudi markings at Dijon, France, where their Egyptian crews trained. Saudi Arabia has financed equipment for neighbouring nations to prevent Soviet encroachment.
Below: 24 Northrop F-5F two-seat combat trainers equip No 7 Squadron RSAF, the F-5 OCU. The F-5F has wing tip-mounted Sidewinders

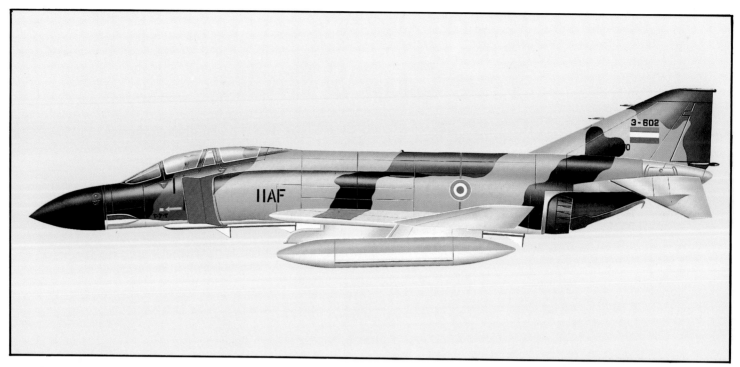

Above: the Iranian air force's Air Defence Command has ten strike-interception squadrons, based at Shiraz, Tabriz and Tehran, equipped with McDonnell Douglas F-4 Phantoms, comprising 32 F-4Ds, 177 F-4Es and 16 RF-4Es. An F-4D is illustrated.

Right: the nose markings of the former Imperial Iranian Air Force, pictured on a Boeing 707-3J9C at the Farnborough Air Show in September 1976. Since the revolution of early 1979, the overthrow of the Shah and the establishment of the Islamic government, the Iranian air force has stagnated and its future is considered uncertain

supply of light transport and training aircraft. Following establishment of a transit base at Dhahran, training and re-equipment were taken over by the USAF, which provided aircraft under MDAP (Mutual Defence Assistance Programme), but the RSAF remained a largely nominal force, comprising by 1962 only one squadron of North American F-86F Sabres for air defence and a unit of Douglas B-26 Invaders for strike duties.

Outbreak of hostilities along the Yemen border demanded a rapid expansion of the RSAF and resulted in renewed ties with Britain through a massive contract involving five BAC Lightning F Mk 52s, two Lightning T Mk 54s, 34 Lightning F Mk 53s, six Lightning T Mk 55s, four Hawker Hunter F Mk 6s, two Hunter T Mk 66s, 25 Strikemasters, 37 Thunderbird SAMs (surface-to-air missiles), radar, airfield equipment and support facilities. Hunters began operations late in 1966, followed by the remainder of the aircraft in 1968–69, the principal Lightning unit being No 6 Squadron at Khamis Mushayt, crewed by seconded Pakistani air force pilots.

Earlier transport types such as the Douglas C-47 Dakota and Fairchild C-123 Provider were withdrawn in 1967 in exchange for an initial batch of 11 Lockheed C-130E Hercules, since augmented by 25 C-130Hs and four KC-130H tankers, although the major US share of the expansion plan was restricted to Hawk SAMs. Further contracts with Britain have primarily concerned infrastructure and training facilities through BAC Military Division, Warton, with the exception of orders for a further 22 Strikemasters. The present UK air defence agreement extends to 1982 when the Lightnings will be nearing the end of their service lives.

In 1971, Saudi Arabia placed a contract for the supply of 30 Northrop F-5E strike aircraft and 20 F-5B trainers, the former with flight-refuelling equipment to operate in conjunction with the KC-130s. As part of a ten-year build-up plan for the RSAF, a further 40 F-5Es and 20 F-5F trainers were ordered late in 1974 and plans formulated for

the establishment of a highly-mobile regional defence army with two assault helicopter battalions, one attack helicopter battalion and two air cavalry battalions with a total of 404 rotorcraft of various types. A ten-year arms-for-oil treaty with France included Shahine mobile SAMs and the funding of 38 Dassault Mirages for Egypt.

Front-line strength of the RSAF currently comprises some 100 combat aircraft (and, apparently, the same number of Saudi nationals trained to fly them), 90 jet trainers with secondary strike capability, and 130 transports and helicopters in 13 squadrons manned by 12,000 personnel. Interceptor duties are undertaken by the Lightnings of No 2 Squadron at Tabuk, following the disbandment of No 6 Squadron, supported by the Lightning Operational Conversion Unit (OCU) at Dhahran. Despite the provision for ground-attack armament, and previous plans for the establishment of a third squadron, Lightnings have remained under-employed in RSAF service, with several held in storage since arrival.

F-5Es now equip No 3 Squadron at Taif and No 10 Squadron at Khamis, following previous deliveries of F-5Bs to No 15 Squadron, Dhahran,

replacing T-33s in the advanced training role; and No 7 Squadron (formerly with the F-86F) at the same base acting as the OCU. Matra R 550 Magic air-to-air missiles permit the F-5 to operate in air superiority roles in addition to its principal strike task, for which it is equipped with Maverick air-to-surface missiles and Shrike anti-radar missiles. Two more F-5 squadrons are to be formed shortly.

Other units with combat potential, but normally used for training, are Nos 9 and 11 Squadrons, equipped with Strikemasters at Riyadh, as part of the King Faisal Air Academy. A 27-month pilot training course at the Academy begins with six months' primary instruction on Reims-Cessna 172s of No 8 Squadron, followed by 14 months' basic training with No 9, and a weapons course with No 11.

Manned interceptors are supplemented by 16 batteries of MIM-23B Improved Hawks, of which six are mobile, together with Thunderbirds under army control. First Thomson-CSF/Matra Shahine SAMs–tracked versions of the Crotale–are due for delivery next year, while in 1981, withdrawal of the Lightning will follow initial deliveries of F-15 Eagles from the US. Israeli objections to the sale of

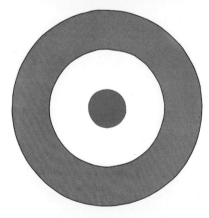

Above: the national marking of Iran.
Centre: the Iranian air force has three Boeing 747s equipped for in-flight refuelling which, together with two Boeing 707 tankers, support the Phantom and F-5 force.
Top: the Iranian air force procured 141 Northrop F-5E Tiger IIs to equip eight fighter-bomber squadrons, based at Buskehr and Tabriz. The F-5Es replaced the original F-5A complement

45 F-15Cs and 15 F-15D trainers prompted Saudi assurances that the aircraft will be based at central or southern airfields and not used for offensive missions. Although in the past, Saudi Arabia has given moral, religious and financial support to the forces opposing Israel, and has joined Arab opposition to the peace treaty with Egypt, it has declined to commit any units to actual combat. Ten Eagles will be delivered in 1981, the balance following in the next two years, all equipped to carry four each of the Sidewinder infra-red and Sparrow radar-homing missiles.

RSAF support elements are now based on the C-130 Hercules, operated by Nos 4 and 16 Squadrons at Jeddah. Two examples are used by the Royal Flight at Riyadh (No 1 Squadron) together with a Boeing 707, two Jet Stars and one AB 206, one AB 212 and two AS-61A-4 VIP helicopters. Remaining rotary-wing equipment is allocated to Nos 12 and 14 Squadrons at Taif, and detachments, comprising 24 Jet Rangers and a similar number of AB 205 Iroquois for training, liaison, crash-rescue and SAR. A small number of AB 212s is in the process of delivery for coastal SAR to supplement two Alouette IIIs, and further transport capability will be added with the reported arrival of 40 CASA 212 Aviocars assembled in Indonesia.

Few are the revolutions which truly succeed in replacing oppression with democracy, as the traumatic events in Iran since February 1979 appear to have proven. Despite the imposition of Islamic law, the new Iranian rulers are fully cognisant of the value of their vast oil reserves to Western industrial society and intend retaining a strong defensive force, albeit smaller than that maintained by the former regime. Current policy of the Islamic Revolutionary Council is international non-alignment and to halve the size of Iranian defence forces, which included an army of some 290,000 men and 300,000 reserves, prior to the exile of the Shah. During the revolution, large-scale desertion considerably weakened Iran's defence structure, and most IAF aircraft have since been grounded

Top: the Iranian air force's tactical transport element has 14 Fokker F-27-400M Troopships (illustrated), including four for aerial survey and four for target-towing duties, and six F-27-600s, three equipped for VIP transport duties.
Above: one of the two CH-47C Chinooks operated by the Iranian air force. The remainder of the CH-47Cs procured by Iran both before and after the 1979 revolution are on army charge. With about 540 helicopters, the Iranian army became one of the world's largest operators of rotorcraft. However, the future of the force is uncertain

through lack of spares and maintenance. Defence expenditure, totalling almost £5,000 million in the year before the revolution, is likewise expected to be drastically reduced, following the dramatic fall in national income due to decreased oil output.

Prior to the revolution, Iran had extensive military orders outstanding with Britain and the United States, but most were then cancelled, including the first 55 of 160 General Dynamics F-16s, seven Boeing E-3A AWACS aircraft, 16 RF-4E Phantoms, 400 Bell 214 helicopters and Phoenix, Harpoon, Standard and Hawk missiles. Britain lost contracts for Rapier SAMs, tanks and construction of a military support complex near Isfahan.

The former Imperial Iranian Air Force (Nirou Hayai Shahahanshahiye Iran) became an independent air arm in August 1955, after which the army and navy also continued large-scale expansion of their aviation branches. Deliveries of modern United States combat aircraft, including 80 F-14A Tomcats, over 200 Phantoms, 141 Northrop F-5Es and 28 F-5Fs gave the IIAF a first-line force of some 22 squadrons, each with 15 aircraft on establishment, plus reserves.

As the principal defensive component, Air Defence Command reorganised, following original UK orders for Rapier and Tigercat low-level SAM systems in 1970. A further order three years later almost doubled the Rapier complement and included Blindfire tracking units for all-weather operation. Further modernisation saw the arrival of semi-automated 3-D radars from the USA through the 'Peace Crown' aid programme.

From a basis of two squadrons with a total of 32 F-4D Phantoms ordered late in 1966, Air Defence Command expanded to the planned figure of ten squadrons with 225 F-4D and F-4E versions, armed with Sparrow and Sidewinder air-to-air missiles. As a counter to MiG-25 Foxbats based in neighbouring Arab countries, the IIAF ordered batches of 30 and 50 F-14A Tomcats, equipped with Phoenix missiles, for squadrons at Khatami, near Isfahan and Shiraz. Since the withdrawal of American technicians, the 77 remaining Tomcats have been largely grounded, and offered for disposal back to the US or Canada.

After going to considerable lengths to retrieve a

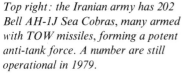

Top right: the Iranian army has 202 Bell AH-1J Sea Cobras, many armed with TOW missiles, forming a potent anti-tank force. A number are still operational in 1979.
Above right: the aircraft equipment of the Iranian navy includes 20 Agusta-built SH-3D Sea Kings normally deployed on anti-submarine warfare and maritime strike duties.
Right: the Iranian air force has six Lockheed P-3 Orions based at Bandar Abbas for maritime reconnaissance and anti-submarine warfare duties. They operate in conjunction with the navy

ditched Tomcat and its secret Phoenix from the depths of the Atlantic, the United States is clearly concerned that these weapons should not be inspected by representatives of foreign powers, or worse, be made available to undesirable nations. Assurances have, however, been given regarding the security of the Tomcat and Phoenix and of Iran's intention to sell them back to the United States if a price can be agreed. Less sensitive in military terms, are 600 of the 1,400 helicopters presently in Iran which have been declared surplus although Western technicians are being invited to return to help operate existing equipment still required.

Tactical elements of the IIAF, originally based on six fighter-bomber squadrons, each with 16

F-5As and F-5Bs from an initial procurement of 104 single-seat and 23 trainers, including 91 through MDAP, subsequently re-equipped with F-5Es following a 1972 order for 169, including 28 F-5F trainers. The latter were supplied to eight squadrons, and most of the F-5As transferred to other air forces under US direction. A squadron of 13 RF-5As was replaced in the tactical reconnaissance role from 1971 by a unit of 16 RF-4E Phantoms, a further 16 remaining uncompleted on the St Louis production line.

Similar expansion was experienced by transport forces, originally comprising a single wing at Shiraz with 16 Lockheed C-130Es and lighter types, which subsequently increased to no fewer than 64 Hercules, although nine examples were

transferred to Pakistan. Jet transports include a dozen Boeing 707-3J9C cargo/passenger aircraft and 11 of the 16 Boeing 747s obtained after consideration had been given to the C-5A Galaxy. Two further 707s are equipped for flight refuelling of the F-5s.

For shorter-range work, Iran undertook large-scale procurement of the Fokker F-27 Friendship 400 and 600 series, of which over 25 were delivered to all three services. The IIAF has operated 14 F-27-400M Troopships, including four for aerial survey, and six Mk 600s, of which three are used as VIP transports. Four Mk 400s were also modified for target-towing, by the installation of a Marquardt system under the port wing. All air force F-27s operate from Dohshan-Tappeh, near the centre of Teheran.

Four C-130Hs were modified for signal-monitoring and collection of electronic intelligence along the borders with the USSR, in connection with the Ibex Elint-gathering network. Having only recently regained access to monitoring bases in Turkey on a limited scale, US intelligence has again been thwarted through the denial of access to Iranian sources.

In addition to lighter types such as Commander 690s, Dassault Falcons and Cessna twins, the IIAF received shortly before the revolution two CH-47C Chinooks, two VIP AS-61A-4s, two AB 212s, two AB 206 Jet Rangers, 11 Bell 212s, two 214Bs and 38 214Cs, for the expansion of its helicopter component.

Maritime reconnaissance and shore-based anti-submarine warfare is undertaken in conjunction

with the navy, by six Lockheed P-3C Orions operated from Bandar Abbas, devoid of certain US avionics but fitted with equipment for flight refuelling from Boeing 707 or 747 tankers. A requirement for further Orions and Harpoon missiles ended with the revolution.

Pilot training is initiated on 49 Beech F33 Bonanzas, followed by a T-33 course, the latter type being now overdue for replacement. Operational conversion courses on the F-5B were expanded with the arrival of 28 more sophisticated F-5Fs in mid-1976.

Expansion of the former IIAF has been exceeded only by the meteoric increase in airborne equipment enjoyed by the Army Aviation forces. From a relatively modest status in the early years of the decade, when it operated 25 Agusta-Bell AB 205s,

Above: about 40 Cessna T-37C basic jet trainers are in PAF service.
Above left and far left: about 140 Chinese-supplied and built Shenyang F-6s (MiG-19s) equip nine Pakistan air force squadrons, deployed in the interceptor and fighter-bomber roles.
Below: the PAF Mirage force comprises 18 IIIEP and 33 5PA fighter-bombers and 13 tactical reconnaissance IIIRPs in four squadrons deployed in strike, interceptor and reconnaissance roles, with three IIIDPs and two 5DPs as operational trainers. A further 32 Mirage III and 5 aircraft were ordered in 1979 for 1981–83 delivery to equip two fighter-bomber squadrons

24 AB 206s, and 14 Kaman HH-43F Huskies as well as a fixed-wing element of 30 Cessna 180/185s and ten O-2As, massive helicopter orders have built Iranian Army Aviation into one of the world's largest military air arms.

Following the IIAA decision to build up a sky cavalry force based on United States' experience in Vietnam, 202 TOW-armed AH-1J Sea Cobras and 287 Bell 214A Iroquois formed the initial equipment pending establishment of local facilities for the production of a further 400 Bell 214s between 1980 and 1985. The latter order has now been cancelled, but Agusta have supplied 91 AB 206B-1 Jet Rangers, over 20 AB 205s and 56 CH-47C Chinooks, augmenting 38 CH-47Cs received directly from the United States. Chinooks have not fallen victim to the widespread cancellations affecting other projects, and deliveries are continuing for civil purposes.

Fixed-wing types include two Fokker F-27s for transport and target towing together with five Shrike Commander liaison aircraft. Air defence missile forces include five squadrons of Rapiers, plus a training unit, backed by a battalion of seven or eight MIM-23A Hawk SAM units in the process of conversion to MIM-23B Improved Hawk standard. Plans for Rapier and TOW licence-production have been abandoned, but the Iranian army operates a few Soviet SA-7 and SA-9 infantry SAMs ordered in 1976.

Iranian Naval Aviation is principally helicopter-equipped for the ASW role, plans for a V/STOL strike force operating from mini-carriers being discarded even before the revolution. Equipment now comprises 20 Agusta-Sikorsky SH-3D Sea Kings delivered in 1976–78 for ASW and strike from shore bases, augmented by at least six AB

Below: some 36 Aérospatiale Alouette IIIs are operated by Pakistan Army Aviation from its main base at Dhamial, where, following initial deliveries from France, local assembly took place. Bottom: Pakistan Army Aviation received its first helicopter type, the Bell 47G, in 1964, of which some 15 remain in service. Plans to replace them with the Hughes 500 were not implemented

212s with provision for AS 12 missiles for shipboard operation. Agusta also supplied five AB 205s and 14 AB 206s to INA, complementing six RH-53D minesweeping helicopters from the United States. A small fixed-wing communications fleet flies two F-27-400Ms, two -600s, six Shrike Commanders and two Dassault Falcons. Harpoon, Sea Killer and Exocet anti-ship missiles, Seacat air defence missiles and Standard anti-radiation missiles equip frigates and fast patrol boats.

Withdrawal of British forces from the Persian Gulf area in 1971 obliged individual sheikhdoms to make their own defensive arrangements. For some, virtually city-states, the obvious solution was a corporate policy, and thus the United Arab Emirates was established, comprising Abu Dhabi, Ajman, Dubai, Fujairah, Ras al-Khaimah, Sharjah and Umm al-Qaiwain. Embryo air forces were already in existence in Abu Dhabi and Dubai, forming the basis of the UAEAF with joint funding from all seven states.

Abu Dhabi's air force was formed with British help in 1968, the main technical support coming from the civilian contractor, Airwork. Many of the initial aircrew and instructors were former RAF personnel, later augmented by Pakistani pilots and, ultimately, a Pakistani commanding officer.

First aircraft to arrive were two Britten-Norman Islanders (later supplemented by two more) for light transport and communications, and two Agusta-Bell Jet Rangers, also joined by a further pair shortly afterwards. Medium transport tasks were allocated to four DHC Caribou received in 1968–69, combat capability following in 1970–71 with the arrival of 12 refurbished Hawker Siddeley Hunters, comprising seven Hunter FGA Mk 76 interception and close-support variants, three FR Mk 76As for recce and two T Mk 77 trainers.

Previously a component of the Abu Dhabi Defence Forces, the aviation section gained air force status in 1972, embarking on an expansion programme with the formation of a second fighter unit with 12 Mirage 5ADs and two 5DAD trainers in 1973–74. A mutual agreement with Pakistan brought additional personnel in return for arrangements to lease Mirages to the PAF if required, and the cancellation of Airwork's contract.

Further purchases from France augmented the helicopter force by three, and subsequently ten Pumas, together with five (later increased to ten) Alouette IIIs. Jet Rangers of the Helicopter Flight were then passed on to Dubai to establish the federally-funded Union Air Force. Mirages replaced Hunters at Abu Dhabi International Airport, the latter then transferring to the ex-RAF base at Sharjah.

In May 1976, all military aviation units within the federation formally amalgamated within the unified armed forces, with headquarters at Abu Dhabi, and main operating base at Dubai incorporating stores and workshops. As wealthiest of the Emirates, Abu Dhabi provides the main financial support and some 80 per cent of the 30,000-strong army. Little unification of purchasing policy has taken place, however, and individual states continue with their own armaments orders, apparently without any effort to standardise.

For Abu Dhabi, further expansion has taken the form of extra Mirages to equip a second squadron with 14 5EAs, three 5RAD reconnaissance variants and one 5DAD trainer. Two C-130H Hercules arrived in 1975 to supplement the Caribou and Islanders, followed in 1978 by five DHC-5D Buffaloes, together with four Agusta-Bell 205As and three AB 212s from Italy. Air defence forces are assisted by BAC Rapier missiles with Blindfire

Below: Pakistan Army Aviation operates about 12 Mil Mi-8 Hip helicopters in the assault and transport roles. Based at Dhamial, they were supplemented by 35 Aérospatiale Pumas in 1979. Army aviation also operates Alouettes, Huskies, UH-19s, Sea Kings and UH-1 Iroquois. Fixed-wing types comprise about 50 Cessna O-1s and a number of Saab MFI-17s

capability and Crotale SAMs for air base and point defence. Seven remaining Alouette IIIs have provision for AS 12 anti-tank missiles, while Vigilant anti-tank missiles are employed by ground forces.

Establishment of the UAEAF has also affected Dubai, the second of the Emirates to form its own air elements. Prior to the 1976 integration, Dubai possessed two aviation services attached to the Union Defence Force and the Police. Evolved in 1971 from the former Trucial Oman Scouts, the Union Defence Force established an Air Wing, commanded by a former RAF officer, with the transfer of three Jet Rangers from Abu Dhabi in 1971. These were again maintained and flown by Airwork with ex-RAF personnel, later assisted by Jordanian and Sudanese pilots and groundcrews. In company with four Bell 205s from Abu Dhabi, they now constitute the Helicopter Squadron of the UAEAF based at Dubai.

A second unit, the Police Air Wing, was formed in 1973 to provide air support for the Dubai Defence Force with two Jet Rangers and a Cessna

Above left: the Sultan of Oman's Air Force has a total of 15 Short Skyvan 3MFs remaining, partly equipping No 2 Squadron based at Seeb, and a detachment with No 5 Squadron at Salalah.
Left: three of the Sultan of Oman's Air Force's four Agusta-Bell AB 206As are operated by No 14 Squadron, based at Salalah, while the fourth is based at Masirah for search and rescue duties

182N, to which were later added a third Jet Ranger and two Agusta-Bell 205s. Italian technical assistance with the formation and training of the DPAW was reflected in a 1974 order for training of a counter-insurgency force equipped with a single SIAI SF 260WD for weapons instruction, one Macchi MB326LD for jet conversion and three MB326KD light strike aircraft. DPAW pilots underwent a conversion course on the MB326 in Italy, and Italian Air Force personnel accompanied aircraft on delivery for training and advisory roles. Later orders have raised totals to nine MB326KDs and three 326LD trainers.

To supplement the helicopter transport squadron, Dubai received its first large aircraft when an Aeritalia G 222 was delivered in November 1976, following an order placed only nine months previously. A second G 222 is on option. VIP aircraft form a special squadron at Dubai Airport alongside the former DPAW helicopters, comprising a Falcon 20 for the use of the Defence Minister and the Presidential Boeing 720. Two Pawnee Brave crop sprayers are also housed at Dubai on UAEAF charge.

With a long-standing background of disputes with India, culminating in two armed conflicts with its neighbour in 1965 and 1971 – the second resulting in the establishment of Bangladesh – Pakistani defences remain geared to confrontation along the Indian border, while recent purchasing policy has turned away from the United States towards a broader source of supply from China, France and the Soviet Union. The recent formation of a Marxist government in Afghanistan, with Soviet assistance, has also presented additional defence problems.

After the 1971 war, the Pakistan Fiza'ya, or air force, was reorganised to upgrade capability and reduce vulnerability to pre-emptive air strikes, with the construction of six small airfields for wartime dispersal. For improved control of operational

Above: No 4 Squadron, Sultan of Oman's Air Force, based at Seeb, operates three BAC One-Eleven 475s. Below: the UAEAF's helicopter transport force, based at Abu Dhabi, operates two AB 205As, while the troop transport force, based at Dubai, operates four

٤٠٤
404

Above left: one of five DHC-5D Buffaloes purchased by the United Arab Emirates Air Force in 1978, pictured in Dubai's national markings at Reykjavik, Iceland, in July 1978 during delivery from Canada. The DHC-5D STOL tactical transports joined the UAEAF's transport force, based at Abu Dhabi.
Above far left: a DHC-4 operated by the transport force of the UAEAF pictured in Abu Dhabi's national markings. Four DHC-4s were delivered to Abu Dhabi in 1968–69, of which three remain.
Left: some 15 BAC Strikemaster Marks 82 and 82A remain in the service of the Sultan of Oman's Air Force. Ten are in front-line combat service, deployed in the ground-attack role against rebel activity. The Flying Training School at Masirah operates the remainder, with two held in reserve

units, the country has been divided into three defence sectors, North, Central and South, with the majority of combat squadrons based in the extremities and Central Sector allocated the principal task of early warning.

Air defences have been augmented by the first Pakistani army procurement of SAMs in the form of nine Crotale batteries, ordered in 1975. Aircraft strength has been increased to over 300 in operational roles, manned by 18,000 personnel, plus 125 support types, compared with 200 combat types and 15,000 officers and men in 1971. Organisation is now based on a three-prong system with flying (operations), administrative and maintenance elements. Subordinate squadrons in fighter, bomber and transport wings have nominal establishments of 16 aircraft each with supporting, training and other units.

After many years as PAF spearhead, the Sabre is now represented only by some 50 of the 90 ex-Luftwaffe aircraft obtained via Iran in 1966. Following suspension of military aid to both India and Pakistan after the 1965 war, the PAF was forced to diversify its sources of supply and gratefully accepted a Chinese offer of 75 Shenyang F-6s (MiG-19s), mainly for air superiority roles, replacing the survivors of 100 F-86Fs received in 1956.

F-6s were modified to take Sidewinder air-to-air missiles in addition to their standard fixed-gun armament of three 30mm cannon, and also have a limited strike capability when fitted with underwing rocket pods. Following the initial re-equipment of Nos 11, 23 and 25 Squadrons, the PAF received a further 60 F-6s from China, the type now additionally serving Nos 14, 15, 17, 18, 19 and 26 Squadrons from bases at Rafiqui, Sargodha and

Masroor. The remaining Sabres of Nos 15 and 16 Squadrons at Chaklala are allocated to training, joined by five Chinese MiG-15UTIs which arrived with the first F-6s. The latter have been modified by the PAF, and are now equipped with Martin-Baker zero-zero ejection seats.

Second main PAF fighter type is the Mirage III/5 for strike-interception duties. An opening order for 18 Mirage IIIEPs, plus three IIIRP tactical reconnaissance types and three IIIDP trainers for No 5 Squadron at Sargodha in 1969 was followed by subsequent deliveries which raised the total Mirage complement to 75, (of which 69 remain) including 28 (later 33) Mirage 5PAs to replace the Starfighters of No 9 Squadron at Sargodha. Mirages are armed with both Sidewinder and Matra R 530 air-to-air missiles and some have been updated by incorporation of the Super Etendard's nav-attack system, including the inertial platform. An order for 32 Mirages 5s was

Below: No 6 Squadron, Sultan of Oman's Air Force operates about 12 of the 31 Hawker Hunters transferred from Jordan in 1975, the remainder being stored. A Hunter conversion unit is attached to the squadron. No 6 Squadron is based at Thumrait, the SOAF's main strike base.
Bottom: a McDonnell Douglas TA-4KU pictured at Prestwick, Scotland, in April 1978 during the delivery flight to the Kuwait air force, still bearing US national markings. As part of a re-equipment programme begun in 1974 and recently completed, the Kuwait air force took delivery of 30 A-4KU strike aircraft and six TA-4KU trainers. These equip two strike squadrons based at one of the two new airfields built by Yugoslav contractors near Kuwait city

placed early in 1979 for the equipment of two more fighter-bomber squadrons between 1981 and 1983, Indian sources having reported a strong Pakistani interest in the Mirage F1 and 2000.

The new Mirages will assume the strike duties of No 7 Squadron at Masroor, which currently flies 11 survivors of the 26 Martin B-57B Canberras supplied in 1958 for night bombing. After losing five Canberras in the 1971 war, No 7 Squadron has been under-strength and in need of a replacement for its ageing fleet for several years. Saudi Arabia offered to finance 110 A-7 Corsairs in 1976, but United States' support for the project was withdrawn following Pakistan's decision to purchase a nuclear re-processing plant from France, thus gaining capability for producing nuclear weapons.

Lack of maritime reconnaissance aircraft enabled Indian patrol boats to launch missile attacks on Pakistani shipping near Karachi in the 1971 war and, to remedy this serious shortcoming, No 29 Squadron formed late in 1975 with the first of three ex-French Atlantics, operating in close conjunction with the small air element of the Pakistan Navy. Naval aviation comprises six Westland Sea King Mk 45s, crews for which were trained by the Royal Navy at Culdrose prior to delivery in 1975, supported by one or two Sikorsky UH-19s and four licence-built Dhamial Alouette IIIs. Four Sea Kings are being fitted with Aérospatiale AM 39 anti-shipping missiles, one helicopter having been detached to Cazaux in France for operational trials.

Lockheed Hercules provide the principal PAF transport force for logistic and tactical air support from Chaklala. Eight C-130Bs were augmented by

nine from Iran, and further deliveries of C-130Es and a Lockheed L-100 civil version. Other transport types include a Falcon 20 and a Fokker F-27 for VIP and government use, plus a few lightplanes and a single Puma, received in 1974. Search and rescue work is allocated to HH-43B Huskies and Alouette IIIs, with additional reconnaissance potential from HU-16A Albatross amphibians.

Pilot training is centred on the PAF Academy at Risalpur, and begins on 30 of the 45 Saab-MFI Supporters obtained as T-6 replacements in 1974–75. Jet conversion then follows on the Cessna T-37C, and after 150–170 hours, students begin advanced flying on the T-33s of No 2 Squadron. Operational conversion is then completed on the Sabre, MiG-15UTI or Mirage III/5DP. Advanced training facilities are also provided for several Arab countries.

Army aviation has recently undergone a dramatic increase in assault transport capability with the delivery of 35 Puma helicopters between 1977 and 1979. The Army Aviation Wing began with a single flight of four Austers in 1947 and was known for the next 11 years as the Air Observation Flight of the PAF. Sixty O-1 Bird Dogs were supplied through US aid programmes for training and the establishment of two squadrons, two-thirds of which continue in service, supplemented by locally-assembled Cessna O-1s. Helicopter equipment arrived in 1964 in the form of some 18 Bell 47Gs, later supplemented by the larger Alouette III. Following the

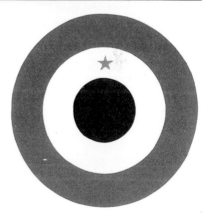

Top: the Sultan of Oman's Air Force has seven Britten-Norman Defenders, which partly equip No 2 Squadron, at Seeb, and No 1 Squadron. Equipped with radar in the nose, the Defender has four underwing hardpoints and can carry two 7·62mm machine guns, bombs, Matra rocket pods, wire-guided missiles and smoke bombs.
Above: the DHC Beaver light transport entered service with the SOAF in 1961

receipt of some early Alouettes from the manufacturer, 503 Workshop at Dhamial initiated a programme of local assembly, Army Aviation presently operating some 20 Alouettes plus a dozen Mil Mi-8s from the main base at Dhamial.

Modernisation plans were earlier formulated for adoption of the Cessna T-41 as the basic training type and replacement of Bell 47s with the Hughes 500, both to be assembled locally. Little progress has been made in achieving this end, and apart from Pumas, Army Aviation has received only 15 Saab Supporters together with a dozen UH-1 Iroquois as United States aid following the 1973 floods in Sind and Punjab. Missile equipment includes MBB Cobra and Hughes TOW anti-tank systems.

Twice the size of Britain and bordered by Pakistan, the USSR and Iran, the inhospitable terrain of Afghanistan is populated by nomadic, principally Islamic, tribes bound to their religion by a fanatical loyalty. Following the overthrow of the monarchy in 1973, the new government was deposed after a military coup in April 1978 which brought to power the pro-Communist Premier Tarakki. Despite Tarakki's professed neutralist policies, considerable Soviet aid began flowing into Afghanistan within two months of the coup, and the Republican Air Force (De Afghan Hanoi Quirah) received much updated equipment during 1979. This year saw the overthrow of the Tarakki regime, which was replaced by another pro-Communist government under President Hafizullah Amin. Amin was in his turn ousted by the Soviet Union's invasion in December 1979. Among the Soviet warplanes involved were MiG-21s, MiG-23s and Su-17s, together with An-22 and Il-76 transports.

Before the Soviet invasion, Russian-built warplanes were flown by Afghan pilots against

of rocket and napalm attacks on rebel strongholds in Kunar and Paktia provinces, and through poor flying and maintenance, or defections to Pakistan. Present strength of the Afghan Air Force is 250–270 aircraft and some 10,000 personnel; 150 of these aircraft are combat types. Three interceptor squadrons centred on Pagram operate 40 MiG-21s, and at Shindand, 30 survivors of 46 Il-28 Beagle jet bombers equip three squadrons, alongside two or three Su-7 units with a total of 30 aircraft. Fifty older MiG-17 Frescos and 24 MiG-19 Farmers equip four additional squadrons. Soviet air defence systems include approximately nine sets of GCI and early-warning radar, operating in conjunction with interceptors and about 140 SA-2 Guideline SAMs, recently augmented by SA-3 Goas.

Stationed primarily at Kabul, the newly-enlarged transport fleet has partly replaced its Il-14 piston-engined twins with the larger An-26 and formed a second squadron of Mi-8 Hip helicopters, although older An-2 Colt biplanes and Mi-4 Hounds form the equipment of a further two units. A flying training academy opened in 1958 at Sherpur accommodates about 400 cadets, training on types such as the Yak-11, Yak-18, MiG-15UTI and L-29.

The Sultan of Oman's Air Force (Al Quwwat Aljawwiya Alsultanat Oman) originally formed with Percival Provost armed trainers and Scottish Aviation Pioneer STOL transports taken from RAF stocks in 1958. DHC Beavers were added in 1961, but not until eight years later did the SOAF gain its first combat types with initial deliveries of BAC Strikemasters in March 1969. By August 1976, 24 had been received for operation by No 1 Squadron from the former RAF base at Salalah. Five were sold to Singapore in 1976 and at least a further two written-off, the remainder serving No 5

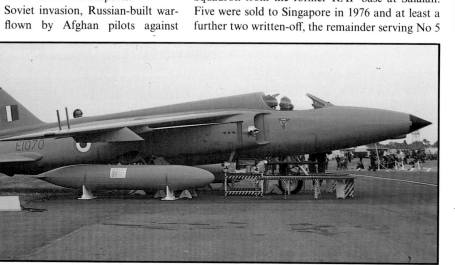

Left: the HAL licence-built Gnat F Mark 1 was, with the Hunter, the Indian Air Force's main fighter until the arrival of the MiG-21. About 150 remain in service, equipping five squadrons. Deliveries of the Gnat F Mark 2, designed and built by HAL, and named Ajit (Unconquerable), began in late 1978. No 9 Squadron became operational in 1979 on the type, and the five F Mark 1 squadrons are scheduled to re-equip with the type. The Ajit has greater internal fuel capacity, making redundant external tanks and freeing the two wing hardpoints to carry ordnance

Islamic insurgents. Closely following the arrival of a small number of L-29 Delfin trainers came ten An-26 Curl transports, possibly as many as 30 Mi-24 Hind assault helicopters, further replacement Su-7s and MiG-21s and 3,000 advisers, all of which had been incorporated into the ARAF by the end of 1978. A squadron of MiG-23 Floggers has been reported in service, possibly with seconded Russian crews, while other unconfirmed sources refer to MiG-27s and Su-20s.

Losses, particularly amongst the MiG-21 Fishbeds and Su-7 Fitters, have been exceptionally high through aircraft being shot down in the course

Squadron at Salalah (ten aircraft) and the newly-established flying training school at Masirah.

Transport links between Salalah and Muscat, some 960 km (600 miles) to the north, were initiated with a single Douglas Dakota obtained in 1968, followed by a further two, and then eight Short Skyvans in 1970–71. Skyvan complement has now been doubled, serving with No 2 Squadron at Seeb, with a detachment of four maintained at Salalah. Five DHC Caribou temporarily entered service with No 4 Squadron at Seeb, but were replaced on the Muscat-Salalah run by three Vickers Viscounts in mid-1971.

Above: the IAF has 32 Antonov An-12 Cub transports, flown by the heavy element of the transport force, Nos 25 and 44 Squadrons. The An-12s are currently undergoing a modernisation programme to extend their lives until 1983–84, as are the IAF's 54 Fairchild C-119G medium transports.

Right: the Indian Air Force's Russian-supplied Ilyushin Il-14s were recently retired from service. Consideration has been given to the DHC-5 Buffalo, to be licence-built by HAL, and the Antonov An-32, as an Il-14 replacement in the medium tactical transport role

Orders for support equipment during 1970 included the first helicopters for the SOAF. Eight Agusta-Bell 205s and four 206s formed the initial equipment of No 3 Squadron at Salalah, later to be joined by a further 17 205s and five Agusta-Bell 214s. A small detachment of helicopters is maintained at Seeb, alongside an AB 212 Twin-Pac of the Royal Flight.

Further Viscounts were added in 1973, together with an additional Beaver and a replacement Caribou, but the apparent stalemate in Dhofar the following year resulted in substantial orders for SOAF expansion. Three BAC One-Elevens replaced the No 4 Squadron Viscounts, but combat elements were provided with a considerable boost with a contract for ten single-seat and two tandem-seat Jaguar Internationals delivered from the British production line in 1977–78. Jaguars were originally intended to replace the Strikemasters of No 5 Squadron, but the upsurge in rebel activity caused their retention, and the Jaguars instead formed No 8 Squadron at Thumrait (formerly Midway).

An unexpected windfall for the SOAF occurred in 1975 when, having failed in its attempts to sell 31 Hunters to Rhodesia, the Jordanian Air Force

presented the aircraft to Oman. A dozen were selected to form No 6 Squadron at Thumrait and the remainder placed in storage. Hunters fly from Khasab on short detachments together with Jaguars armed with over-wing Matra Magic air-to-air missiles, although principal role of the latter aircraft is ground attack.

Operating in a pseudo-civilian capacity are the aircraft of the Oman Police Air Wing, with a Learjet, two Turbo Porters, two Swearingen Merlins, two Buffaloes, four AB 205 Iroquois, two Jet Rangers and a Hughes 500. The Royal Flight has a VC 10, a Falcon 20, a Gulfstream II and two AS 202 Bravos both units being based at Seeb.

The Qatar peninsula and its neighbouring islands of Bahrain chose to retain their independence following the British withdrawal in 1971, rather than join the United Arab Emirates as recommended by the Willoughby report. Accordingly, defence forces were established in both sheikhdoms, although Qatar's deterrent potential is nominal, and Bahrain's all-but non-existant. Made prosperous through its oil deposits and extensive port facilities, the Bahrain archipelago presently maintains a small Defence Force of 2,300

Designed as a replacement for the IAF's de Havilland Vampire trainer, the HAL HJT-16 first flew in 1964. IAF Training Command, whose HQ is at Bangalore, and the Indian Navy jointly require some 270, but deliveries have been slow, amounting to about 160. After a course on the HAL HT-2 piston-engined trainer at the Elementary Flying School at Bidar, fixed-wing IAF pilot trainees proceed to the Air Force Academy at Dundigal, near Hyderabad for 180 hours basic and advanced training on the Kiran, or to the TS-11 Iskra-equipped Fighter Training Wing at Hakimpet, used in parallel pending the delivery of the full complement of Kirans

men on an annual defence budget of some £20 million. British equipment was initially chosen, the ground units operating Saladin armoured cars and Ferret scout cars.

Three government agencies fly a total of eight helicopters, and first to become airborne were the Bahrain State Police, following the arrival of two Westland Scouts in August 1965. In 1977 an air wing of the BDF was formed with two MBB-Bölkow 105s, soon followed by a third, while Bahrain Public Security operates a Bell 205 and two Hughes 500s. Attempts have recently been made to obtain more effective armament in the form of Northrop F-5Es, together with TOW anti-tank missiles, Harpoon anti-ship weapons and Redeye infantry SAMs, but the United States Government has declined to authorise delivery.

Qatar received one of several RAF missions seconded to the sheikhdoms in 1972 to advise and supervise the formation of military aviation organisations and to provide air and ground crews pending the availability of qualified national personnel. From a population of 205,000, some 4,000 Qataris are in military service, all three branches coming under nominal army control. Annual defence expenditure is currently about £30 million per year.

First aircraft to arrive in Qatar were two Westland Whirlwind helicopters delivered to the Air Wing of the Public Security Forces in March 1968. In mid-1969, Short Tigercat SAMs were ordered for air defence and plans announced for the purchase of six refurbished Hawker Hunters to provide the first combat element. Requirements were later revised, and the Hunter requirement raised to 12, but financial constraints resulted in only four being delivered from late 1971 onwards, comprising three Hunter FGA Mark 78s and one T Mark 79 trainer.

In 1974, the re-titled Qatar Emiri Air Force ordered three Westland Commando Mark 2As for assault transport, together with a Mark 2C, with VIP interior, deliveries beginning in October 1975. Two Gazelle helicopters were obtained by the Qatar Police in October 1974, to be flown by military pilots until police personnel could be trained. The gradual build-up has continued with an Islander light transport and three Lynx helicopters, the latter delivered in June 1978.

As one of the first Gulf States to form its own air arm, Kuwait received two DH Doves, a Heron and eight Austers during the 1950s to form an extension of the government Security department. Requests for further British aid resulted in the arrival of an advisory mission to supervise KAF formation in 1960, followed in 1961 by an order for six Jet Provost T Mark 51 armed trainers to be flown by Kuwaiti nationals trained in the RAF.

A helicopter component was then added with two Westland Whirlwinds, and subsequently, in 1965–66, initial deliveries of an eventual total of four refurbished Hunter FGA Mark 57s and five T Mark 67 trainers gave the KAF its first combat equipment, albeit flown by seconded British personnel. Two Caribou STOL transports were added in 1966 together with six Agusta-Bell 205 Iroquois in 1968, although both these types were withdrawn in the past year.

Brief use was made of a loaned RAF AW Argosy in 1969 and, in the same year, Kuwait took delivery

BHARATIYA VAYU SENA – INDIAN AIR FORCE		
Unit	Aircraft	Role
No 1 Sqn	MiG-21	Fighter
No 2 Sqn	Gnat F Mark 1	Fighter
No 3 Sqn	MiG-21	Fighter
No 4 Sqn	MiG-21	Fighter
No 5 Sqn	Canberra	Strike
No 6 Sqn	Canberra, Super Constellation	Anti-shipping, transport
No 7 Sqn	MiG-21	Fighter
No 8 Sqn	MiG-21	Fighter
No 9 Sqn	Ajeet	Fighter
No 10 Sqn	Marut	Strike
No 11 Sqn	Dakota	Transport
No 12 Sqn	C-119G, HS 748	Transport
No 14 Sqn	Hunter	Strike
No 15 Sqn	Gnat F Mark 1	Fighter
No 16 Sqn	Canberra	Strike
No 18 Sqn	Gnat F Mark 1	Fighter
No 19 Sqn	C-119G	Transport
No 20 Sqn	Hunter	Strike
No 21 Sqn	MiG-21	Fighter
No 22 Sqn	Gnat F Mark 1	Fighter
No 23 Sqn	Gnat F Mark 1	Fighter

of 12 BAC Lightning F Mark 53s and two T Mark 55 trainers. These proved too sophisticated for the KAF and flew infrequently until their retirement in 1977–78. More effective mounts from the same stable were six Strikemasters for both instruction and light strike duties received in 1970, a similar quantity following in 1971, together with two L-100 Hercules (the civil version of the C-130) to reinforce the transport fleet.

Border clashes with Iraq in 1973 prompted the Kuwaiti government to expand the KAF, principally for the protection of the oilfields so vital to its economy. After extensive international evaluation, 18 Mirage F1CK interceptors and two F1BK trainers were chosen and, following pilot and groundcrew training in France, the aircraft were delivered in 1977, complete with armament of Matra Super 530 and 550 Magic air-to-air missiles.

Modernisation plans also included 30 A-4KU Skyhawks and six TA-4KU trainers received in 1977–78 and a French package of 12 Pumas and 24 Gazelles, plus HOT, SS 11 and Harpon missiles. Two new airfields were built at Jakra and Ahmadi by Yugoslav contractors, to supplement Kuwait International Airport, while the USSR supplied SA-7 infantry SAMs and other army weapons. Transport capability was doubled following the arrival of two DC-9s in late 1976, but all single-seat Hunters have now been withdrawn, leaving the two-seat T Mark 67s to continue in the training role.

Born out of a six-year civil war between 1962 and 1967, in which Egyptian and Soviet intervention defeated supporters of the former Imam, the Yemen Arab Republic (North Yemen) later sought non-alignment and re-opened its contacts with the West. Reconciliation with Saudi Arabia brought increased economic aid, but expansion of the Republican Air Force was restricted through funding problems and a switch of Soviet interests towards South Yemen.

Earlier Russian equipment, comprising MiG-17s and MiG-21s for strike and interceptor roles, An-24 and Il-14 transports, plus Mi-4 and Mi-8 helicopters were supplemented by purchases of Agusta-Bell Iroquois and two Skyvans following the Soviet departure from bases allocated to them at Janad, Sana'a and Hodeida. Continued friction between North and South Yemen prompted Saudi Arabia to transfer a small number of BAC Vigilant anti-tank missiles to the Republic in 1976, followed by two Hercules and four surplus F-5B trainers. Seeking replacements for the MiG force, which was reportedly down to two MiG-17s and a dozen MiG-21s by early 1979, North Yemen requested 12 Northrop F-5E strike aircraft from the United States, backed by financial assurances from Saudi Arabia. Deliveries were effected early in 1979, most of the aircraft being from the batch originally destined for Egypt. Military airfields are located at Hodeida, Sana'a and Taiz.

Whereas North Yemen has recently added Western equipment to its military inventory, the South Yemen People's Republic has reversed the process and entirely disposed of its former British aircraft, retaining only four examples of the ubiquitous Dakota as a reminder of its previous pro-Western inclinations.

Assumption of power by the National Liber-

ation Front accelerated British withdrawal from what was formerly known as Aden, this being complete by November 1967. Plans for the transfer of No 43 Squadron's Hunter FGA Mark 9s and considerable military assistance were shelved in favour of a more moderate aid programme, although the extensive infrastructure at Khormaksar airfield was handed over intact.

Formation of the air force had, however, begun in the previous August with the arrival of four Dakotas modified by Aviation Traders. Six Beavers were obtained from Canada, together with a similar number of Westland Sioux helicopters. Delivery of eight ex-RAF Jet Provost T Mark 4s converted to armed T Mark 52s began in late 1967, followed by four Strikemaster Mark 81s in August 1969.

As Soviet influence increased, MiG-19s and later MiG-21s equipped two new squadrons, training being undertaken with Russian or Cuban guidance. Some of the last British types to depart were the four Strikemasters, sold to Singapore in 1976, by which time the SYAF was flying Il-28 light bombers, Il-14 and An-24 transports and Mi-4 and Mi-8 helicopters. Soviet base facilities formerly in Somalia have now been re-established in South Yemen, but expansion remains limited

No 6 Squadron, Indian Air Force operates two Lockheed L-1049C Super Constellations as freighters. Five were transferred to the Navy in 1976, to equip INAS 312 maritime reconnaissance unit, based at Goa, when it was re-numbered from the original No 6 Squadron, IAF

No 24 Sqn	MiG-21	Fighter
No 25 Sqn	An-12	Transport
No 26 Sqn	Su-7	Strike
No 27 Sqn	Hunter	Strike
No 28 Sqn	MiG-21	Fighter
No 29 Sqn	MiG-21	Fighter
No 30 Sqn	MiG-21	Fighter
No 31 Sqn	Marut	Strike
No 32 Sqn	Su-7	Strike
No 33 Sqn	Caribou	Transport
No 35 Sqn	Canberra	Strike
No 37 Sqn	Hunter	Strike
No 41 Sqn	Otter	Transport
No 43 Sqn	Dakota	Transport
No 44 Sqn	An-12	Transport
No 45 Sqn	MiG-21	Fighter
No 47 Sqn	MiG-21	Fighter
No 48 Sqn	C-119G	Transport
No 49 Sqn	Dakota	Transport
No 59 Sqn	Otter	Transport
No 101 Sqn	MiG-21	Fighter
No 106 Sqn	Canberra	Strike
No 108 Sqn	MiG-21	Fighter
No 220 Sqn	Marut	Strike
No 221 Sqn	Su-7	Strike
No 222 Sqn	Su-7	Strike

An Indian Air Force BAC Canberra T Mark 4 pictured in 1975 on Malta. Following advanced training at the Air Force Academy, Dundigal, bomber pilots are streamed to the Jet Bomber Conversion Unit at Agra, to fly the Canberra T Mark 4. Canberras were introduced to the IAF in 1957

through continued financial constraints, holding defence expenditure to some £25 million per year. The most recent addition to the inventory has been 30 Su-22 fighter-bombers, giving the South a considerable edge in strike capability compared to their Northern neighbours.

Organised on RAF lines, with at one time predominantly British front-line equipment, the Bharatiya Vayu Sena, or Indian Air Force, maintains a constant alert along the country's northern frontier areas. Boosted to a peak of 45 squadrons by a five-year plan instituted between the two wars with Pakistan, the IAF is further upgrading its deep-penetration strike capability with BAC Jaguars, the first of which were delivered in July 1979 to begin conversion of No 14 Squadron. Forty British-built Jaguars, including some ex-RAF aircraft, will be joined by 110 built under licence for the eventual re-equipment of five squadrons.

Present IAF strength is in the region of 1,300 aircraft of all types, operated by some 36 first-line units and two dozen transport and helicopter squadrons. In addition, about 30 SAM squadrons are equipped with SA-2 and SA-3 missiles supplied by the USSR, but India is now tending to turn away from the Soviet Union towards a wider source of arms supply, with self-sufficiency as the ultimate goal.

Of completely independent status since 1 April 1966, the IAF underwent a major change in 1972 following the establishment of No 1 (Operational) Group, Jodhpur, as an offshoot of Western Air Command, Delhi. Remaining operational formations comprise Central Air Command at Allahabad (which also controls all bomber, transport and maritime aircraft from Bengal to southern India) and Eastern Air Command, Shillong. In addition to these three operational commands and an Independent Group, IAF organisation is completed by Training and Maintenance Commands.

Mainstay of the IAF fighter force for many years was the diminutive, indigenously-built Folland Gnat interceptor, over 200 being produced for eight squadrons. Nos 21 and 24 Squadrons have recently converted to the MiG-21bis, and No 9 has received the first of 100 Gnat Mk 2s, or Ajits, which will eventually equip Nos 2, 15, 18, 22 and 23 Squadrons. First Indian-assembled MiG-21FLs were accepted in late 1967, after delivery of 14 MiG-21Fs and 42 MiG-21UTI trainers from the Soviet Union. Hindustan Aircraft went on to complete 198 MiG-21FLs and 150 MiG-21M/MFs and is now beginning assembly of up to 200 MiG-21bis variants. First examples of the MiG-21bis came from the USSR for Nos 21 and 24 Squadrons,

following the conversion of Nos 7 and 108 Squadrons with the MiG-21PFMA. Remaining MiG squadrons – including the Operational Conversion Unit – forming the main combat element, are gradually re-equipping from earlier day fighters to all-weather variants, each with 16 fighters and two MiG-21UTI trainers.

From over 100 Canberras of various marks obtained since 1958, the IAF continues to operate about 80 in strike roles with Nos 5, 16 and 35 Squadrons, together with 12 reconnaissance versions in No 106 Squadron. No 6 Squadron has the unlikely complement of anti-shipping strike Canberras and two Super Constellation freighters, from nine maritime reconnaissance versions now partially transferred to the navy.

Strike forces comprise four squadrons (Nos 14, 20, 27 and 37) with the survivors of 252 Hunters delivered to India from 1957 onwards, backed by some 150 Su-7BMK Fitters received in 1967–68. Heavy losses against Pakistan in 1971 have reduced Fitter units from six to four (Nos 26, 32, 221 and 222 Squadrons). Engine shortcomings prevented the HF-24 Marut from fulfilling its intended role as primary strike weapon, 125 Mk 1s and 15 Mk 1T trainers equipping only Nos 10, 31 and 220 Squadrons, although a Soviet power plant has been suggested for a re-launched version in the 1980s.

Himalayan airlift operations dominate transport aircraft commitments, undertaken by Nos 25 and 44 Squadrons with the remainder of 34 An-12s, and 54 jet-boosted C-119Gs operating with Nos 12, 19 and 48 Squadrons. Lighter transports include the 22 Caribou of No 33 Squadron, scheduled for replacement by HAL-built Buffaloes or Soviet An-32s, if necessary funding can be found. Locally-assembled HS 748s equip No 12 Squadron, plus navigational training units and the VIP squadron and Communications Flight, together with BAC (de Havilland) Devons and Dakotas. Other Dakotas serve with Nos 11, 43 and 49 Squadrons, while Nos 41 and 59 share some 25 Otters.

Helicopters have become increasingly important in supply roles, and nearly 400 have been delivered or ordered in the past few years. Original mainstay of the helicopter fleet was the Mi-4, of which about 125 were received for Nos 105, 107, 110, 111, 115 and 116 Squadrons, followed by 40 Mi-8s for Nos 109, 118 and 119. HAL built 143 SA 319 Alouette IIIs to augment the 30 received from France for all three services, these going to Nos 104, 112 and 114 Squadrons. Army liaison squadrons (Nos 659, 660, 661 and 662) have replaced Auster AOP 9s and HAOP-27 Krishaks with 120 SA 315B Lamas, including the first 20 from France.

Flying training for students from all three services begins with 40 hours' instruction on some 70 HT-2 primary trainers, planned for eventual replacement by the HT-32 in 1981–82. Army officers convert to the Alouette III at the Helicopter Training School, Hakimpet, while naval helicopter students proceed to No 561 (Indian Navy) Squadron at Cochin. Remaining pilots transfer to the HJT-16 Kiran for 180 hours' basic and advanced training at the Air Force Academy, Dundigal. Premature retirement of the Vampire through structural problems brought the purchase of 50 TS-11 Iskras from Poland, and a twin-seat

supply of eight Sea Harriers, including two trainers, to begin upgrading the naval strike force with a view to obtaining an eventual total of 24 initially for INAS 306.

Shipborne ASW is performed by the turboprop Bregnet Alizes of INAS 310, formed at Hyères in 1961 with 12 aircraft, supplemented by a further three in 1968. Only six aircraft remain serviceable, but these are to be joined by 12 refurbished examples from Aéronavale in the near future. Two batches of six Westland Sea Kings provided for the establishment of INAS 330 in 1971 and INAS 336 at Cochin four years later, the latter for ASW duties

Ajeet is projected as a Hunter Trainer replacement.

Combat pilots complete the full course at Dundigal before continuing to the Hunter OCU at Kalaikunda or Jet Bomber CU at Agra on the Canberra T Mark 4. Transport and helicopter students transfer in mid-stream to the HS 748 Transport Training Wing at Yelahanka or Alouette III-equipped Helicopter Training School, Hakimpet. Selected pilots receive a 12-week Tactics and Combat Development Establishment course at Ambala on MiG-21s, Su-7s and Gnats, the Armament Training Wing at Jamnagar providing two squadrons (Nos 3 and 31) with target-towing Hunters. Other training facilities include the Navigation and Signals School, Begumpet with HS748s.

Indian Naval Aviation centres on the aircraft carrier INS *Vikrant* (the former HMS *Hercules*), obtained and modernised in 1957 prior to the formation of Indian Naval Air Squadron (INAS) 300 at Brawdy with Sea Hawks in 1960. An initial batch of 24 Sea Hawk FGA Mk 6s was followed by 12 refurbished Mks 4 and 6 the following year, plus 28 Mks 100 and 101 from Germany in 1966. Replacement of the 22 surviving aircraft in INAS 300 and training unit INAS 551 at Dabolim is now

overdue, and authorisation has been given for the along the western coast. Alouette IIIs are allocated to INAS 331, formed in May 1972 for deployment aboard six *Leander*-class frigates with torpedo armament, and INAS 321, commissioned at Dabolim for SAR and liaison tasks in 1969.

A protracted wrangle over responsibility for land-based maritime reconnaissance aircraft was resolved in 1976 with the transfer of five Super Constellations from the IAF to form INAS 312 at Dabolim. These were supplemented in mid-1977 by four Ilyushin Il-38 -ay ASW patrollers delivered from the USSR to equip INAS 315 at the same base, two more being expected when the Constellations are withdrawn in about 1981. Naval Aviation is completed by three second-line units: INAS 330, Cochin, has five Islanders, originally for training and communications, but now modified with nose-mounted radar for coastal patrol, and a small number of HJT-16 Kiran jet trainers. Kirans, together with Sea Hawks, equip INAS 551, while INAS 562 is the Cochin-based helicopter school with four Hughes 300s and some Alouettes. A further helicopter operational squadron (INAS 333) is scheduled to form with five Ka-25 Hormones.

The Hindustan HF-24 Marut (top) and Mikoyan MiG-21 (above) represent two of India's main sources of front-line equipment. The locally-designed and built Marut was the intended successor of the Hawker Hunter, which still equips four IAF units, but failed to emulate that type's success due to power plant deficiencies. The tried-and-tested MiG-21, however, proved more effective and was licence-produced by Hindustan Aircraft. Various improved versions have seen service, a MiG-21PF being illustrated

CHAPTER THIRTEEN

North-East Asia
China/Taiwan/Japan/South Korea/ North Korea

It would be difficult to name another five nations so geographically close and yet politically, culturally and otherwise so far apart, than China, Taiwan, Japan, North Korea and South Korea. Within the last half century there have been wars between Japan and China, North and South Korea, and a Communist take-over in China which has seen the last vestiges of the old regime exiled to the island of Taiwan. Cold wars continue. The battle between China and Taiwan is perpetuated by the propaganda loudspeaker and leaflets attached to toy balloons, while the two Koreas continue an arms race on each side of the 38th Parallel.

Despite these two long-standing stalemates, the strategic position continues to change. China, idealogically estranged from its former Soviet ally, moves cautiously towards the West and seeks large arms purchases to replace obsolete equipment dating from before 1960. Having recently suffered a humiliating encounter with Vietnam, the Sino-Communists would appear to be in no mood to attack Taiwan, which is well armed with American arms and equipment.

Diplomatic recognition of China by the United States on 1 January 1979 has undoubtedly contributed to world peace, but the corresponding break with Taiwan left the Nationalists with few friends. A transitionary period of five years will follow the curtailment of the Taiwan-US military treaty in January 1980, during which time sufficient armament to dissuade a Communist invasion will be assured.

In South Korea, however, US force reductions were halted in February 1979, with the realisation that the North's army had been greatly increased.

Despite this apparently aggressive move, the Soviet position in North Korea is uncharacteristically restrained, and the air force has received none of the advanced types such as MiG-23/27s, MiG-25s, Su-7s or Tu-22s normally issued to other friends of the USSR. This omission has not gone unnoticed in Washington, for the South has yet to receive approval for the F-16s and A-10s it has requested.

Japan, too, relies heavily on the United States for aircraft. The majority of these are licence-built in this heavily-industrialised nation, but a strong left-wing influence in the Diet precludes armament deals with South Korea. Despite outward signs of buoyancy, the Japanese economy is not as robust as might at first be indicated by car and transistor radio sales, and purchases of new aircraft are being delayed through financial stringency and massive cost inflation.

As China cautiously emerges into the light of a slightly more liberal regime following almost three decades of doctrinaire Communism under the awe-inspiring figure of Mao Tse-Tung, it finds itself in the frustrating predicament of Gulliver as he awoke on the shores of Lilliput. By reason of its immense size, China should be in a more commanding position in the world, but tied-down by obsolescing technology and near-subsistence agriculture, it is denied both the means to produce, or the finance to buy, the modern equipment so urgently needed to put it on a par with other world-leaders.

Nowhere is this deficiency more keenly felt than in the sphere of aviation, where the most up-to-date technology is demanded to produce aircraft for both civil and military roles. Copy and adaptation

Left: as with the Soviet airline Aeroflot, China's civil carrier CAAC acts as a backup to the military transport force. A Boeing 707 is pictured.

Below left: 18 Hawker Siddeley Tridents were acquired for government transportation duties, representing one of the few Western types in the Chinese aviation inventory.

Below: despite the design's obsolescence, the MiG-19 continues to provide the backbone of Chinese fighter strength as the indigenous Shenyang F-6.

Bottom: the F-6bis carries avionics in a pointed nose, but is otherwise broadly similar to its precursor. It has been allocated the code-name Fantan-A by the West

have been the twin watchwords of Chinese aviation, and the most advanced indigenously-designed aircraft yet to appear is the Pinkiang Y-11, a light piston-engined utility transport and crop sprayer, similar in size and concept to the Australian GAF Nomad currently in production.

Close relationships with the Soviet Union after the communist take-over in 1949 brought aid and industrialisation initially to China, although doctrinaire differences in the interpretation of Marxist-Leninism soon resulted in a parting of the ways. When new equipment was required for the Air Force of the People's Liberation Army (Chung-kuo Shen Min Taie-Fang-Tsun Pu-tai), it came directly from Soviet production, or from drawings supplied

to Chinese industry. After the two nations severed relations, China found itself with little or no design or research facilities on which to base the next generation of aircraft.

Present-day assembly plants at Shenyang (formerly Mukden) and Pinkiang (formerly Harbin) were originally established during the period of

Top: the ubiquitous Lockheed F-104 Starfighter equips three squadrons of Taiwan's 5 Fighter Wing. Above: standard Taiwanese fighter type is the Northrop F-5, with over 200 ordered or delivered. Below: the T-CH-1 trainer combines the Lycoming T53 turboprop and the T-28 Trojan's airframe. Opposite: a Japanese F-4 Phantom at Chitose

Left: McDonnell Douglas F-4EJ Phantoms complement F-104J Starfighters in the JASDF's interceptor force, with five squadrons operating the type in 1979. Similar to the USAF's F-4E, the F-4EJ is built under licence in Japan by Mitsubishi. The pictured Phantoms serve with No 302 Squadron. Below: the F-4EJ first entered service in 1973 with No 301 Squadron

Japanese occupation in the 1930s. These modern and well-equipped facilities were dismantled during the Soviet take-over of Manchuria, but the installation of a Sino-Communist government and the establishment of close ties with Moscow following the return of Manchuria to China, found the Soviets refurbishing these very same factories for the fledgling Chinese aviation industry.

Shenyang played a vital role in the Korean War, assembling and repairing MiG-15s supplied by the USSR, and simultaneously preparing for licence-production of other Soviet types. With work proceeding at the highest priority, production rights for the relatively simple Yakovlev Yak-18 primary trainer and its radial engine were granted in November 1952, the first production aircraft, designated Type 3, flying in July 1954. Engine production was based on a longer time-scale, and not until 1956 did Chinese units replace those imported from Russia.

As the production base expanded, with the extension of Shenyang and Pinkiang and construction of further plants at Chungking and Sian, China looked towards greater self-sufficiency. Licences were acquired in October 1954 for the MiG-17F Fresco C fighter, Il-28 Beagle light bomber, MiG-15UTI Midget trainer, Mil Mi-4 Hound helicopter and Antonov An-2 Colt utility biplane together with their associated power plants. Shenyang delivered its first examples of the MiG-17F (known as the F-4) and MiG-15 (F-2) in the autumn of 1956, following with the MiG-17PF (F-5) limited all-weather interceptor. The MiGs were eagerly welcomed by the AFPLA, and by mid-1959, some 25 per month were joining interceptor regiments. Light bomber units were issued with the Il-28 (B-5), while the Mi-4 (H-5) and An-2 (C-5) took up both civil and military roles from early 1958 onwards.

Seeking even more advanced weaponry, including a strategic bomber force, to replace the small number of Tu-4 Bulls (a copy of the B-29 Superfortress) received from the Soviet Union, China initiated plans for the manufacture of the Tupolev Tu-16 Badger and the new MiG-19 Farmer during 1958. The programme was less than complete two years later when ideological differences precipitated

an irrevocable division of the two Communist camps. Over 1,000 Soviet technicians and advisers packed their bags and their blueprints and left.

Repercussions for China and its aspirations in both the long and short terms were catastrophic, although it is conceivable that had the relationship continued, the West would now be faced with a much stronger Soviet bloc incorporating well-equipped Chinese forces. Presented with this *fait accompli*, China had little alternative than to persevere with its original plans, although it would now have to duplicate the jigs and tools awaiting delivery in Russia, and supply items of equipment such as hydraulics and ejection seats previously obtained from Soviet sources.

Through a prodigious effort of adaptation and improvisation, the first MiG-19S (F-6) was delivered to the AFPLA in mid-1962, joining earlier examples supplied by the USSR. First Tu-16s (B-6s) were not to arrive until 1968, the production line closing in 1973 after some 60 had been delivered, while Il-28 assembly also terminated in the same year after a run of about 600 aircraft.

Even as the first F-6s were being test-flown, the AFPLA was looking towards the time when a replacement would be required, a task made the more complex by China's acute lack of friends. Europe's historic neutrals, the Swedes and Swiss, offered Saab Drakens and FFA P-16s, and France put forward the Mirage III, but omnipresent foreign-exchange shortages as well as political problems precluded any firm undertakings. Although China was now installing research facilities and training engineering students, the fruits of this investment would take too long to ripen, and only one solution presented itself. A small number of MiG-21F Fishbed-Cs had been received from the Soviet Union prior to the break in relations, and these would be copied down to the last rivet as the next generation fighter.

Flown for the first time in late-1964 and designated F-7, the MiG-21 copy showed a disappointing performance. Having staked all on the F-7, China had no more cards to play, and thus the F-6 was put back into full-scale production, remaining so in 1979. Modified variants have been produced, and output is continuing at the rate of at least 60 aircraft per month while a completely new fighter is now on the drawing board at Shenyang. For the moment, however, the AFPLA continues to make the best use it can out of obsolescent equipment.

Avowed aim of China's present rulers is complete modernisation of the armed forces by the year 2005, and interest has been expressed in a wide range of aircraft and avionics from Britain, France and Japan. Chinese military posture is essentially defensive and tactical, although some changes may accompany the development of nuclear weapons and a new political stance stemming from recognition by the United Nations as a major power.

Nuclear testing began at Lop Nor in the Sinkiang Desert during October 1964, and China is estimated to have stockpiled at least 300 nuclear warheads. Strategic missile delivery systems are, however, restricted to 30 or 40 CSS-1 medium-range ballistic missiles with a 1,600 km (1,000 mile) range and a similar number of CSS-2 intermediate-range ballistic missiles with up to 4,000 km (2,500 mile) capability. Development of the CSS-3 limited-range intercontinental ballistic missile began in 1976, although this has yet to be operationally deployed, while a full-range CSS-4 is in prospect for the early 1980s to give China a comparatively modest ICBM force, comparable in numbers with that of France. Emulating the other nuclear powers, a submarine-launched missile is also believed to be under development, but the single People's Navy submarine fitted with launch tubes is presently known to be unarmed.

Remaining nuclear-delivery forces are restricted to the 50 or so Tu-16 Badgers, which are capable of covering a sizeable portion of the Soviet Union and Asia from their home bases, while MiG-19/F-6s are available for tactical nuclear delivery. Continued testing, involving at least 23 explosions of fission and fusion devices, indicates the Chinese ambition to become a significant nuclear power, given sufficient delivery systems.

Growing reliance on the nuclear arsenal as protection against the USSR and Western powers has seen a reduction in spending on conventional weapons and aircraft, although it may be argued that there are few worthwhile indigenous types on which finance could logically be expended. Since 1976, however, arms spending has again risen. In an unprecedented revelation of its finances, China has made public the figure of 20,230 million Yuan (about £6,000 million) for 1979 defence spending, but this falls far short of the CIA's estimates of around £18,000 million per year. This may be explained by the fact that it is not uncommon for Communist countries to quote only the actual cost of their hardware purchases, while research and development, pay and pensions, and the like, are attributed to other categories.

With an estimated strength of 4,500 aircraft, the AFPLA is claimed to be the third largest in the world, and the only one at present in East Asia with nuclear capability. Personnel total around 400,000 including over 100,000 in the separate air-defence organisation and some 15,000 pilots. Apart from strategic bombers, the remaining nuclear forces are controlled by the ground forces,

through the 2nd Artillery Corps, but China also has a sizeable naval air arm with 450–500 land-based aircraft and 30,000 personnel.

Air defence duties are assigned to a force of some 900 or more Shenyang F-6s, equivalent to the PF and PM variants of the MiG-19, backed by 200 older F-4s, the 50–60 remaining F-7s and a small number of the new F-6bis Fantan A interceptors. Causing much confusion to Western observers, the F-6bis was originally identified (erroneously as it later transpired) as the F-9, which was duly allocated the reporting name Fantan-A. Representing a major re-design of the basic MiG-19, the aircraft is fitted with a pointed nose, and air-intakes re-located in bifurcated configuration astride the cockpit. With justification, it was assumed the drastic modification was to permit the installation of updated fire-control radar, as in times past similar surgery transformed the Republic F-84F Thunderstreak into the camera-nosed RF-84F Thunderflash reconnaissance fighter.

Far left: a Lockheed F-104J Starfighter of No 203 Squadron, Japanese Air Self-Defence Force crosses the runway threshold at Chitose in late 1976. Together with the Phantom-equipped No 302 Squadron, the unit comprises 2 Air Wing. Production in Japan of the McDonnell Douglas F-15 Eagle will permit the Starfighter's withdrawal in the early and mid-1980s.
Below left: 3 Air Wing is composed of two squadrons flying the Mitsubishi F-1 fighter, together with a support flight. Aircraft of No 3 Squadron (pictured) and No 8 Squadron are based at Misawa, having entered Japanese service in April 1978. A development of the T-2 trainer, the F-1 had an order book of 69 aircraft in 1979. No 3 Squadron was the first unit to form with the type in March 1978

It now seems, however, that the F-6bis is employed as a strike-fighter. As it would be unreasonable to conclude the re-design was merely undertaken to further the belief in Chinese inscrutability, it must be assumed that the intended avionics have not yet been perfected, or that they have been tried and found wanting. This apparent symptom of technological poverty is clearly expressed by the Chinese themselves through the attempts which have been made to acquire modern equipment for study. Whilst defection in Communist terms, normally implies a one-way movement, China has offered considerable rewards to any Taiwanese defecting with military equipment. Gold bars to the value of $5 million await the crew of the first Chinese Nationalist warship to sail into a mainland harbour, and a publicised scale of proportionally smaller inducements invite the presentation of lesser equipment.

More fruitful has been the recent liaison with Egypt, after it, too, suffered disillusionment with the Soviet system. Chinese technicians have assisted in the repair and overhaul of grounded Egyptian MiGs, following the withdrawal of Russian spares supplies. In return for this, and a generous gift of 40 Shenyang F-6s, China has received single examples of latest Soviet weaponry, including a MiG-23, a Sagger anti-tank missile and armoured vehicles, all of which have no doubt been subjected to searching examination and assessment of their individual production prospects.

After a protracted gestation period, the Chinese aviation industry is at last showing signs of delivering the AFPLA its first indigenously-designed high-performance fighter, albeit with a transplanted British heart. Late in 1975 Rolls-Royce undertook to supply China with 50 Spey 202M afterburning turbofans (similar to those powering the Buccaneer and F-4K/M Phantom) together with facilities for the establishment of an engine plant, and it subsequently became known that the Spey would power a new fighter designated F-12. Trial running of the first Chinese Spey, made from British castings machined to final shape after delivery, is scheduled to take place this year. Although it has not been officially admitted that the Mach 2.4 F-12 will be fitted with Speys, first flight of the new warplane will probably take place in 1980–81, for eventual issue to all air defence units.

Several types of Western aircraft have been, or are being, considered for the AFPLA, but as yet, no firm contracts have been placed, due to the ever-present foreign exchange shortage. China has expressed an interest in the Mirage 50, Mirage 2000, P-3 Orion, Gazelle, Transall and Hercules together with Crotale, Exocet, HOT and Milan missiles, although it is the Harrier which has come closest to gaining the all-important signature. Fiercely opposed by the Soviet Union, the Harrier deal, if it comes to fruition, could involve up to 40 aircraft from British production and further quantities assembled in China, for the equipment of some half-dozen squadrons or more. Planned Harrier deployment will be along the Soviet borders, and around China's air-fields. While the Harrier has no strategic capability, obvious Soviet discomfiture at the prospect of a Chinese order highlights the sensitive nature of these areas.

Present AFPLA organisation is associated with ten military air regions, and district headquarters situated at Chengtu, Fu-chow, Junan, Kuangchow, Kunming, Lanchow, Nanking, Peking, Shenyang and Wuhan. Largest operational unit is the air division, comprising three regiments, of either three or four squadrons, each with 12 aircraft. Operational conversion is undertaken at regimental level and air regiments are allocated trainer versions of operational types.

Each division has a tactical bomber unit, 12 regiments operating some 400 Il-28 Beagles. All military forces are geared for confrontation with Soviet divisions along the Chinese border, although at least six air divisions are installed in the coastal belt covered by the military regions of Fu-chow, Nanking and Tainan, facing Taiwan under the control of Eastern Air Region. Air defence of Peking is the responsibility of 38 Air Division with over 100 SA-2 Guideline SAMs backed by 10,000 AA guns of varying calibres.

Western Air Region, centred on Nagchuka, in what was formerly Tibet, flanks India and Pakistan, while North-West Air Region includes the nuclear test base at Lop Nor and four regiments of Tu-16 bombers. Northern Air Region controls operational

Above: the F-86F Sabre was one of the JASDF's main combat types throughout the 1960s. It is still operated by the JASDF's No 6 Squadron, 8 Air Wing, based at Tsuiki, the Air Defense Command Flight (Headquarters Squadron) at Iruma and as an advanced trainer by No 1 Squadron, 1 Air Wing. It was not until the early 1970s that the F-86 was substantially replaced by the locally-built and imported McDonnell Douglas F-4EJ. The remaining front-line F-86s are scheduled to be replaced by the Mitsubishi F-1 by late 1980

units along the Mongolian and Soviet borders, but North East Air Region in Manchuria incorporates large training bases with Yak-18s (BT-5s) and the indigenous BT-6 development, which entered service in 1977. Southern Air Region controls AFPLA units along Vietnamese and Laotian borders, and having provided advisers and equipment for Communist elements in the former, prior to Hanoi's transfer of allegiance to Moscow, SE Region recently found itself involved in the punitive expedition into Vietnam early in 1979.

Chinese veneration of its senior citizens was clearly reflected in the AFPLA's choice of a 73 year old overall commander, and a tactical commander only four years younger. Beginning on 17 February, the 16-day invasion, supported by a force of 1,000 F-6s stationed in the immediate area, soon proved to be a classic case of the pupil beating the master. Poor communications prevented the F-6s from giving their full weight to the task; transmissions from the infantry's walkie-talkie radios refused to disobey the basic laws of short-wave electromagnetic radiation and curve themselves round the hilly terrain; while tanks attempted to communicate with each other by flag signals.

The severe mauling inflicted by the Vietnamese left China in no doubt as to the outcome of any future confrontation and has resulted in the formulation of a crash '1,000 Day' re-equipment programme to remedy some of the worst shortcomings. For this, China may relax some of the stipulations attached to other projects under negotiation with the West, which involve provision for industrial offsets to ease the balance of payments problem. This condition has in the past presented a major stumbling block to military programmes, for a while it is normal practice in the West for a nation to produce components for the F-16 in return for its purchase of the type, Chinese aircraft are virtually hand-made with resultant spares-interchangeability problems.

A few direct purchases from Europe have, nevertheless, found their way into the AFPLA inventory, in the form of 13 Aérospatiale Super Frelons to augment the helicopter fleet and 18 Tridents transferred from civil aviation. Not as yet tried in combat, the aviation forces of the People's Navy,

appear equally in need of modernisation. Tasked with shore-based defence of naval bases, maritime patrol and anti-shipping operations, the navy additionally contributes about 300 MiG-17s and MiG-19s to the Chinese air defence system. Strike forces comprise up to 100 Il-28 torpedo bombers, together with Beriev Be-6 Madge flying boats for patrol and 50 Mi-4 helicopters for ASW and SAR, backed by transport and communications types.

Anti-shipping missiles are restricted to SS-N-2 Styx installations in 23 destroyers and frigates and some 150 lighter vessels. Interest in the Lockheed Orion may prove difficult to translate into a firm order through the Carter administration's reluctance to provide China with any military equipment, defensive or otherwise. Even should there be a change of heart, it is unlikely that the latest mark of Orion will be offered.

Beset with financial difficulties, China will thus be torn between producing its own military hardware on a long timescale, or buying the best which the West is willing to provide to patch the mantle of its defences. Until the F-12 fighter emerges as a production item and is supplied to Chinese frontline units in quantity, the long-term effects of the Soviet departure will continue to be apparent.

While not entirely unexpected, the United States' decision to give diplomatic recognition to Communist China as the only China came as a severe blow to Taiwan. In the wake of Mao Tse-Tung's take-over of the mainland in 1949, Nationalist China was established on the island of Formosa (Taiwan), and between 1950 and 1969 received more than $3,000 million in military assistance from the USA. Originally committed to the 'liberation' of the mainland through military action, Taiwan has been forced by US pressure to modify its ambitions in the light of the *rapprochement* between America and the People's Republic, and a corresponding reduction in offensive-equipment deliveries. Taiwan has thus diversified its sources of supply, and established the Aero Industry Development Centre at Taichung for the licence assembly of US aircraft.

On 1 January 1979, the United States formally transferred recognition to Peking and simultaneously gave one year's notice of concluding the

Right: a formation of F-86E Sabres of the JASDF's 'Blue Impulse' aerobatics team pictured in November 1976 over Tsuiki air base, the home of 8 Air Wing. From the JASDF's inception the F-86 was envisaged as its main combat type, and following initial deliveries from the United States, Mitsubishi undertook the licence production of some 300 F-86Fs. About 120 F-86D all-weather interceptors were supplied by the United States from 1957. The JASDF's first fighter unit, 2 Fighter Wing, was activated in 1956 with F-86's and T-33s. Left: a 'Blue Impulse' F-86F

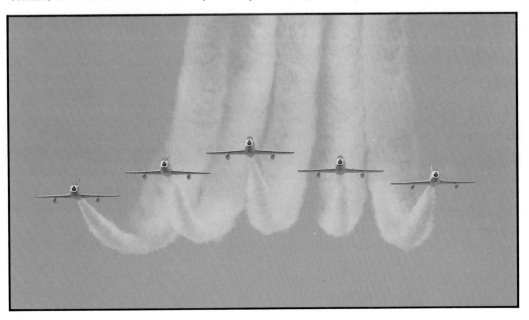

mutual defence treaty in force since 1954. Taiwan will be permitted to obtain sufficient military hardware to dissuade a Communist invasion during the next five years, after which the supply position will be reviewed. Purchases of equipment from the United States, provided it is defensive in nature, will be allowed, although Taiwan has been demoted in status from favoured credit-client, to a cash-and-carry customer.

Nationalist China's air force (Chung-kuo Kung Chuan) formed under American tutelage through a USAF Military Assistance Advisory Group which advised on air tactics, training, doctrine and formulation of long-term planning. Not unnaturally, American aircraft predominated, and by the early 1970s more than 750 combat types had been supplied under the Military Assistance Programme. Until recently, air defence of Taiwan and Pescadores was the joint responsibility of the CNAF and Air Task Force 13 of the US Pacific Air Force (13th AF).

Although Taiwan's military stance is now purely defensive, the armed forces are efficient and held at a high state of readiness. Defence spending for 1978 amounted to no less than 48% of the national budget, at some £800 million, and with cessation of US aid, will in all probability exceed that sum by a considerable margin in the years to come. CNAF strength includes over 300 combat aircraft and 70,000 personnel.

Operational forces are spearheaded by five interceptor and fighter-bomber wings, plus a single squadron for tactical and maritime reconnaissance. Each wing is subdivided into three squadrons of between 18 and 25 aircraft, dependent on function.

Taiwan is now standardising on the Northrop F-5, having received 92 F-5As and 23 F-5B trainers from the US for the re-equipment of 1st Fighter Wing. Numbers have been reduced through transfers to South Vietnam before the 1975 collapse, the wing now having three squadrons, each of 15 F-5As and three F-5Bs.

Sabres are being progressively withdrawn from Nos 2 and 3 Wings following deliveries of the latest batches of F-5Es. Taiwan received an eventual total of 320 Sabres, although by the mid-1970s, strength had been reduced to six squadrons of 25 aircraft each. Direct deliveries of 60 F-5Es began the re-equipment of two squadrons of each wing, while the Taichung factory is assembling further single-seat and two-seat versions, sufficient for all six Sabre units.

Consistent with its anti-invasion policy, the United States authorised the co-production of a further 39 F-5Es and nine F-5Fs beyond the 187 F-5Es and 21 F-5Fs previously approved, a few days before the discontinuation of formal recognition. These will be delivered by the end of 1983 for the conversion of a further two squadrons, although the CNAF had been hoping for more potent aircraft from several sources. All of these, however, failed to materialise.

During 1978, unconfirmed reports mentioned Taiwanese interest in 48 IAI Kfirs from Israel for two 20-aircraft squadrons, plus attrition replacements. Sales of the Kfir are effectively controlled by American approval for the export of its J79 engine, and it rapidly became apparent that no J79s would be permitted to go to Taiwan. This

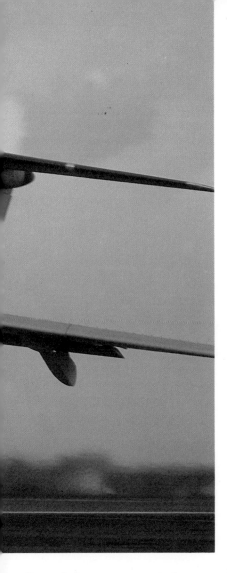

disclosure was followed by denials that the CNAF had ever considered the Kfir and an announcement to the effect that 60 Phantoms would be obtained for the same price.

At the same time, the Nationalists were pressing for supply of the Northrop F-5G, a version powered by a single General Electric F404 in place of two J85s, together with the advanced Sparrow air-to-air missile to complement stocks of Sidewinders. The F-5G foundered through Congressional opposition and a lack of further customers, as Taiwan required only 30 of the 300 production-run needed to break even on the project, after which the Phantom request was also turned-down. Despite a subsequent US statement professing no objection to the Kfir in principle, the supplementary F-5E/F order was ultimately agreed.

The problem of obtaining a suitable replacement for two wings of interceptors continues to cause anxiety in Taipei. Super Sabres and Starfighters are rapidly becoming obsolete, but the F-16 which is replacing these aircraft in Europe, together with the F-18 and Phantom, are specifically barred to Taiwan by White House decree.

From 80 Super Sabres supplied to Taiwan in 1960, heavy attrition–to which this aircraft appears to have proved particularly prone–resulted in 4 Fighter Wing having only two squadrons of 18 aircraft each by 1970. In July of that year, 34 F-100As from surplus USAF stocks permitted the re-formation of a third squadron, all aircraft having been refurbished to F-100D standard. Replacement with Sidewinder-armed F-5Es is expected soon.

Once NATO's standard interceptor and fighter-bomber, the Starfighter is now nearing the end of its career in Europe, but appears destined to continue in CNAF service for some years. Formed with 24 ex-USAF F-104As and 18 new F-104Gs, 5 Fighter Wing later expanded to three squadrons and standardised on the F-104G during 1964–66 after receiving additional deliveries of 26 aircraft, plus six TF-104G trainers and 21 RF-104G reconnaissance variants. The last-mentioned replaced RF-101C Voodoos as the sole air force tactical reconnaissance element, while later deliveries have involved trainer versions, comprising two TF-104Gs in 1969 and six surplus F-104Ds in 1976. About 60 Starfighters of all types remain in service.

Equipment of the military transport fleet is similarly beginning to show its age. Wartime-vintage types comprise 35 Dakotas and 25 C-46 Commandos, supported by 30 C-119G Flying Boxcars and ten C-123 Providers. Commandos and Boxcars have been employed on clandestine missions involving shallow penetration of mainland airspace for dropping agents, but their contribution towards the overthrow of international Communism remains minimal. In this connection, however, it is worth noting that the wreckage of three Lockheed U-2s in CNAF markings was exhibited at an open-air display in Peking some years ago. Modern fixed-wing transport equipment is restricted to a single Presidential Boeing 720B.

A substantial helicopter force stems from licence production of the UH-1H Iroquois, of which 118 were assembled at Taichung for all three armed

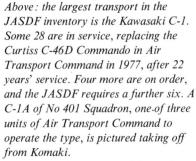

Above: the largest transport in the JASDF inventory is the Kawasaki C-1. Some 28 are in service, replacing the Curtiss C-46D Commando in Air Transport Command in 1977, after 22 years' service. Four more are on order, and the JASDF requires a further six. A C-1A of No 401 Squadron, one of three units of Air Transport Command to operate the type, is pictured taking off from Komaki.
Right: a NAMC YS-11 of the Flight Check Squadron, based at Iruma. The YS-11 is operated in the personnel, freight and logistic transport roles by Air Transport Command JASDF, as an instrument trainer by No 205 Squadron JMSDF, and as an ECM trainer.
Below right: a Fuji T-1B of the Air Proving Wing JASDF, based at Gifu

services. Garrisons on the offshore islands are equipped with small detachments of Iroquois and Grumman HU-16A Albatross amphibians. Programmed replacement type for the older transports is the indigenous C-2, resembling a Provider with fin-mounted tailplane. Aimed at both civil and military markets, the aircraft was flown in prototype form as the XC-2 in May 1979, and may be quickly converted from passenger to cargo configuration.

A second, more ambitious project from Taichung concerns a new trainer and attack aircraft, derived from the Northrop T-38 Talon. Known as the T-3 and A-3 for its two versions, or alternatively under the joint designation AT-3, the aircraft is expected to begin trials in 1980. Apparently, the AT-3 will be a re-engined T-38, as the latter's J85 engine has not been made available for licence production in Taiwan. Source of the AT-3's new engine has not yet been revealed.

This will not be the first time the Aero Industry Development Centre has fitted a new engine to an existing design. A turboprop basic trainer has recently been produced by installation of a Lycoming T53 (manufactured under licence at Kang Shan for the UH-1 programme) in a North American T-28 airframe, 30 examples being on order as the AIDC T-CH-1B. It would seem logical for the AT-3 to assume the strike roles of the F-5E, releasing the latter for air defence.

Pilot training for the CNAF begins on 55 Pazmany PL-1B two-seat light aircraft, built under licence, following by a basic course on the T-28. Advanced instruction on 30 T-38s then precedes operational conversion to the designated aircraft type via TF-104Gs, F-100Fs or F-5Bs attached to fighter and fighter-bomber wings.

The Chinese Nationalist Army has an expanding aviation element, based mainly on the bulk of Iroquois deliveries between 1970 and 1975, but also including seven Sikorsky CH-34s and two Kawasaki KH-4s. Air defence is also partly an army responsibility through equipment with two battalions of Nike-Hercules SAMs and a third operating 24 MIM-23B Hawks. A fourth Hawk unit is in prospect, although earlier attempts to obtain the Rapier were defeated by a British government veto.

Naval forces have been more successful in buying the Sea Chaparral system for patrol boats (although the supply of Harpoons was blocked by the US government) and the IAI Gabriel ship-to-ship missile ordered from Israel as a Harpoon alternative, under the local name of Hsiung Feng (Male Bee). Airborne maritime reconnaissance and anti-submarine warfare are delegated to nine S-2A Trackers, supported by a few HU-16s for SAR. A recent innovation has been the order for 12 Hughes 500MD/ASW helicopters for shipboard operation, deliveries of which began in mid-1979.

In common with Germany, its former Axis partner, Japan was prohibited from forming military forces until the mid-1950s. It was not, therefore, until July 1954 that the Japanese Air Self-Defence Force was established with American equipment and guidance, remaining since that time purely defensive in nature. Presently in the middle of its Fifth Build-up Programme, extending from 1977 to 1982, the JASDF currently comprises 14 operational fighter squadrons in four regional commands with 420 first-line aircraft and 460 trainer, transport and helicopter types, supported by a 1979 air defence budget of 482,653 million yen. While the majority of combat types are of American origin, licence-production has accounted for the greater proportion of Starfighters, Phantoms and Sabres in the current inventory, and Mitsubishi is shortly to begin deliveries of the F-15 Eagle to update air defence capability still further.

The JASDF, or Koku Jiei Tai, is administered by Air Defence Command (Koku Sohtai) from headquarters at Fuchu, alongside those of the USAF 5th Air Force, controlling SAM groups as well manned interceptors and associated radar sites. Five Starfighter squadrons operate 167 F-104J versions from 210 originally supplied, together with 17 of the 20 F-104DJ trainers. Replacement by the

No 21 Squadron JASDF took delivery of its first Mitsubishi T-2 advanced jet trainer in March 1976, since which time the type has assumed an important role in the pilot training syllabus. The T-2 flew initially in July 1971 and was intended to replace the North American F-86F Sabre with the two squadrons of 4 Air Wing. Similar both in mission and appearance to the Sepecat Jaguar, the T-2 likewise provided the basis for a strike fighter in the Mitsubishi F-1

F-15 has been delayed, and the proposed order for 123 reduced to 100, of which 12 will be two-seat F-15Ds and the remainder F-15Cs.

Two Eagles are due to be delivered from McDonnell Douglas in July 1980, followed by four in the Spring of 1981 and two in April 1982, after which indigenous production will allow the gradual disbandment of remaining Starfighter units over the next four years. To alleviate the shortage of fighters as a result of the Eagle delays, a further 12 F-4EJ phantoms are to be added to the present production line bringing the total on order to 140, including two supplied from the United States. Phantoms are armed with AIM-4D Falcon and AIM-9F Sparrow III air-to-air missiles, to be superseded by the Mitsubishi AAM-2 which is, of course, indigenously produced

Japan has had its own programme for a support

fighter type, resulting in the Mitsubishi F-1 development of the T-2 trainer, unkindly referred to as the 'Japanese Jaguar' through its remarkable similarity to the Anglo-French strike aircraft. The F-1 entered service with No 3 Squadron on 1 April 1978, and the total of 69 on order or projected may be increased to 86 given sufficient funding. F-1s will be ultimately equipped with the indigenous Mitsubishi ASM-1 missile, now undergoing trials, for anti-shipping strike.

Of the four operational area commands, Northern Air Defence Force, headquartered at Misawa (Hokubu Koku Homen-tai) comprises 2 Air Wing (Koku-dan) at Chitose with 203 Squadron (Hiko-tai) equipped with Starfighters and 302 Squadron with Phantoms. At Misawa, 3 Air Wing has two squadrons (Nos 3 and 8) operating F-1s, aided by a support flight of T-33s and T-34 Mentors. Northern

forces are completed by 3 Air Defence Missile Group at Chitose with three units of Nike J SAMs and the nine radar sites of Northern Aircraft Control & Warning Wing (AC & WW) at Misawa.

Central Air Defence Force, Iruma, (Chubu Koku Homen-tai) has a similar complement, but increased missile forces, comprising 1 Air Defence Missile Group (ADMG), Iruma, with four Nike squadrons and 4 ADMG, Gifu, with a further three, plus eight radar sites of the Central AC & WW. Interceptor squadrons are operated by 6 Air Wing, Komatsu (205 Squadron–Starfighter and 303 Squadron–Phantom) and 7 Air Wing, Hyakuri (301 and 305 Squadrons–Phantom).

Western Air Defence Force (Seibu Homen-tai) at Kasuga has 5 Air Wing (202 Squadron–Starfighter and 204 Squadron similarly equipped) based at Nyutabaru; 8 Air Wing, Tsuiki, (6 Squadron –F-86F Sabre and 304 Squadron–Phantom); 2 ADMG, Kasuga with three Nike J units; seven radar sites of Western AC & WW, headquartered at Kasuga, and a support flight of T-33s and T-34s at the same base.

Following the Japanese return to Okinawa in 1972, South-West Composite Air Wing (Nansei Koku Kansei-dan) formed two years later and now comprises 83 Air Unit with a single squadron (No 207) of Starfighters, three Nike J units of 5 ADMG and four radar sites of South-West AC & WW, all stationed at Naha.

Although within the area of Central ADF, 501 Squadron at Hyakuri, equipped with 14 RF-4EJ Phantoms for tactical reconnaissance, remains an autonomous unit. In the same region come the similarly independent Air Defense Command Flight with F-86s and T-33s at Iruma and the associated Electronic Warfare Training Unit (Koku Sohtai Denshi Kunren-tai) with three specially converted NAMC YS-11s.

Logistic support for operational units is administered by Air Transport Command (Yusoh Koku-dan) at Miho. Three attached Air Transport Units each have one squadron equipped with a mixed complement of Kawasaki C-1s and YS-11s. No 1 ATU is assigned 401 Squadron at Komaki; No 2 ATU–402 Squadron, Iruma; and No 3 ATU–403 Squadron, Miho. The ASDF has received 28 C-1As, the final four with additional fuel capacity, and has hopes of obtaining a further ten. Three have been granted in the last defence budget and a further one in Fiscal 1979, but financial stringency is now being felt even in the apparently prosperous Japan. From 65 aircraft of all types requested for 1979 funding, only 36 were approved.

Pilot training within Air Training Command, headquartered at Hamamatsu, begins with 40 hours' basic instruction on T-34 Mentors of 12 Air Training Wing at Hofu, followed by an advanced course of 90 hours on Fuji T-1A/Bs in 13 Air Training Wing, Ashiya. After 120 hours with 1 Air Wing at Hamamatsu flying T-33s with 33 and 35 Squadrons or the Sabres of No 1 Squadron, operational conversion is completed on Mitsubishi T-2s of Nos 21 and 22 Squadrons, 4 Air Wing.

Deliveries of T-2s to No 21 Squadron began in March 1976, and with the conversion of 11 Air Training Wing from T-34s to the first of 60 Fuji T-3s from March 1978, training schedules are currently being revised. In future, pilot training will consist of 70 hours on the T-3, 70 hours T-1A/B, 100 hours T-33A and 140 hours T-2A. A turbofan successor to the T-1 and T-33 is being sought through the MT-X programme, with production of the selected type to begin in 1987.

Above: despite its considerable age, the Lockheed T-33 continues to perform a useful training function with the JASDF.
Top right: the turboprop-powered Kawasaki P-2J Neptune will remain in the Maritime Self-Defence Force inventory until finally replaced by the Orion in the early 1990s.
Above right: another long-serving JMSDF type is the Grumman S-2 Tracker anti-submarine hunter/killer.
Right: pictured beached at their Iwakuni base, the Shin Meiwa flying boats of No 31 Squadron are among the few exclusively water-borne aircraft types still operational.
Below right: the Sikorsky S-61 Sea King was licence-manufactured for the Japanese navy by Mitsubishi, 81 examples being ordered

Three further flying units complete the ASDF organisation. SAR facilities are provided by nine detachments of the Air Rescue Wing (Koku Kyunan-dan) headquartered at Iruma and equipped with 29 Kawasaki-Vertol V-107 and seven Sikorsky S-62 helicopters together with 21 Mitsubishi MU-2Es and a few T-34 Mentors. Also at Iruma is the Flight Check Squadron with YS-11s, T-33s and MU-2Js for inspection and calibration of radionavaids. Three T-2 trainers, modified to ET-2 standard are expected to join the unit in late 1979 or early 1980.

At Gifu is the Air Proving Wing (Jikken Kokutai) tasked with evaluation of prototypes and trials aircraft. Present equipment includes one or two examples each of the Phantom, C-1, Starfighter, F-1, T-2, T-1, T-33 and T-34. In addition, Maintenance Training Command has five schools for the instruction of ground personnel.

One of the less gratifying aspects of the well-publicised defection of a Soviet pilot and his MiG-25 to Japan in September 1976 was the performance of the ASDF air defence system. When the MiG approached at low altitude over the sea, it was only momentarily picked-up on the Air Defence Command radar screens and two Phantoms were scrambled to intercept. These completely failed to make contact and were forced to return to base, while the MiG, finding it impossible to land at the airfield originally chosen, flew down the coast to Hakodate. As the Soviet pilot made practice approaches to the short runway, much to the delight of the local aviation enthusiasts, the ASDF was left to puzzle what had entered its airspace, and where it was at that time.

Clearly, such an event caused acute embarrassment to Air Defence Command, and steps were taken to provide an airborne early warning aircraft capable of locating low-flying aircraft. Having formulated a requirement for up to 15 Grumman E-2Cs, the ASDF was refused funding in 1978 and was forced to wait until the following year for the first four aircraft to be ordered. Planned deployment is detachments of three aircraft to each of five bases.

Japan is also considering a further increase in air transport capability by the purchase of further American types. A longer-range and greater capacity aircraft to complement the Kawasaki C-1A will be ordered in 1982/83 and may well result in the delivery of a dozen C-130 Hercules, although a stretched version of the C-1 is also in the running. The CH-X programme for a new cargo helicopter currently favours the Sikorsky CH-53, of which 30 would be required, while three DC-10s are to be ordered for a special VIP flight.

Of equal status to the more traditional maritime elements under control of the naval Self-Defence Fleet at Yokosuka, MSDF Air Command (Koku Shuh-dan) headquartered at Atsugi controls 11 operational squadrons together with support units and SAR flights. A second command, also directly subordinate to MSDF HQ, administers air training, while two district commands each have an attached air squadron. The MSDF presently comprises 140 operational aircraft, backed by 170 support types and helicopters, and is shortly to embark on a major re-equipment programme which will replace older models of Neptune with Lockheed Orions. Naval budget for 1979 is 454,004 million yen, while personnel in all branches of the service total 43,000.

Mainstay of the maritime patrol fleet is the Lockheed P-2 Neptune, which in various forms has served the MSDF since its earliest days. A dozen SP-2Hs are scheduled for replacement in 1981, and of these, four have been converted for target towing to replace S-2 Trackers used hitherto. Principal variant is, however, the P-2J Turbo-Neptune which entered production at Kawasaki's Nagoya plant in 1969. The last of 83 aircraft was completed in March 1979, concluding deliveries of a basic type which was first produced in 1944. Turbo-Neptunes will remain in MSDF service until 1992 when further Orions are likely to be obtained, but delays and postponements in military programmes will undoubtedly see some Neptunes flying on the type's fiftieth birthday.

After consideration of 13 maritime patrol aircraft, the P-3C was chosen in December 1977 to update submarine detection capability in response

1 Air Group, with No 3 Squadron (P-2J), No 14 Squadron (Tracker) and Rescue Squadron (S-62).

All-helicopter 21 Air Group at Tateyama is equipped with the SH-3 Sea King in Nos 101 and 121 Squadrons, plus the Komatsujima Air Squadron. Over 60 Sea Kings are presently in service, and an eventual total of 81 will be delivered.

One of the world's few flying boat units is 31 Air Group at Iwakuni, operating the Shin Meiwa PS-1 in No 31 Squadron, (together with a few S-2A-U Trackers) and the US-1 amphibian in No 71 Squadron. The 23rd and last PS-1 has been ordered in the 1979 defence budget, but four of the type have already been lost in accidents. While the PS-1 is employed on anti-submarine duties, three US-1s are used for SAR and a further four are on order.

Four independent squadrons of Air Command comprise the flight trials unit, No 51 Squadron at Shimofusa with P-2J, P-2H, Sea King, Mentor and PS-1 aircraft; No 61 Squadron at Atsugi, for transport tasks, equipped with YS-11Ms and S-2A-Us; No 111 Squadron at Shimofusa, flying

The Boeing KV-107-IIA is operated by the Japanese Air, Maritime and Ground Self Defence Forces. A KV-107-IIA of No 101 Squadron, 1 Composite Air Command based at Naha in JGSDF's Western Air Command is illustrated. Other JGSDF operators of the type are Western Air Command's Helicopter Squadron at Takayuhara, the Training Support Squadron at Akeno and the Kasumigaura School. Nos 1 and 2 Helicopter Units and 1 Helicopter Wing based at Kisarazu currently fly 58 in the transport role. No 111 Squadron JMSDF, based at Shimofusa, operates 11 KV-107-IIAs on minesweeping duties, while the JASDF's Air Rescue Wing has some 30

to recent advances in submarine technology which render the Neptune less effective. Between 1981 and 1987, 45 Orions will be handed over to the MSDF, with Kawasaki as prime contractor. First three aircraft will come from Lockheed, and the next six will be assembled from knocked-down components, prior to full-scale indigenous production of the type.

Little publicised is Japan's interest in the Harrier, which at present covers two aircraft for evaluation and a possible follow-up order for 30 to equip one of the three close-support squadrons scheduled for formation, as an alternative to an upgraded Mitsubishi F-1. The two trials aircraft were to have been funded under the 1979 programme, but have now been delayed until next year.

Present organisation of the MSDF Air Command comprises five Air Groups with attached squadrons and four independent units. At Kanoya, 1 Air Group has No 1 Squadron with P-2Js, No 11 Squadron with Trackers and a Rescue Squadron with Sikorsky S-62s. At Hachinoe, 2 Air Group consists of No 2 Squadron (P-2Js), No 4 Squadron with a mixed complement of P-2H and P-2J Neptunes, and an S-61A-equipped Rescue Squadron. Atsugi-based 4 Air Group duplicates

Vertol 107s for minesweeping; and Okinawa Air Squadron with P-2Js at Naha. Funding for two Sikorsky RD-53Ds for evaluation with a view to re-equipment of No 111 Squadron has been postponed until 1980, although the MSDF has a requirement for 12 Hercules for use as minelayers. Units attached directly to Self-Defence Fleet HQ, comprise Sea King flights of Omura Air Squadron (Sasebo District Command at Sasebo) and Ominato Air Squadron (Ominato District Command).

Air Training Command (Kychiku Koku-shudan) at Shimfusa controls six second-line squadrons with both training and combat equipment. After ground training with No 221 Squadron, basic pilot training is begun on 26 of 38 Fuji KM-2s on order for No 201 Squadron, Ozuki Air Training Group at Ozuki, together with about five remaining Mentors. For twin-conversion, students progress to the Tokushima Air Training Group (which additionally has an attached Rescue Squadron with S-62s) to fly on 28 Queen Airs and six King Airs, alongside trainees from the ASDF. Operational conversion is undertaken within Shimofusa Air Training Group.

The 180,000 strong army is supported by aviation elements comprising 66 fixed-wing aircraft and 341

helicopters for support and liaison, rather than offensive roles. However, six UH-1s in each of the five district commands are scheduled for equipment as anti-tank helicopters, probably armed with TOW missiles. Two Bell AH-1S Cobra gunship helicopters are being delivered for operational trials at Utsunomiya, the first arriving on 1 June 1979, and subject to satisfactory evaluation, a further 54 will be assembled by Fuji from 1981 onwards for three new squadrons. Plans are also in hand for the replacement of Vertol 107s with Chinooks from 1983, and two examples will be obtained for evaluation within the next one or two years.

In addition to an independent helicopter wing and Air Training School, GSDF elements are organised into five regional commands, each with an HQ air squadron (Homen Hiko-tai); a helicopter squadron (Homen Hërikoputa-tai); several divisional helicopter units (Hiko-tai) and groups of Hawk missiles for air defence. Northern Air Command (Hokubu Homen Koku-tai) at Okadama comprises Northern Air Squadron with LR-1s (MU-2s), Fuji LM-1s, OH-6Js and L-19s; Northern Helicopter Squadron with UH-1B/H Iroquois and four detached squadrons–No 2 Squadron, Asahigawa, and No 5 Squadron, Obihiro both with

Top: helicopters outnumber the fixed-wing contingent of the JGSDF by five to one. Although due for replacement, the Bell UH-1 Iroquois (pictured) soldiers on in the support and liaison role, assisted mainly by the Hughes OH-6.
Above: the venerable Cessna L-19/O-1 Bird Dog spotter monoplane also survives in numbers, being deployed alongside helicopters in each of the five regional air commands

OH-6s, UH-1s and L-19s and Nos 7 and 11 Squadrons, both at Okadama, flying the OH-6, UH-1 and H-13 (Bell 47), although the latter has an additional flight of L-19s. Air defence Missile Groups are 1 ADMG, Chitose and 4 ADMG, Nayoro.

In a similar organisation, North-Eastern Air Command (Tohuku Homen Koku-tai) at Kasuminome has similarly equipped Air and Helicopter Squadrons at the main base, together with two units of OH-6s and L-19s (6 Squadron, Jinmachi and 9 Squadron, Hachinoe), plus 5 ADMG's Hawks at Hachinoe. Eastern Air Command (Tobu Homen Koku-tai) at Tachikawa comprises the Air Squadron (LM-1 and OH-6J); Helicopter Squadron (UH-1); 1 Squadron, Tachikawa and No 12 Squadron, Utsunomiya (both OH-6J and L-19); and Hawks of 2 ADMG at Matsudo. This is duplicated by Central Air Command (Chubu Homen Koku-tai) at Yao with divisional units 3 Squadron, Yao, 10 Squadron, Akeno; 13 Squadron, Hofu; and 8 ADMG, Aonogahara, with the exception of LR-1s replacing LM-1s in the Air Squadron.

Western Air Command (Seibu Homen Koku-tai) at Takayubaru has an enlarged complement including elements at Naha, Okinawa and comprising the Air Squadron (LR-1, LM-1, OH-6J and H-13); Helicopter Squadron (UH-1 and Vertol 107); No 4 Squadron at Metabaru and No 8 Squadron at Takayubaru (both OH-6J and L-19); 1 Composite Command, Naha with No 101 Squadron (Vertol 107, UH-1, LR-1 and LM-1); 3 ADMG, Izuka, 6 ADMG, Naha and 7 ADMG, Takematsu.

Helicopter transport facilities are undertaken by 1 Helicopter Wing (Herikoputa-dan) at Kisarazu with Nos 1 and 2 Helicopter Units, each equipped with Vertol 107s, plus a small number of OH-6Js. Aircrew training is administered by the Air Training School (Koku Gakko).

Under the terms of the 1953 armistice agreement, introduction of new weapons to both North and South Korea was banned, and a prohibition placed on any increase in forces. However, even while talks were in progress, the Korean People's Army Air Force was being rapidly expanded with Soviet assistance to five Air Divisions, comprising Air Regiments operating Tu-2 and Il-10 bombers, plus Fighter Regiments with MiG-15s, Yak-9Ps and La-11s. Over 200 aircraft, including a few Il-28 jet bombers, were ferried to Uiju and Sinuiji early in July 1953 before final signing of the truce.

In the next five years, KPAAF strength was doubled with Soviet and Chinese assistance, and obsolete piston-engined aircraft replaced with jets. During 1957 the Army Air Force was integrated with the Sino-Communist and Soviet Far East forces in a joint operational command, but following Russian signing of the nuclear test-ban treaty, North Korea moved into the Chinese camp. Five fighter regiments re-organised along Chinese lines with MiG-17s built at Shenyang and by 1963, the KPAAF had received 380 MiG-15bis and MiG-17 fighters for three Fighter Air Divisions, each of two Fighter Regiments with 35 aircraft apiece.

Following the 1965 *rapprochement* with the Soviet Union, an initial batch of MiG-21Fs was supplied, while further aircraft have been assembled locally from about 1975 onwards. Front-line forces currently comprise nine units with some 170

MiG-21s, and an additional interceptor element with 50 or so MiG-19s. About 350 MiG-17s and 30 Su-7s are allocated to ground attack regiments, backed by a strike force of 70 Il-28s. Transport forces number about 40 aircraft, including An-2s, An-24s, two Il-18s and a single VIP Tu-154B. A similar number of helicopters is evenly divided between Mi-4 and Mi-8 types, while about 70 trainers comprise Yak-18s, MiG-15UTIs and a few examples of the MiG-21U.

Current overall strength is thus 670 operational types and 150 support aircraft, manned by 45,000 personnel including conscripts serving between

three and four years (army and navy conscripts serve 5 years, and the former was, until recently, seven years).

Celebrating the 30th anniversary of its foundation on 1 October 1979, the Republic of Korea Air Force (Hankook Kong Goon) has already entered a period of modest expansion to compensate for the previously-planned American withdrawal of 34,500 ground troops, although elements of the USAF would have been unaffected. This expansion is superimposed on a nationally-launched plan in 1976 to build up the combat strength of the Korean forces over a five-year term.

Having given consideration to the purchase of 40–50 Fairchild A-10s and 70 Vought A-7 Corsairs, South Korea is instead believed to be in the process of negotiating licence rights for the Northrop F-5E, covering some 70 aircraft including F-5F trainers. These will augment the 126 F-5Es and nine F-5Fs already delivered under earlier aid programmes. South Korea has for some time been developing a major military industrial complex aimed at achieving self-sufficiency in the production of all military equipment except combat aircraft and nuclear weapons by the early 1980s.

Work on the assembly of Hughes 500MD Defender helicopters is already in hand at Hanjin, where 66 of the 100 on order will be produced, the remaining 34 coming direct from the United States. Most of the Defenders will be delivered to the RoK Army for anti-tank roles, armed with TOW missiles.

Allocated principal tasks of air defence, tactical reconnaissance, logistic support of ground forces and SAR, the RoKAF is subordinate to the Ministry of National Defence through the Commander, Air Forces Korea (who also commands 314 Air Division, US Pacific Air Force). Tactical elements are controlled by Combat Air Command, with

The South Korean air force operates about 20 Curtiss C-46 Commandos, the majority employed by the transport force with a few used for liaison duties. Heavily reliant on support and equipment from the United States, the South Korean air force has ordered six Lockheed C-130s to increase the effectiveness of its present transport force, which operates C-46s, Douglas C-54s, Fairchild C-123Ks and Rockwell Aero Commanders

headquarters at Osan, alongside the USAF's 51 Composite Wing (Tactical).

Main combat strength comprises 16 fighter squadrons, armed principally with the F-5 and the McDonnell Douglas F-4 Phantom. In 1957–58, 112 North American F-86F Sabres re-equipped 10 and 11 Fighter Wings with two squadrons each, present disposition being only half that number. In 1965 10 Wing converted to the F-5, the RoKAF receiving 87 F-5As and RF-5As, plus 35 F-5B trainers for 102 and 105 Squadrons. About 45 of these were subsequently transferred to South Vietnam in 1972 and replaced by later F-5E/F deliveries. Nine squadrons currently fly the F-5, together with ten RF-5As forming a tactical reconnaissance unit.

In 1965, 1 Fighter Wing began operating the first of 50 ex-USAF F-86D Sabres, armed with Sidewinders for all-weather interception. These were augmented, and ultimately replaced, from 1969, following the delivery of 18 F-4D Phantoms. A second batch of 18 Phantoms arrived in 1972 after the transfer of F-5s to South Vietnam, and a few others were loaned from the USAF. Later deliveries included 37 F-4Es in two blocks, plus 341 AIM-7E Sparrow air-to-air missiles.

Plans to form a fifth Phantom squadron equipped with RF-4Es for reconnaissance as RF-5E replacements have yet to be realised. Unless ex-USAF RF-4Cs are substituted, South Korea's only alternative will be to take-over the cancelled Iranian order for 16 aircraft, which were to have been RF-4Es. Further equipment stemming from the build-up programme includes 24 Rockwell OV-10G Broncos and 733 more Sidewinders for F-5 armament. The Broncos will be used for COIN operations, but may also function as forward air control aircraft augmenting 14 Cessna O-2As delivered in 1975. Support of ground units is becoming increasingly important to the RoKAF, and although the substantial forces of the US 2nd Division will now remain in South Korea at least until 1981, 1,800 TOW anti-tank missiles have been requested to help combat the 2,000 or so tanks which the North is estimated to possess.

Air Transport Group is similarly engaged in modernisation, some 20 C-46 Commandos, ten C-54s and ten C-123 Providers being due for replacement. Six Hercules are on order, while two BAe 748s and a Bell UH-1N form the VIP flight. A few Aero Commanders are used for liaison flights. Rescue facilities are provided by two Sikorsky H-19s, five UH-1D Iroquois and six Bell 212s. Six CH-47C Chinooks are on order to provide a heavy-lift helicopter capability.

Pilot training is initiated on about 20 Cessna T-41Ds, followed by a basic course with the North American T-28A Trojan. Advanced training is flown on the T-33A, prior to operational conversion on the F-5B/F.

Naval aviation is confined to a single land-based squadron of some 20 Grumman Trackers, but it is expected to receive some anti-submarine versions of the Hughes Defenders on order. The remaining Defenders will go to the RoK Army and will substantially increase its present aviation force of Cessna L-19s and DHC Beavers plus four Kawasaki KH-4 (Bell 47) helicopters. An initial batch of 25 US-built Defenders is already in service.

CHAPTER FOURTEEN
South-East Asia & Australasia
Australia/New Zealand/Indonesia/Malaysia/ Philippines/Singapore/Thailand

The General Dynamics F-111C entered service with the Royal Australian Air Force in 1973. Twenty-one examples of the variable-geometry fighter are operated by Nos 1 and 6 Squadrons, based at Amberley, Queensland and they comprise the RAAF's long-range strike force. Four aircraft are being converted to the reconnaissance role and the remainder are to be updated

Australia's air force, which received its 'Royal' prefix in June 1921, was then already nine years old, having been brought into being as the Army Aviation Corps in September 1912. In January 1913 it was renamed the Australian Flying Corps, possessing at that time 43 personnel and five aircraft. Its introduction to combat came in April 1915, when an AFC contingent became a part of the Royal Flying Corps' No 30 Squadron in Mesopotamia. Thereafter, the Australian Flying Corps took an increasing part in World War I, particularly in the Middle East theatres of operations. The AFC was disbanded in 1919, but a new Australian Air Corps was formed in 1920, equipped with ex-Royal Air Force aircraft under the 'Imperial Gift' scheme offered to all of Britain's major Dominion forces.

When retitled Royal Australian Air Force, its subservience to Army control ceased and it became a fully independent service, continuing to employ aircraft of British origin. These were supplied directly by the UK or built in Australia under licence, the latter creating the nucleus of a domestic aircraft industry which made healthy growth during the period between the two world wars. By late 1939 the RAAF had grown to a strength of about 3,500 officers and men, with some 165 operational aircraft organised into 12 squadrons. This was, of course, massively augmented during 1939–45 by British and American aid, resulting in a strength by the war's end of over 180,000 personnel plus nearly 3,200 first-line aircraft. A postwar cutback was inevitable, though its severity was lessened to a small extent by Australia's commitment to provide

Left: No 10 Squadron operates the Lockheed P-3C Orion from Townsville, Queensland, on maritime reconnaissance duties. Its sister unit, No 11 Squadron, has flown the earlier P-3B Orion since 1968 from Edinburgh, SA.
Below: the RAAF operates the F-111C in the land and maritime strike roles.
Inset below left: the RAAF's fighter-bomber wing operates 100 Dassault Mirage IIIO As. The wing comprises the home-based No 77 Squadron and Nos 3 and 74 Squadrons in Malaysia.
Inset below right: No 1 OCU operates 16 Mirage IIIDO trainers from Amberley

a proportion of the occupation force in Japan. The service also made an important contribution on the opposite side of the globe during the Berlin airlift of 1948–49.

The RAAF was back in action again in mid-1950 when its sole remaining occupation squadron in Japan (No 77) was drawn into the war in Korea, while the newly-formed No 90 Wing operated with the RAF against the Communist terrorists in Malaya. By this time several of its units were flying jet aircraft, locally-built de Havilland Vampire fighter-bombers being supplemented by Gloster Meteor fighters from England in 1951. In 1953 the Melbourne-based Commonwealth Aircraft Corporation flew the first of many locally-built North American Sabres, having developed its own version of this excellent US fighter, powered by a Rolls-Royce Avon turbojet engine. Similarly, the medium bomber squadrons were re-equipped at about the same time with English Electric Canberra jet bombers, built under licence by the Government Aircraft Factories.

Administered since early 1976 by a unified Ministry of Defence, the RAAF is now under the operational direction of its Chief of Air Staff, Air Marshal J. A. Rowland, from headquarters in Canberra. It has only two Commands: Operational Command, with headquarters at Glenbrook, near Sydney, and Support Command, centred upon Melbourne. Of these, the former is responsible for all RAAF flying within Australia except training; flying outside the sub-continent is the duty of Support Command, which also looks after training requirements, maintenance, supply, recruitment and other aspects. Manpower strength is some 21,500.

There are 16 first-line squadrons in Operational Command, the two most important of these (Nos 1 and 6) being equipped with the General Dynamics F-111 variable-geometry aircraft for strike and reconnaissance. Nos 3, 74 and 77 Squadrons fly the Dassault Mirage III-OA for fighter/ground attack duties. Canberras still remain with No 2 Squadron and are now used for strike/reconnaissance plus secondary duties as target towing and photographic survey. The F-111s and Canberras are based at Amberley (Queensland) and the Mirages at Williamstown NSW and Butterworth, Malaysia. Over the past two or three years the RAAF has evaluated various potential replacements for its Mirages, most of which were built in Australia under licence from Dassault, their French manufacturer.

Two maritime patrol units Nos 10 and 11 Squadrons are operated by the RAF, based respectively at Townsville, Queensland and Edinburgh, South Australia. Both fly the Lockheed Orion, some of which now carry the Australian-designed Barra AQS-901 acoustic (sonar) detection system. Detachments of these two squadrons also fly from Darwin or Pearce air force bases in Northern Territory and Western Australia.

There are five fixed-wing transport units: Nos 36 and 37 Squadrons with Lockheed Hercules, Nos 35 and 38 Squadrons with Caribou STOL transports, plus No 34 (VIP) Squadron, which flies a mixture of BAe One-Elevens, Dassault Mystère/Falcon 20s and BAe HS 748s. The Hercules and Caribou are based at Richmond, near · Sydney, while No 34 Squadron operates from Fairbairn, near Canberra. The RAAF transport element also includes No 12 Squadron with the Boeing-Vertol Chinook heavy-lift helicopter at Amberley, while Nos 5 and 9 Squadrons are equipped with the Bell UH-1 Iroquois at Fairbairn and Amberley respectively.

Overseas deployment of RAAF units is under the jurisdiction of Support Command. Generally speaking, units are based outside Australia for fixed-term tours on a rotational basis. In 1978–79, these deployed units included two squadrons of Mirage III-OA fighter-bombers and one flight of C-47 transports based at Butterworth, Malaysia, with detachments at Tengah, Singapore.

Basic training, under the aegis of Support Command, takes place primarily at the Central Flying School at East Sale, Victoria, No 1 Flying Training School at Point Cook, Victoria and No 2 FTS at Pearce, Western Australia. The CFS and No 1 FTS have Aerospace CT-4 Airtrainer piston-engined primary trainers, a New Zealand-built type which was originally designed in Australia. At No 2 FTS, jet training is given on the Aermacchi MB 326H, some of these aircraft also being used at the CFS to train flying instructors. Additional training establishments include the School of Air Navigation at East Sale (HS 748s for training other aircrew and the pilots of multi-engined types), plus No 1 Operational Conversion Unit at Amberley, where combat pilots are trained on the two-seat Mirage IIID.

Military aviation on behalf of the Royal Australian Navy has been the responsibility of a separate service only since 1948. Prior to that date the Air

Force had undertaken all flying commitments required by the other two services. Its first naval aircraft were six Fairey IIID floatplanes which formed part of the RAAF's initial equipment on its formation in 1921. Even today, Royal Australian Navy aviation is confined primarily to sea-going operations, such tasks as maritime patrol, surveillance and anti-submarine warfare from shore bases still being undertaken by the RAAF.

The original basis for the formation of a separate naval air arm was the purchase, in 1948, of two light fleet carriers from Britain. Entering service as HMAS *Sydney* and *Melbourne*, the former went to sea initially with Fairey Fireflies (anti-submarine) plus Hawker Sea Furies (fighter-bombers), with which it saw action in the Korean war theatres in

1951 and again in 1953. In mid-1954 the Navy received its first jets (Vampires for training), followed by Sea Venom all-weather fighters; turbo-prop-engined Fairey Gannets replaced the carrier-based Fireflies in the ASW role. HMAS *Melbourne*, fitted with angled deck, steam catapults and mirror landing aids, at last joined the fleet in 1956. *Sydney* was then redesignated for the training role, but was put up for disposal in 1973, going to the breakers two years later.

Thus, *Melbourne* now remains the RAN's only serving carrier, but a series of refits, and the re-equipment of her embarked squadrons with more modern aircraft, have ensured her continuing viability beyond the 1980s. Having a standard displacement of 16,000 tons, she has a 40,000 shp

power plant which gives a top speed of 24 knots, and a principal armament of twelve 40 mm anti-aircraft guns. Her normal complement comprises 14 McDonnell Douglas A-4G Skyhawk attack bombers of No 805 Squadron, six Grumman S-2G Tracker anti-submarine aircraft of No 816 Squadron, and six Westland Sea King Mk 50 ASW helicopters from No 817 Squadron.

These three squadrons are the operational element of a Fleet Air Arm which totals six squadrons in all (about 60 aircraft altogether) and has a manpower strength of around 1,500. The Skyhawks and Trackers began to replace the earlier Sea Venoms and Gannets in the late 1960s; they are also to be found ashore at the RAN's only current land base, HMAS *Albatross* at Nowra, south of Sydney, NSW. This is the home of the three second-line units, Nos 723, 724 and 851 Squadrons. No 723 Squadron undertakes fleet support, search/rescue and training duties with Wessex, Iroquois and JetRanger helicopters, while No 724 is a fixed-wing unit equipped with A/TA-4G Skyhawks and Macchi MB 326Hs and fulfilling fleet support and training roles. No 851 Squadron gives fixed-wing training in ASW (with Trackers) and electronic warfare (with HS 748s), providing also a fleet transport element with a handful of Douglas C-47s.

Youngest of Australia's three aviation services, the AAAC only became an independent arm as recently as July 1968, although it still has the status of a regiment within the overall Army structure. Its headquarters are at the former RAAF base at Oakey, Queensland, and about three-quarters of its 100 or so aircraft are helicopters. The Corps was originally created in 1960, as the Army Aviation Branch of the RAAF, becoming the Air Regiment six years later with a strength of three operational units, Nos 16, 17 and 18 Squadrons, equipped with Bell 47G helicopters. By 1968 it had almost 80 of these, plus a smaller number of Cessna 180 light-planes; some of its aircraft and personnel saw action in the war in Vietnam.

A major change in equipment began during the early 1970s, when the AAAC began to receive the first of just over 50 Bell JetRanger II light helicopters, and these remain in service along with about two dozen of the earlier Bell 47s. Main fixed-wing type, in terms of numbers, is the Pilatus Turbo-Porter single-engined STOL aircraft, about 15 of which are in service; these have taken the place of the Cessnas since the mid-1970s and are used for liaison, light transport and artillery observation. The Army's largest aircraft is the domestically-designed Nomad Mission Master, produced by the Government Aircraft Factories at Fishermen's Bend, Melbourne, which began to enter service in 1975 and will eventually replace the Turbo-Porters. Eleven were ordered, most or all of which had been delivered by 1979.

Oakey, which has been the headquarters base of No 1 Air Regiment of the AAAC since late 1973, now also accommodates the Army School of Aviation plus the AAAC's maintenance and repair depot. Army pilots complete their training at the School, although primary flying training is given by the RAAF's No 1 FTS at Point Cook, Victoria. Operational units, known as Air Cavalry Flights, are deployed with various regiments of the

Far left: No 723 Squadron, RAN has four Westland Wessex HAS Mark 31Bs among its mixed complement of helicopters. Based at Nowra Naval Air Station, New South Wales, the unit operates in the fleet requirements, communications, search and rescue and training roles.

Left: No 817 Squadron, RAN operates Westland Sea King Mark 50 anti-submarine warfare helicopters, six of which are normally deployed aboard HMAS Melbourne as part of the aircraft carrier's Air Group.

Below: McDonnell Douglas A-4G Skyhawks are operated by No 805 Squadron, forming the strike and ground-attack element of HMAS Melbourne's Air Group. No 724 Squadron, based at Nowra, has a few A-4Gs and two-seat TA-4Gs for fighter and ground attack training, alongside the Aermacchi MB 326H. Skyhawks entered RAN service in the late 1960s and some 16 are currently in service

Below: six Grumman S-2E and S-2G Trackers of No 816 Squadron form HMAS Melbourne's fixed-wing anti-submarine warfare element. The RAN's remaining S-2s are based at Nowra, New South Wales, where the type flies with No 851 Squadron for ASW training. The Tracker replaced the RAN's Fairey Gannets in the late 1960s.
Inset left: a hangar fire on 5 December 1976 decimated the RAN's S-2 force, destroying seven Trackers. The single undamaged and two repairable Trackers were joined by 16 replacements from US Navy stocks. An S-2G variant, the ultimate production model for the US Navy, is pictured.
Inset far left: an RAN S-2G is displayed with its tail-mounted magnetic anomaly detector boom extended. The type also carries a search radar and sonobuoys are housed in the engine nacelles

Australian Army at home, and at such overseas bases as Indonesia, Malaysia and Papua.

The Hon Henry Wigram, after whom one of the RNZAF's principal air bases is named, was the first New Zealander to urge (in 1909) the creation of a military flying corps in the country; three years later New Zealand sent its first pupil to Great Britain for flying training. However, one pilot does not make a flying corps; consequently, the Blériot monoplane donated to the Dominion by Britain in 1913 was returned upon the outbreak of World War I, and New Zealanders who became aviators flew instead with the Royal Flying Corps during that conflict, although they had been trained in their own country. Efforts made by these pilots after the war eventually succeeded, by 1920, in securing government acceptance of 30 or so war-surplus aircraft from the UK. Even then, it took a further three years, and additional persuasion by Mr Wigram, before a Permanent Air Force, with a somewhat larger Territorial Air Force, were set up by the New Zealand authorities. Expansion of these was curtailed by the poor economic climate of the late 1920s, and this state of affairs was not relieved until the middle of the following decade, when, in common with most of the rest of the world, New Zealand began to take serious note of Germany's rearmament.

Re-established as an independent service in April 1937, the Royal New Zealand Air Force was still modest in size, its manpower totalling only 339 officers and men; but during the next two years, several expansion programmes were put in hand, and by the outbreak of war in Europe in 1939, this figure had more than trebled. While RNZAF crews flew with RAF squadrons over Europe, production at home of de Havilland Tiger Moths was laying the foundations for one of New Zealand's great contributions to the Allied war effort: as a breeding ground for pilots, as a part of the huge Empire Air Training Scheme. In all, more than 13,000 pilots and other aircrew passed through New Zealand flying schools for service with the RAF, RNZAF and other Commonwealth air forces. New Zealand's participation in the 'shooting war' of 1939–45 took a sharp upward turn when Japan plunged the south-west Pacific into chaos; in addition, six RNZAF combat squadrons fought in Europe and a Short Sunderland equipped squadron operated from West Africa.

A post-war regular establishment, after the inevitable demobilisation and run-down, took a few years to become settled, but a 4,000-strong force was finally decided upon in 1948, with an operational strength of five front-line squadrons. A four-squadron Territorial Air Force (also later increased to five) was reintroduced in 1948.

After the war, the RNZAF had re-equipped some of its squadrons with de Havilland Mosquito fighter-bombers, but a more significant programme began in 1950, when the first DH Vampire jet aircraft (fighter-bombers and trainers) arrived from Britain. Short Sunderlands replaced the Catalinas in the sole flying boat squadron, and the two transport units had a mixture of British types to replace its American Douglas C-47s. Before long, the RNZAF was caught up in the Korean War, and was involved in the anti-terrorist campaigns in Malaya. One result of this level of operational activity was a decision, in the late 1950s, to disband the Territorial Air Force and expand the capability of the full-time RNZAF.

Since 1971 New Zealand, like Australia, has administered all three of its armed services through a unified Ministry of Defence, with its headquarters in Wellington, North Island. Operational command of the RNZAF is exercised by the Chief of Air Staff, Air Vice-Marshal C. L. Siegert; like the RAAF, the RNZAF has two principal elements: an Operations Group and a Support Group. The

Top left: No 36 Squadron, with 12 Lockheed C-130Hs (illustrated) and No 37 Squadron, with 12 C-130Es form the RAAF's heavy transport force, operating from Richmond, New South Wales.
Above left: No 35 and 38 Squadrons, RAAF, based at Richmond fly DHC-4 Caribou transports.
Left: the 171st Air Cavalry Flight, Australian Army Aviation Corps, operates the Pilatus PC-6 Turbo Porter from Holsworthy

Top right: No 5 Squadron, operating five Lockheed P-3B Orions from Whenuapai is the RNZAF's only maritime reconnaissance unit.
Right: six Bell 47G/H-13G Sioux are among the helicopters which equip No 5 Squadron for its diverse duties. The Sioux are operated in the battlefield support role on behalf of the NZ Army.
Below right: DH Devons are operated by RNZAF Support Command's Flying Training School, at Wigram. The two Devons of No 42 Squadron, RNZAF Operations Group, and the single example of No 40 Squadron are scheduled to be replaced.
Below: the Bell UH-1 Iroquois is flown by two RAAF squadrons in search and rescue, support and training roles

former, with headquarters at Whenuapai near Auckland, North Island, is responsible for the strike, maritime and transport squadrons, while the latter group, centred upon Wigram air base, South Island, co-ordinates and controls RNZAF ground as well as flying training.

Operations Group comprises eight squadrons in all, of which the only recognised combat unit is No 75 Squadron at Ohakea, North Island, equipped with late-model McDonnell Douglas Skyhawk attack aircraft. However, a second unit–No 14 Squadron, also at Ohakea–flies BAe Strikemaster armed trainers, which can provide a back-up force for the light strike unit if required. The other five squadrons are made up of one maritime patrol, one helicopter, two transport and one communications/liaison unit. No 5 Squadron, at Whenuapai, is the MR (maritime reconnaissance) unit, flying P-3B Orions purchased from the USA. The two medium transport squadrons are No 1 at Whenuapai, with ex-RAF Andovers and No 40, at the same base, flying Lockheed C-130H Hercules. The helicopter squadron is No 3, based at Hobsonville, near Auckland. This operates on behalf of all three New Zealand services, flying a mixture of Sioux, Wasp and Iroquois rotorcraft. Its work includes battlefield support for the Army with the Sioux, anti-submarine duties with the Royal New Zealand Navy, deploying its Wasps aboard RNZN frigates and search/rescue and Army support with the Iroquois. One flight of Iroquois also operates from Tengah, Singapore, as a part of New Zealand's contribution to the ANZUK force in South-East Asia. At Ohakea, No 42 Squadron is a composite unit which provides VIP transport, communications and liaison with Andovers and Devons.

Flying training is conducted by Pilot, Navigation and Electronic Training Squadrons, which make up the Wigram-based Flying Training Wing of the RNZAF's Support Group. Basic pilot training is given on the Aerospace (NZ) Airtrainer, with about six Devons and four Sioux helicopters sharing the other tasks. The Strikemaster is used for advanced training, and two-seat TA-4K Skyhawks provide the necessary operational conversion for combat pilots. Tactical training for paratroops and assault forces is given on Hercules or Andover aircraft.

Below: the five Lockheed C-130Hs of No 40 Squadron, RNZAF are deployed as long-range and heavy logistic transports and assault trainers.

Inset right: No 14 Squadron, RNZAF has 16 BAe Strikemaster Mark 88s for light strike and basic jet training.

Inset below right: delivery of ten ex-RAF HS Andover C Mark 1s to the RNZAF was completed in 1977. Six are used by No 1 Squadron as troop/freight transports and four by No 42 Squadron in roles including VIP duties and pilot and tactical assault training.

Inset opposite below: the Aerospace (NZ) CT-4B Air Trainer replaced the last North American AT-6 in the basic training role at the RNZAF's Flying Training School, Wigram in 1977. In 1979, 13 CT-4Bs are operated

The United States of Indonesia was created as a republic at the end of World War II; in April 1946 an aviation division of the People's Security Force was set up, equipped initially with about 50 ex-Japanese wartime fighters, bombers, reconnaissance, transport and training aircraft. In its previous incarnation as the Netherlands East Indies, Indonesia had been part of a considerable Dutch colonial empire in the Far East, and one in whose history aviation had played a considerable part. Before the war it was a major staging point for KLM Royal Dutch Airlines and its local subsidiary KNILM; during the war, both the NEI Army and Navy possessed small but well-equipped air arms.

After the official transfer of sovereignty in December 1949, a Dutch military mission provided assistance to the new republic by expanding and re-equipping its armed forces. Known initially as the Angkatan Udara Republik Indonesia (AURI), the Indonesian Republican Air Force became responsible for the air defence of the republic, taking over all existing NEI Air Force airfields and equipment when the latter was disbanded in June 1950. The republic was divided into seven 'air districts', while the AURI was organised into two operational elements–one for combat and the other for transport. Equipment at this time was virtually all of United States origin, including North American F-51D Mustang fighters and B-25 Mitchell medium bombers, Consolidated Catalina amphibian flying boats, the ubiquitous Douglas C-47 for transport, and North American Texans for training.

The first jet aircraft appeared in late 1955; this was a small batch of de Havilland Vampire T Mk 55 trainers. For the next few years Indonesia continued to shop in western markets for the comparatively few additional aircraft needed by the AURI. Eastern bloc influence was first introduced in the spring of 1958 when, after an insurrection on the main island of Sumatra, 60 MiG-17 jet fighters and 40 Ilyushin Il-28 twin-jet bombers were purchased from Czechoslovakia.

This was only the beginning of a considerable influx of Soviet-designed military aircraft over the next eight years, at the end of which time Indonesia possessed the largest air force in the whole of Southeast Asia. Additions to the fighter inventory included 35 supersonic MiG-19s and 18 MiG-21Fs, with 15 MiG-15UTIs for training; the Il-28s were augmented by two dozen Tupolev Tu-16 Badger strategic jet bombers; transports included 28 Avia (Ilyushin) 14s and 10 An-12s; Indonesia received its first helicopters in the form of about three dozen Mil Mi-1s, Mi-4s and Mi-6s.

A large proportion of these aircraft remain in the country, though they are now grounded and in store through lack of spares since the present anti-Communist government took over after the end of the Malaysian conflict in the mid-1960s. With the severing of trade relations with the Soviet Union, Indonesia turned once again to the West for its military requirements; most of its current equipment has come from the United States and Australia, backed up in recent years by a modest output from its own domestic aircraft industry.

Known today as the Tentara Nasional Indonesia –Angkatan Udara (TNI–AU or Indonesia Armed Forces–Air Force), the former AURI no longer occupies the position of strength which it possessed in the early 1960s, largely as a result of the grounding of its Soviet-supplied equipment and the lack of sufficient funds to replace this with the numerical equivalent of western aircraft. The TNI-AU is nominally organised into five commands, though only three of these have an effective air strength. In Operations Command only three front-line combat squadrons remain; one of these flies the Rockwell OV-10 Bronco in a COIN (counter-insurgency) role, while the other two fly the Northrop F-5E. The latter type recently began to replace Commonwealth Avon Sabres supplied from Australia in 1972.

Logistic Command operates two transport squadrons, the heavy-lift No 130 Squadron at Halim, with Lockheed C-130 Hercules, the medium-lift No 2 Squadron in which the Douglas C-47s are assisted by eight Fokker Friendship Mk 400Ms, six CASA C 212 Aviocars and a single Short Skyvan. A third squadron is equipped with an extremely mixed bag of aircraft, ranging from modern business twins to Lockheed Super Constellations, Scottish Aviation Twin Pioneers and a pre-war Lockheed 12. The primary functions of this unit are VIP transportation and general com-

munications/liaison duties. Completing the transport element is the helicopter force, comprising two squadrons with a mixture of sizes and types from France, Germany, Italy and the USA.

The principal unit in Training Command of the TNI-AU is the flying school at Jokjakarta, in central Java. Here, propeller-driven training is given on Japanese-built Beech T-34 Mentors and US-built T-34C-1 Turbo-Mentors, plus the remnants of the ancient T-6 Texans, with Lockheed T-33As for jet training. A small batch of British Aerospace Hawks is now on order.

Also operated by the TNI-AU are two squadrons which support the work of the Army and Navy air arms. No 4 Squadron flies Piper L-4 (Cub) and Cessna 180/185 lightplanes, and a few examples of the larger de Havilland Canada DHC-3 Otter, on Army support, while No 5 Squadron augments naval coastal patrol work with Grumman Albatross amphibians.

Above left: eight Lockheed C-130Bs, one C-130H and an L-100-30 (illustrated) equip the Indonesian Air Force's heavy transport squadron.
Above far left: one Indonesian squadron flies Rockwell OV-10F Broncos on COIN duties.
Below: Nos 11 and 12 Squadrons, Royal Malaysian Air Force operate Northrop F-5E fighter-bombers from Butterworth. Fourteen were received in 1978

Top: Indonesia operates 12 GAF Nomad Search Master Bs.
Centre: Nos 1 and 8 Squadrons' 17 DHC-4 Caribous form the Royal Malaysian Air Force's medium transport element.
Above: 16 Canadair CL-41G armed trainers serve with the RMAF.
Opposite top: the 5th Fighter Wing, Philippine air force, operates Northrop F-5A fighter-bombers from Basa.
Opposite above: delivery of 25 LTV F-8H interceptors to the 5th Fighter Wing, PAF was completed in 1979

Earlier attempts to foster a domestic aircraft industry, to provide at least part of Indonesia's internal needs, foundered a few years ago with the cessation of work at the former Lipnur factory at Bandung. This concern built more than 30 examples of the Polish PZL-104 Wilga under licence as the Gelatik, and had designed as well as flight-tested an attractive little piston-engined trainer, the LT-200. Plans to build about 30 of these were, however, frustrated by the inevitable lack of funds.

Nevertheless, the domestic scene looks more promising in the hands of the P. T. Nurtanio company of Bandung, which is now assembling the CASA C 212 Aviocar twin-turboprop transport and the MBB Bö 105 helicopter under licence. This concern was set up in the summer of 1976 by merging the Lipnur factory with the Indonesian oil company Pertamina, thereby providing both additional finance and a wider home-based market for the domestic industry's new products, since Pertamina operates a small fleet of Aviocars for inter-island transport. The Bo 105 is an excellent general-purpose helicopter which had already demonstrated its capability for offshore oil rig support in the North Sea and the Baltic; the larger Aérospatiale Puma helicopter from France is among the types which Nurtanio has rights to assemble in the future. Tooling for the LT-200 trainer remains available should it be decided to reinstate this programme.

Like the air force, the Indonesian Navy has an aviation history dating back to the days of the Netherlands East Indies, the island nature of the area being particularly conducive to the development and use of waterborne aircraft. After the postwar change to republic status, the naval air arm took longer to become re-established than the air force, emerging in 1958 with the title of Angkatan Laut Republik Indonesia (ALRI, or Indonesian Republican Naval Air Force). Among its first aircraft were Fairey Gannets from Britain, in two versions: the AS Mk 4 for anti-submarine warfare and the T Mk 5 for radar observer training. Fighter and transport elements were later added to the original patrol/reconnaissance force, the fighters being Soviet-supplied MiG-19s and MiG-21Fs; like their air force counterparts, these are now in store.

Compared with the 28,000-strong air force, the naval air arm has a personnel strength of only about 1,000. Now known as the Tentara Nasional Indonesia-Angkatan Laut (TNI-AL or Indonesian Armed Forces-Naval Air Force), its primary function is still that of coastal patrol. For that role its main aircraft are about half a dozen Grumman Albatross amphibians – now no longer young – and twice that number of Australian GAF Nomad Search Master B twin-engined landplanes. As mentioned earlier, the TNI-AU also provides an Albatross squadron to assist the Navy in the coastal patrol role. Helicopter backup for these duties and for general transport is provided by a mixture of French and American types, while fixed-wing transports still include a handful of the unfailing Douglas C-47, plus a trio of Rockwell Grand Commanders. The main air bases are at Gorontalo, Kemajoran and Surubaya; the last-named also accommodates the headquarters of the TNI-AL.

In addition to the Army co-operation function

Coastal patrol is, in fact, a term with considerable meaning for Indonesia, composed as it is of more than three thousand islands, of varying sizes, strung out in a line stretching for nearly 4,800 km (3,000 miles) across the south-west Pacific. Administratively, the air force is still divided into geographical regions, now numbering six instead of the original seven. A key region is that of West Java, with the TNI-AU headquarters at Jakarta, the main combat aircraft base at Iswahyudi and the home-based aircraft industry centred upon nearby Bandung. The other five regions comprise East Java (with Central and Lesser Sunda), Sumatra, the Celebes group, Kjalimantan and West Irian.

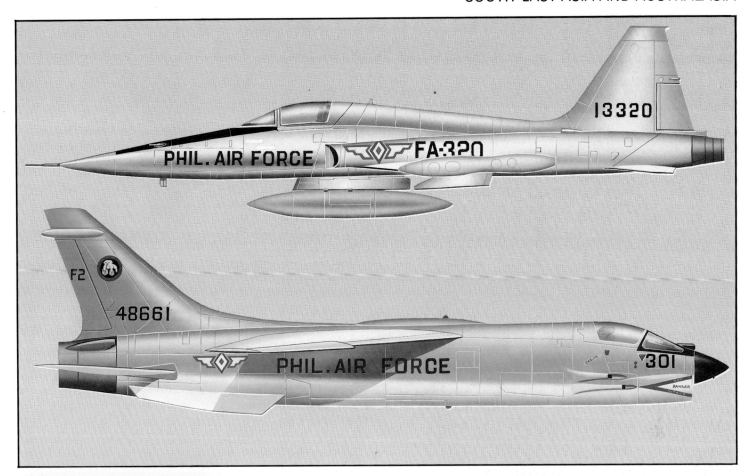

performed by the air force's No 4 Squadron, the Indonesian Army has its own aviation element, known as the Tentara Nasional Indonesia–Angkatan Darat (TNI-AD, or Indonesian Armed Forces-Army Air Force). This is a somewhat gradiose title, since the force has only somewhere in the region of 50 aircraft, nearly half of which are lightplanes. Most of the latter are locally-built Gelatik (PZL-104) lightplanes, while about half a dozen twin-engined aircraft are employed for transport duties.

The history of military aviation in this part of the world goes back through several incarnations of the Malayan peninsula, to the days before World War II when it was known as the Straits Settlements. It then had a Volunteer Air Force, formed in March 1936, which was equipped with Hawker Audax general-purpose biplanes; its personnel were a mixture of British and Malayan nationalities. A year after the outbreak of the war in Europe, this service was reorganised under the title of Malayan Auxiliary Air Force, its numerical strength being increased by commandeering unarmed de Havilland Rapides, Tiger Moths and other light aircraft from the civilian flying clubs in the country. During the three months of fighting before Malaya was overrun by the advancing Japanese forces, these aircraft with their crews played a valiant part in the defence of the peninsula by carrying out reconnaissance, communications/liaison and jungle rescue duties in support of the Allied ground forces. When Malaya fell to the Japanese, most of them succeeded in escaping to the island of Sumatra, in what is now known as Indonesia, from where they could continue to fight until the end of the war in the Pacific.

It was, however, an uneasy 'peace' that returned to Malaya in 1945. During the late 1940s and early 1950s the jungle provided a breeding ground and hiding place for some 5,000 Communist insurgents, against whom the national security forces and the British armed services were to fight for many bitter years before the terrorists were finally defeated. The Malayan Auxiliary Air Force was revived, still on a volunteer basis, in 1947, and the flying training of new would-be pilots by the Royal Air Force began in 1950. For approximately a decade, Malayan nationals in the MAAF and the RAF Regiment (Malaya) supported the Royal Air Force in its struggle to rid the Federated Malay States of the Communist guerillas.

The country gained independence, though not total peace, in the summer of 1957 with the creation of the Federation of Malaya. Great Britain continued to be responsible for the country's defences, but a new Royal Malayan Air Force came into being, whose primary duty was to provide air support for the ground forces. With Air Commodore A. V. R. Johnstone as its first leader, the RMAF was inaugurated in June 1958. Its first aircraft were light STOL types–four examples each of the Scottish Aviation Pioneer and Twin Pioneer–used for such varied tasks as light bombing, coastal patrol, reconnaissance, multi-purpose troop/cargo/casualty transportation, communications/liaison, and the psychological warfare operation of 'sky-shouting' propaganda messages to the terrorists in the jungle. The new RMAF continued to depend heavily upon RAF assistance for some years. Its next aircraft to arrive were half a dozen ex-RAF DHC Chipmunks, on which flying training began at Kuala Lumpur in late 1958 and early 1959, although many Malayans continued to receive their training in the United Kingdom for some time after this.

From this modest start the postwar RMAF was able to expand relatively quickly, especially after the national state of emergency declared in May 1948 came to an end in the summer of 1960. By the end of the following year, piston-engined Hunting Percival Provost trainers had replaced the Chipmunks; the original Pioneer/Twin Pioneer strength had been almost trebled, de Havilland Doves and Herons forming the basis for a more up-to-date transport element. Two years later, in August 1963, the RMAF received its first helicopters–Sud Alouettes from France. A month later, Malaysia replaced Malaya in the title of both the country and its air force when the former formed a new Federation with the neighbouring states of Sabah (North Borneo), Sarawak and Singapore. (Two years afterwards, Singapore withdrew from the Malaysian Federation to become an independent state.)

Almost immediately, the new Federation found itself in political and military confrontation with Indonesia, again necessitating RAF support pending completion of the RMAF expansion/modernisation programme. One other result of this was that the Malaysian Provosts were fitted with armament, and more of the same were ordered. In 1964, 20 examples of a more capable ground-attack aircraft, the Canadair CL-41G, were ordered from Canada, and these began to arrive in 1967; they were known as the Tebuan (Wasp) in Malaysian service.

By that time, plans had been made for the British forces to be withdrawn from Malaysia; this made an expansion of the RMAF's combat capability both necessary and urgent. After considering various alternatives, the indecision over the choice of a suitable fighter type was solved by the donation of 10 ex-RAAF Avon-Sabres by the Australian Government; these entered service in 1969. Sikorsky S-61A-4 tactical assault helicopters had been introduced some two years earlier, while fixed-wing transport was provided by a squadron of DHC-4 Caribou acquired in 1965. From 1971, the piston-engined training role has been fulfilled by Scottish Aviation Bulldogs replacing the now-aged Provosts, while the Sabres have given way to Northrop F-5E Tiger IIs in the fighter/strike role.

In 1979, the strength of the Royal Malaysian Air Force (Tentara Udara Diraja Malaysia), under the command of Air Vice-Marshal Mohammed Taib, is about 6,000 personnel and 175 aircraft, some three-quarters of the latter being combat types. The Northrop Tigers equip Nos 11 and 12 Squadrons, both accommodated at Butterworth alongside the two overseas Mirage squadrons of the Royal Australian Air Force. Operational conversion training is given on the two-seat Northrop F-5B. The CL-41G Tebuans are still in service, currently equipping Nos 6 and 9 Squadrons, for advanced training as well as ground attack duties; they are based at Kuantan.

The transport element of the RMAF now comprises approximately 30 fixed-wing aircraft of six widely different types, the largest being the Lockheed Hercules, flown by No 14 Squadron at Kuala Lumpur. Squadron Nos 1 and 8 at Labuan each have DHC-4 Caribou STOL transports, while No 2 Squadron (Kuala Lumpur) is a communications/

VIP transport unit with a mixture that includes Fokker F-28 Fellowships, British Aerospace HS 125s plus a few Doves and Herons. Additional airlift capability is provided by the Sikorsky S-61A-4 helicopters, which serve with Nos 7 and 10 Squadrons, based at Kuching and Kuantan respectively, plus one other unit.

Several other helicopter squadrons form part of the RMAF. Nos 3 and 5 Squadrons fly the other principal type, the Alouette III, from Kuala Lumpur and Labuan; helicopter training is given at the latter base, using the Bell 47G. Gazelles are on order to augment the helicopter force.

Fixed-wing pilot training, as mentioned, is given on the Bulldog Mk 102. The flying training school is at Alor Star, near the northern border with Thailand; here about a dozen Cessna 402 light twins are available for multi-engine training, plus such other incidental duties as communications, liaison and photographic survey.

Created in May 1935 as the Air Corps of the Philippine Army, the air force of this large Pacific island republic was, at that time, an extremely modest organisation, whose main task was to assist the national police force by using light aircraft to track down criminals from the air. However, between then and the outbreak of World War II in the Pacific, the Philippines became a major overseas base for the US Army Air Corps; with General Douglas MacArthur as military adviser, all branches of the Philippine armed services were considerably reorganised, expanded and re-equipped. Aircraft included such relatively modern types as Boeing P-26 fighters and Martin B-10 bombers, but most of the Philippine Air Corps' material was quickly destroyed or overrun in the early campaign successes of the Japanese advance. Philippine resistance ended with the fall of Corregidor in May 1942, and a new air arm was not created until after the end of the war when the Philippine Air Force came into being in July 1947.

Equipped once again with US military aircraft–notably North American Mustang fighters and Douglas C-47 transports–the PAF was given comparable status with the Army and Navy in 1950. At the time of writing it is a modest but well-equipped force with a manpower strength of some 16,000 and well over 100 combat aircraft. Known officially as the Hukbong Himpapawid ng Pilipinas, the headquarters is at Nichols Air Force Base, near Pasay City, with other airfields at Basa (Pampanga), Clark Field (Manila), Cubi Point (Bataan), Gozan, Mactan (Manila), Paranal, Paredes, San Fernando (Luzon), Sangley Point (Mindanao) and Zamboanga (Mindanao). It is organised into five major functional wings, plus a number of miscellaneous formations.

The 5th Fighter Wing at Basa comprises the air defence element of the PAF, with three combat-ready units and a Combat Crew Training Squadron. The Northrop F-5, which had largely supplanted the North American Sabre as standard equipment, now serves alongside the LTV F-8 Crusader. Twenty-five of these aircraft were bought from the United States Navy, with a further ten being used as spares.

Four operational squadrons (Nos 16, 17, 18 and 19) plus a light transport group make up the 15th

Above right: six McDonnell Douglas TA-4S Skyhawk conversion trainers are flown by the Singapore air force at Tengah, an aircraft of No 143 Squadron being pictured. They have been modified from ex-US Navy A-4B single-seaters by the addition of a second cockpit. All Singapore's Skyhawks have updated avionics and more-powerful engines.
Right: this insignia is carried by the Skyhawks of No 142 'Gryphon' Squadron at Tengah.
Below: A-4S Skyhawks of No 143 Squadron photographed in 1978

Strike Wing, whose headquarters are at Sangley Point AFB on the island of Mindanao. Of these, No 18 is equipped with Bell UH-1H Iroquois gunship helicopters and No 19 with Douglas AC-47 armed transports. The other two squadrons fly North American T-28Ds (No 16) and SIAI-Marchetti Warrior light aircraft (No 17) for COIN and light attack duties. About 10 de Havilland Canada DHC-2 Beavers equip the 290th Special Air Missions Group for light transport and liaison work.

The major transport organisation of the PAF is the 220th Heavy Airlift Wing, based at Mactan. This has three squadrons, flying respectively Fairchild Providers (No 221), Lockheed L-100-20/C-130H Hercules (No 222) and Australian GAF Nomad Mission Masters (No 223).

Transportation is also one function of the PAF's fourth command. This is the Nichols-based 205th Composite Wing, whose duties also include the other important functions of reconnaissance, maritime patrol and search/rescue. A mixture of Grumman Albatross fixed-wing amphibians, Bell UH-1H and Sikorsky UH-34 helicopters carry out these last three duties, all serving with the 505th Air Rescue Squadron. No 204 Squadron (Fokker F-27s) and Nos 206 and 207 (C-47s) provide the main transport capability of this command, with No 601 Squadron fulfilling the air liaison requirement with Cessna lightplanes.

San Fernando Air Force Base, near Lipa City on the northernmost island, Luzon, is the home of the 100th Training Wing, where primary and basic flying training is given on Cessna T-41D Mescaleros, North American T-28As and SIAI-Marchetti SF 260s–all piston-engined types. The

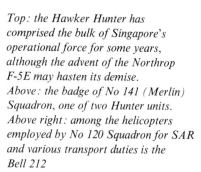

Top: the Hawker Hunter has comprised the bulk of Singapore's operational force for some years, although the advent of the Northrop F-5E may hasten its demise.
Above: the badge of No 141 (Merlin) Squadron, one of two Hunter units.
Above right: among the helicopters employed by No 120 Squadron for SAR and various transport duties is the Bell 212

T/RT-33As, F-5Bs and Sabres are the only jet trainers used by the PAF.

Miscellaneous units include the 506th Composite Tactical Wing at Mactan; a Composite Air Commando Squadron at Nichols AFB and the 533rd Air Base Squadron (Zamboanga). A general support unit known as the Reserve Airlift and Tactical Support Service flies Beech Mentors, Hughes light helicopters and assorted Cessna types.

Until recently a domestic aircraft industry in the republic was non-existent, but over the past few years steps have been taken which indicate an intention to remedy this situation. The leading organisation in this field is the National Aero Manufacturing Corporation (NAMC) at Pasay City, which in 1974 signed an agreement with Britten-Norman of England to assemble the Islander light utility transport in the Philippines. Over the years, as production know-how and experience accumulated, NAMC has progressed from simple assembly of British-built component kits to gradually increasing parts manufacture, and eventually to the construction of major sub-assemblies. Over approximately the same length of time, NAMC has been following a similar programme with the German MBB Bö 105 helicopter, and it currently has in the development stage a four-seat light utility/agricultural aircraft of all-Philippine design. The

Philippine Air Force, too, has made tentative inroads into the design and manufacturing field in the past few years, first by redesigning an Italian SF 260 trainer to its own requirements – the XT-001, of which the prototype has flown. More recently, the prototype and manufacturing rights of the American Jet Industries (originally Temco) Super Pinto tandem two-seat jet trainer/light attack aircraft have been acquired. No decision has yet been taken (mid-1979) to put the latter into production, but it could fill a definite need for a PAF jet trainer in the not-too-distant future.

Singapore, formerly a British settlement and colony since the early 19th century, became one of the 14 states in the Federation of Malaysia in September 1963, at which time the United Kingdom's defence agreement with Malaya was extended to the members of the new Federation. After only a brief membership, Singapore withdrew from the Federation in August 1965 and became an independent republic, but entered into a new external defence and mutal assistance treaty with Malaysia, introducing compulsory military service of 24 to 36 months some two years later. In November 1971, following a 1968 decision by the British government to withdraw its military presence from bases in the Far East, a new five-nation defence treaty came into force between Malaysia, Singapore, Australia, New

Zealand and Great Britain, with the main purpose of defending Malaysia and Singapore against external attack.

As a result, the Republic of Singapore Air Force (RSAF) has a current strength of approximately 3,000 officers and men, and a total of about 150 aircraft, of which just over 100 are combat types. All armed services in the republic are administered by a single Ministry of Defence, having its headquarters at Tanglin. The RSAF also has its headquarters there, with other major air bases at Bukit Gombak, Changi, Paya Lebar, Seletar and Tengah. The early warning radar station at Bukit Gombak is considered to be one of the most advanced installations in south-east Asia.

All of these bases were former RAF stations, and were inherited when the republic's air force began to be formed in 1968. At that time it was known as Singapore Air Defence Command, and founded its new organisation on the Flying Training School at Seletar. To begin with, students were sent abroad from here for training, but local instruction began in the following year using eight Cessna 172 lightplanes, later joined by a pair of AESL Airtourers donated by New Zealand. Now known as No 150 (Falcon) Squadron, the unit is currently equipped with SIAI-Marchetti SF 260MS trainers from Italy, some of which have been converted to

SF 260W Warrior armed configuration. In late 1969 the first British Aerospace (BAC) Strikemasters arrived to equip No 130 (Eagle) Squadron at Tengah, where from the following January they began to provide jet and weapon training. Later in the same year advanced training was introduced on two-seat Hunter T Mark 75s.

This reflected the fact that the main operational element of SADC then comprised an interceptor/strike and tactical reconnaissance squadron of Hunter FGA Mark 74s and FR Mark 74As: No 140 (Osprey) Squadron, also based at Tengah. Additional Hunters arrived in 1971 to form a second combat unit – No 141 (Merlin) Squadron. Most of the 47 single and two-seat Hunters acquired were still in service in 1979, although one of the two squadrons will re-equip with the Northrop F-5E Tiger II.

The other major combat aircraft in the Singapore inventory is the McDonnell Douglas Skyhawk, the RSAF being one of the small number of air forces to operate a version improved and modernised by Lockheed Aircraft and Service Company in recent years. Beginning life as US Navy A-4Bs, they were refurbished to an advanced standard by LAS and redesignated A-4S. Improvements include an uprated power plant, installation of 30mm cannon with a Ferranti lead-computing gunsight, additional

Top left: the British Aerospace Strikemaster, an armed development of the Jet Provost T Mark 5 trainer, equips No 130 Squadron RSAF.
Top: the badge of No 143 (Phoenix) Squadron, seen on a Changi-based A-4.
Above left: fulfilling a dual role as both a transport and search and rescue type is the Short Skyvan, flown by No 121 Squadron. The unit also operates the C-130 Hercules
Above: the badge of No 140 (Osprey) Squadron

wing control surfaces, and a braking parachute in the tail. Starting in 1974, Singapore received 40 single-seat A-4S Skyhawks and six tandem two-seat, operations-capable TA-4S aircraft, which now equip Nos 142 and 143 Squadrons.

In view of Singapore's small size, the airlift requirements of its air force are quite modest. No 121 Squadron's main workhorses are half a dozen Short Skyvan short-range STOL transports, with a small number of larger Lockheed Hercules to meet any heavy-lift demands. This unit is backed up by No 120 (Condor) Squadron, which performs search/rescue and general VIP/utility transport roles with Alouette, Iroquois and Bell 212 helicopters; some of No 121's Skyvans also double up in the search and rescue role.

It comes as a slight surprise to learn that military aviation in Thailand can be traced back as far as 1911, only eight years after the birth of powered flight. French-trained Army officers from Siam, as the country was known until 1939, helped to form a Royal Siamese Flying Corps in early 1914 and, for the final 16 months of World War I, a contingent of this service fought in Europe on the side of the Allies. When they returned to Siam in the summer of 1919, one in five of the pilots had received either the Légion d'Honneur or the Croix de Guerre. Postwar, as the Royal Aeronautical Service, it remained an air arm of the Army, but was raised to the status of a fully separate and independent arm, with the title of Royal Siamese Air Force, in April 1937. The RSAF's first Commander-in-Chief, Phya Chalerm Akas, was one of the three original Siamese pilots who had learned to fly in France in 1911.

By the late 1930s the Air Force was equipped almost exclusively with aircraft of American design, but its pilots were trained in Britain, France and Italy as well as in the USA. Thailand was occupied by Japanese forces for most of World War II, but the RTAF followed a policy of non-cooperation and successfully avoided making any active contribution to the Japanese war effort. At the war's end it had accumulated an extremely mixed bag of US, French and Japanese aircraft, most of which by that time were more or less obsolete.

In 1947, however, with Royal Air Force assistance plus the purchase of replacement aircraft from Britain and the USA, the RTAF began to reorganise and modernise itself. Further expansion was prompted six years later when the neighbouring state of Laos was invaded by Communist forces. As the troubles in the Indo-China region increased in subsequent years, Thailand also became a major platform for US forces in the area, a maximum of about 80,000 US troops being stationed in the country at the peak of the American involvement in the war in Vietnam. Since the ending of that war, with the consequent withdrawal of the US forces, Thailand has run down the number of US bases within its borders.

Currently organised into four operational groups, the RTAF has a manpower strength of some 43,000 and about 150 first-line combat aircraft. In addition, there is a single Naval air squadron, which flies Grumman S-2F Trackers for maritime patrol and reconnaissance, a fairly substantial Army aviation section, equipped mainly with helicopters and the Air Wing of the Thai Border Police, which has both helicopters and fixed-wing aircraft.

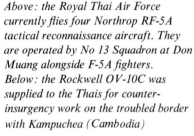

Above: the Royal Thai Air Force currently flies four Northrop RF-5A tactical reconnaissance aircraft. They are operated by No 13 Squadron at Don Muang alongside F-5A fighters. Below: the Rockwell OV-10C was supplied to the Thais for counter-insurgency work on the troubled border with Kampuchea (Cambodia)

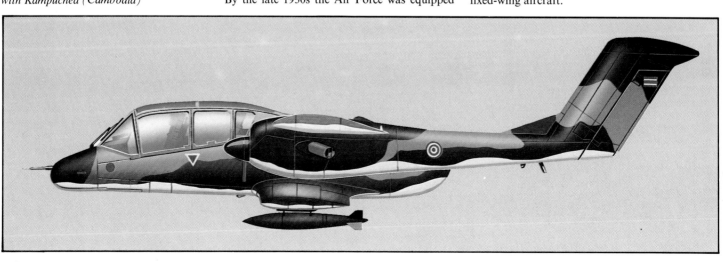

Headquarters of the Royal Thai Air Force are Don Muang air base, near Bangkok; other major bases include Chiengmai, Khorat, Lop Buri, Nakhon Phanom, Nam Phong, Prachuap, Satahip, Takhli, U-Tapao, Ubon and Udon. Nine tactical squadrons, in the 1st and 2nd Wings, make up the strike element of the combat group. These units fly a mixture of Cessna Dragonfly, Fairchild Peacemaker, North American T-6G and T-28D, Northrop F-5A/E and Rockwell Bronco fighter or attack aircraft. A number of Lockheed RT-33As and RF-5As are used for tactical reconnaissance.

Logistics support and tactical training is the responsibility of the 6th Wing, with two large squadrons (Nos 61 and 62) of C-47 and Provider fixed-wing transports; another (No 63) has about 100 helicopters of varying sizes for assault troop and tactical transport duties, or search/rescue. More recent acquisitions include a small number of CASA C 212 Aviocars, from the Indonesian assembly line of this Spanish transport aircraft. A few fixed-wing HS 748s and Bell 212 helicopters are operated by the Thai Royal Flight.

Four basic types of piston-engined primary or basic trainer are used, the most numerous being the Aerospace (New Zealand) Airtrainer, followed in descending order by the SIAI-Marchetti SF 260MT, de Havilland Canada Chipmunk and Cessna T-41D Mescalero. Advanced training, on jets, is given on Cessna T-37s and Lockheed T-33As; operational conversion for combat pilots on two-seat F-5B/Fs

and multi-engined conversion on C-47s. Helicopter training is on the Kawasaki KH-4, a Japanese-developed version of the Bell 47. Most of the RTAF's training is undertaken at Khorat.

A miscellany of utility and light aircraft operates with the 'special services' group of the RTAF, including Pilatus Turbo Porters and Britten-Norman Islanders. Liaison, communications and general 'hack' transport duties are performed by these, as well as by a mixture of Beech and Cessna light twins, Helio Couriers and Bell OH-13 helicopters. There are even a few Airtruk agricultural aircraft, of Australian origin.

Army aviation is quite considerable. Most of it is helicopter-mounted, with the Bell UH-1 Iroquois being by far the most numerous type–there are about 90 in service. Much smaller numbers of Bell and Hiller rotorcraft are also operated; four Boeing-Vertol CH-47 Chinooks provide a heavy-lift capability. A few fixed-wing Beech 99 and Swearingen Merlin aircraft fill the staff transport requirement; the Army's other major type is the Cessna O-1 Bird Dog, which is used for liaison and observation. This is in service in about the same numbers as the Bell UH-1 helicopter.

The Border Police Department, which has about 14,000 personnel, also has a fairly substantial fleet of nearly 50 aircraft. About half of these are helicopters (Bell 204/205/206s), the remainder being a very mixed bag, ranging in size from the Turbo Porter to the DHC-4 Caribou.

Above: the inventory of the Royal Thai Air Force is cosmopolitan by nature, including aircraft of American, Canadian, Italian, Japanese and Indonesian manufacture. The former country is represented, among others, by the Fairchild-Swearingen Metro.
Below: the Thai Navy supplemented its squadron of Grumman S-2 Trackers in late 1978 by the acquisition of two Canadair CL-215 multi-purpose amphibians. The first of these aircraft is pictured shortly after delivery

CHAPTER FIFTEEN

Sub-Saharan Africa
Ethiopia/Kenya/Nigeria/Rhodesia/ South Africa/Sudan/Tanzania/Zaire

Heavily committed to anti-guerrilla operations in the Eritrean region, and fresh from combat against neighbouring Somalia, the Ethiopian Air Force (Ye Ethiopia Ayer Hail) has received considerable Soviet support in the past two years and is now one of Black Africa's strongest air arms.

Until recently, the EAF was reliant on the United States, receiving its first combat equipment in July 1960 with the arrival of a squadron of North American F-86F Sabres, augmented ten years later by a few surplus examples from Iran. Transition training was handled by the Lockheed T-33, while North American T-28 Trojans were received for both basic instruction and COIN

operations against dissident rebels.

In addition to three de Havilland Doves, the transport wing obtained Douglas C-47s and a couple of C-54s, supplemented in later years by some 18 Fairchild C-119 Boxcars converted to C-119K standard by the addition of underwing jet pods, the last two arriving from Belgium as late as 1972. In addition to providing paratroop transport facilities, the C-47s were also used for civilian roles, serving areas not covered by internal airline routes.

First deliveries of the Northrop F-5A and F-5B trainer were made in 1966, resulting in the formation of a new fighter-bomber squadron. With the addition of a further three aircraft in 1972, the EAF had a total strength of 13 F-5As and two trainers. Four refurbished Canberra B Mark 2s were added in 1969 for light bombing missions, giving Ethiopia four operational squadrons, equipped with the Sabre, F-5, Trojan and Canberra.

At the main instructional base of Harar Meda near Addis Ababa, a dozen or more Saab Safirs provided primary training facilities for students from the EAF and other African countries, including Nigeria, Somalia, Kenya, Uganda, Tanzania and Sudan. Basic and advanced courses on the Trojan and T-33 were subsequently established. Final T-33 deliveries were made in late 1972 with the arrival of two aircraft obtained from Holland.

Diversifying its sources of supply, Ethiopia also received an Ilyushin Il-14 transport and two Mil Mi-8 helicopters from the Soviet Union. An Aérospatiale Puma and some Alouettes were obtained from France, and Agusta-Bell 204B

Iroquois via Italy, together with six UH-1H versions from the USA. Three DHC Twin Otters were acquired from de Havilland Canada for use as patrol aircraft.

In 1974, the political status of Ethiopia deteriorated with the overthrow of Emperor Haile Selassie in August, but faced with a growing threat from Soviet-supplied Somalia, plans for further American aid remained unaltered. Some 22 million dollars in military equipment had been promised to the armed forces in the previous year, including 16 F-5E fighter-bombers, 12 Cessna A-37B light strike aircraft and 15 Cessna 310s for training, together with tanks and other army weapons.

Three months later, 59 government officials, including the new head of state, were massacred in a communist take-over, and the United States ceased all military aid to Ethiopia, although eight F-5Es were later obtained by cash transaction. A secret arms agreement, not revealed for six months, was concluded with the USSR in December, and in May 1977 all American personnel in Ethiopia were expelled.

Scarcely had the first Soviet aircraft been unpacked, before they were committed to a bitter civil war which rapidly escalated in the spring of 1977. With Somali support, Eritrean separatists in the coastal strip bordering the Red Sea had been mounting a guerrilla campaign since 1960, in an attempt to secure an independent Eritrean state. Over the same period, Somalia had laid claim to the Ogaden region of Ethiopia, and in concert with increased guerrilla activity, launched an invasion of the disputed area on 13 July.

The situation was not without its irony, as both antagonists claimed a Marxist-Leninist government and both were being supplied by the Soviet Union. Ethiopian F-5s bore the brunt of the early fighting, attacking Somali armoured columns as they advanced further into the Ogaden; within a month, Ethiopia was claiming the destruction of eight MiGs, against counter claims of 25 EAF aircraft shot down.

During September, giant Antonov An-22 transports delivered 48 MiG-21s to Ethiopia, along with SA-3 and SA-7 SAMs and Sagger anti-tank

missiles, although by the time the MiGs entered service the following month, half of Somalia's 56 MiG-17s and -21s had been claimed as destroyed. Throughout the campaign, EAF fighters, aided by Canberra bombing missions mounted from Dire Dawa, were able to maintain air superiority and prevent the Somalis from exploiting their undoubted successes on the ground.

A second airlift from the Soviet Union in December reportedly brought MiG-17s, 21s and 23s in 225 cargo flights by An-22s and Ilyushin Il-76s, while aid to Somalia was reduced to a trickle of SAMs when its offensive faltered. As Ethiopian forces took the offensive in February 1978, missile-armed Mi-4 and Mi-8 helicopters took a heavy toll of Somali tanks and vehicles, while the following month, Mi-6 'flying cranes' with Russian crews airlifted 70 tanks for the final and decisive assault.

By 14 March, after receiving 500 million pounds in Soviet aid, Ethiopia had driven out the last of the invaders. Estimates of the number of aircraft delivered vary considerably, but are believed to total 48 MiG-21s and 24 MiG-23s, although Eritrean guerrilla leaders state 160 MiG-21s, 40 MiG-17s and 24 MiG-23s were received in the early months of 1978 alone. Somalia claimed to have shot down at

Opposite below: the ageing Fairchild C-119 transports of the Ethiopian Air Force have been bolstered by the thrust of two wing-mounted turbojets. Opposite bottom: deliveries of the Northrop F-5 to Ethiopia ceased abruptly when a communist-inspired revolution occurred in the mid-1970s. Eight further examples have, however, since been obtained.

Below: prior to the revolution, Western designs featured strongly in the Ethiopian ranks. The de Havilland Twin Otter, pictured in army markings, was one such type.

Bottom: despite the design's undoubted obsolescence, the English Electric Canberra medium bomber only entered Ethiopian service in 1969

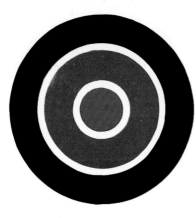

MiG-23s were involved in the re-taking of Keren from the guerrillas in December 1978, and the EAF continues to launch periodic attacks on known Eritrean positions in the north of the country. In early 1979, guerrillas shot down an Antonov transport as it was approaching Asmara and have destroyed at least one other aircraft in a similar attack. Although fragmented, the guerillas are still harassing government forces in a type of warfare demanding the counter-use of COIN-type aircraft and attack/troop-transport helicopters, of which the EAF has only small numbers.

Although under one-party rule, Kenya is one of Africa's more liberal states. Its air force is equipped throughout with Western aircraft and continues to look to Britain and the United States for its future requirements. Pro-Western leanings have earned Kenya some animosity from fellow African states, not least through the use of Nairobi as a staging post in the famed Israeli raid on Entebbe. Consequently, the KAF is undergoing a period of re-equipment, bringing more potent strike aircraft into service as a counter to Soviet supplies to nearby countries.

After attaining independence from Britain in December 1963, Kenya sought the help of a Royal Air Force mission to assist formation of the KAF at Eastleigh, near Nairobi. Inaugurated on 1 June 1964, the KAF began pilot training on six British-supplied DHC Chipmunks, while ground crews and technicians were also instructed by the RAF.

Shortly afterwards, the first operational squadron formed with 11 DHC Beavers supplied by Canada for light transport and liaison duties. In late 1965, a second unit was established following the arrival of four DHC Caribou, some Beaver pilots being transferred to the new squadron after a three month conversion course at Eastleigh under RAF supervision. Navigators for the Caribou were among some of the first Kenyans to be trained in Britain.

By mid-1967, the KAF had 40 African pilots, of which several were later sent to the Central Flying School at Little Rissington for RAF training as instructors. On return, they were able to relieve the British personnel still engaged in pilot training and complete Africanisation of the air force. From some 700 personnel on strength in 1976, the KAF has recently expanded to 1,200 within the framework of a typical defence expenditure of about £40 million per year.

During its early years, the air force engaged solely in transport and support work, including trooping, freighting, paratrooping, supply dropping and casualty evacuation. Some Beavers were additionally fitted with spray-bars for anti-locust operations. Caribou undertook regular services to supply Kenyan garrisons at Carissa, Mandera, Moyale and Wajir.

least 12 F-5s, a figure which is probably exaggerated, in addition to the two aircraft reported to have been flown to Sudan by defecting pilots in July 1977. The loss of three transport aircraft, including one civil and one military C-47, was admitted by Ethiopian spokesmen.

Further Soviet-supplied aircraft will no doubt be incorporated in the EAF in the next few years as existing equipment is due for replacement. The USSR intends strengthening its hold on the Horn of Africa. It already has a considerable number of advisors installed, and is training Ethiopian pilots in Russia, while some 14,000 Cuban troops assist the Ethiopian army.

Top left: the Kenyan air force's ten Northrop F-5E fighter-bombers and two F-5F trainers operate in the air defence and strike roles from Nanyuki.
Middle left: the Kenyan air force acquired six Dornier Do-28D Skyservant transports in 1978.
Above left: six DHC-4 Caribou are operated by the Kenyan air force for transport. They are likely to be replaced by the DHC-5 Buffalo

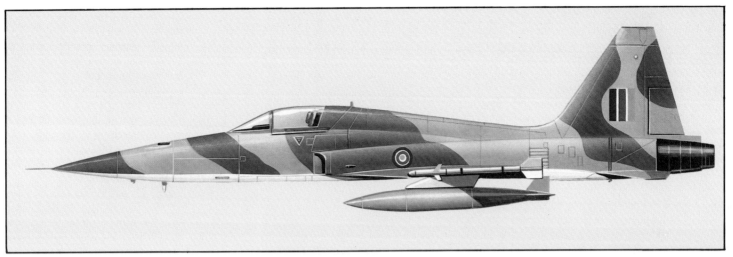

Combat potential was introduced in the early 1970s with the arrival of six Strikemaster Mk 87s to serve in the dual role of advanced trainers and light strike aircraft. At the same time, basic training transferred from the Chipmunks to five new Bulldogs, ordered from Scottish Aviation in October 1969. Following completion of an airfield at Nanyuki in 1973, this having been built with the assistance of RAF engineers, all flying units were moved into the new base, leaving only the administrative elements at Eastleigh.

It was to Nanyuki that the first of six refurbished Hawker Hunters was delivered in June 1974, the order comprising four F Mark 80 fighters and two T Mark 81 trainers to increase further combat potential. It had originally been proposed to maintain KAF operational strength at six Strikemasters and six Hunters; an offer of surplus F-5As and F-5B trainers from Iran as part of a United States aid programme, then totalling about 5 million dollars per year, was turned down in 1976. Orders in that year included nine more Bulldogs for a further expansion of the training programme. Helicopter operations continued to mount, however, and following two Bell 47Gs, France was to supply small numbers of Aérospatiale Pumas, Alouette IIIs and Gazelles, principally for liaison and army support.

As the Soviet presence in Somalia built up, Kenya re-assessed its defensive capability in the light of the changing situation. It transpired that the Somalis were preoccupied with Ethiopia and were later to lose their Russian backing, but events in Uganda were a secondary cause for concern. These threats, too, were later to recede with the ousting of the infamous President Amin and his regime, but air defences were strengthened in the meantime by the addition of 10 Northrop F-5Es and two F-5F trainers bought outright from the USA and based at Nanyuki. Apart from their armament of Sidewinder air-to-air missiles, the

Top: the Kenyan air force's F-5Fs are armed with Sidewinders for air defence duties.
Top right: about 12 MiG-17s remain in Nigerian air force service. The Soviet Union supplied 41 during the Biafran civil war of 1967–70.
Above right: the Federal Nigerian air force received five Ilyushin Il-28 bombers from the Soviet Union during the Biafran conflict

F-5s may also be used for strike missions.

While tension between Kenya and Somalia and Uganda has lessened recently, a new problem has emerged with the growing Soviet and Cuban presence in Ethiopia. If, or perhaps when, the Eritrean guerillas are defeated, Ethiopia may look south under the leadership of Lt Gen Mengistue Haile Mariam, the self-styled 'Castro of Africa'.

Kenya has therefore made plans to increase its defences and begun the construction of two new airfields in the north of the country, each with a 2,750m (9,000ft) runway. As a Strikemaster replacement, 12 BAe Hawks were ordered early in 1978 under conditions of some secrecy, and the

first of these is expected to fly before the end of 1979.

Lacking specific anti-armour weapons, Kenya has turned to the United States for the supply of 32 Hughes 500MD Defender helicopters, together with a batch of 2,100 TOW anti-tank missiles, agreed in March 1979. The Defender order will include 17 'Quiet Armed Scout' versions fitted with the revolutionary mast-mounted stabilised sight, which enables the helicopter to shelter behind natural features with only the laser-seeker exposed, while acquiring targets for missile-equipped counter-parts. Army units will be provided with BAeGW Swingfire ATMs to be mounted on armoured cars, from a further unannounced order placed in the United Kingdom. These moves have doubt-less been prompted by the operations of some of the several hundred tanks recently shipped to Ethiopia and used in Soviet-directed attacks on Eritrean strongholds.

France, too, has promised economic and tech-nical assistance, following a visit to Paris by Presi-dent Daniel Moi in November 1978. The form of this aid has yet to be made public, but it reportedly includes some helicopters as well as armoured vehicles and communications equipment. Signifi-cantly, Kenya presently has no surface-to-air missiles to support its F-5s, and an order for Cro-tales or similar weapons may result.

KAF second-line equipment includes two Piper PA-31P Pressurised Navajos and one Aero Com-mander 680F, for communications, while the tran-sport force was further strengthened in 1978 with the delivery of four DHC Buffaloes and six Dornier Do-28D Skyservants to supplement the Caribou and Beavers. The Buffalo complement has recently been increased to eight, indicating possible re-placement of the Caribou.

Although granted sovereignty in 1960, the Federation of Nigeria did not establish an air force for more than three years. A former British protectorate, in which English remains the official language, Nigeria chose to accept military assist-ance from several sources in the creation of its armed forces.

Initial pilot training was begun in Canada, West Germany and Ethiopia, and after some participation from India, Germany supplied 42 Luftwaffe personnel, 14 Piaggio P149Ds and 20 Dornier Do-27s in mid-1963 to provide the nucleus of the Federal Nigerian Air Force. Officially formed in January 1964, the FNAF began operations from Kaduna, training its pilots with 120 hours on the P149s, followed by conversion to the Do-27.

A major expansion resulted from the civil war which accompanied the secession of the eastern states on 30 May 1967, under the collective name of Biafra. While the breakaway forces attempted to gain what military equipment they could via illicit world sources, the Federal Government received its first combat equipment from the Communist Bloc in the form of 16 MiG-17 strike-fighters and MiG-15UTI trainers, delivered through Egypt and Algeria.

As the bitter war continued, five Ilyushin Il-28 light bombers were received, and were subsequently flown by Egyptians and white mercenaries as well as FNAF pilots. Czechoslovakia also sold a dozen or so L-29 Delfin strike-trainers, while ten ex-

airline C-47s were impressed for service as transports and 'bombers'. The harrowing campaign ended with the defeat of Biafra in January 1970, at which time the Federal forces had lost some 20 jet air-craft, although only one was actually shot down in aerial combat.

Operation of Soviet aircraft continued on a limited basis after the civil war, but Western equipment was again made available. As replace-ments for four Nord Noratlas aircraft supplied earlier, the FNAF obtained six Fokker F-27s and four Do-28D Skyservants for transport and com-munications. Transport capability was further enhanced with the arrival of the first of six Lockheed C-130 Hercules at Ikeja Airport, Lagos in Septem-ber 1975. In the previous year, an initial batch of 20 Bulldog Mk 123s was supplied to replace the remaining P149Ds in the primary training role, while three Piper Navajos joined the communi-cations flight.

A coup of July 1975 brought a new military regime to power with plans to modernise the armed forces. These aspirations were rapidly realised with the delivery of about a dozen MiG-21MF fighters to equip a new squadron based at Kano, before the end of the year. They were joined by a further batch of 12 aircraft together with two MiG-21UTI trainers, also believed to be based at Kano alongside the remnants of 41 MiG-17s originally received. The Il-28 Beagles have now been with-drawn. Helicopter strength has been increased from the original three Westland Whirlwinds and ten Alouette IIIs by the addition of six, and later a further seven Pumas for medium-lift work and four

Top: the Nigerian air force's four Nord Noratlas transports were replaced in the mid-1970s by more modern types. The transport force in 1979 comprised six Lockheed C-130Hs, two Fokker F-27 Series 400 and one Fokker F-28.
Middle: a total of 20 Scottish Aviation Bulldog Mark 123 primary trainers was supplied in two batches in 1974 and 1978 to the Nigerian air force.
Above: the national insignia is not generally carried by Zimbabwe-Rhodesian combat aircraft, both for security reasons and to disguise their source of supply

Top: a Hawker Hunter FGA Mark 9 of
No 1 Squadron ZRAF, which flies the 9
survivors of 12 delivered to Rhodesia in
1963. They are deployed in the tactical
fighter and strike roles.
Middle: No 24 Squadron, South African
Air Force operates the Hawker
Siddeley Buccaneer S Mark 50 in the
maritime strike role from Waterkloof.
Sixteen examples were acquired in 1965.
Above: despite the efforts of the United
Nations' arms embargo, the South
African air force has modern equipment,
much of it built under licence. The
country's insignia is illustrated

MBB Bö-105s for SAR. The last-mentioned were
augmented by a subsequent order for 20 locally-
assembled examples in 1978–79.

Because of capacity shortages and accident
losses incurred by Nigeria Airways, the FNAF
leased two F-27s to the national airline, replacing
these by a presidential F-28 Fellowship and a VIP
Grumman Gulfstream II, delivered in late-1976.
C-47s and C-54s have also been retired, while
further light transport work has been assumed by 12
more Skyservants ordered from Germany in 1976,
resulting in the disposal of the four Do-28Bs.
In late 1978, the FNAF was reportedly operating
five Aermacchi AM 3C liaison aircraft which are
assumed to have come from the South African
production line.

First Nigerian missiles were the K-13 Atoll
air-to-air rounds equipping MiG-21s in the air
defence role, but the armed forces are now taking
delivery of several other systems. These include
Short Seacats on two corvettes being built in
Britain for the navy, MM-38 Exocet anti-ship
missiles on three fast patrol boats under construction
in France and OTOMAT ship-to-ship rounds for
three German-built patrol boats.

Links with West Germany have resulted in an
order for 12 Alpha Jets, placed in December 1978
for delivery in 1981–82. Unlike other export orders,
the aircraft will be built on the German assembly
line and will thus presumably be versions optimised
for the strike role. The Alpha Jet will replace the
L-29 Delfin in FNAF service and was apparently
chosen after the evaluation of comparable types,
including the Delfin's successor, the L-39 Albatross.

As a major oil exporting nation, Nigeria is capable of devoting considerable financial resources
to arms purchases, which totalled in the region of
£1,400 million in 1977–78, about a third of all
government spending. Envisaging a dominant position in African affairs in the years to come, Nigeria
continues to expand its armed forces and has recently
strengthened relations with the United States,
becoming the second largest external supplier of
US oil.

US hopes of military equipment sales to Nigeria
have yet to be officially confirmed, but ties with the
Soviet Union were weakened with the announcement that the MiG-21 training mission of 40
instructors and technicians is to be reduced to
five by 1980. Citing condescending attitudes and
inefficiency (the reasons for which President Sadat
expelled his advisors from Egypt), the FNAF is
clearly dissatisfied with Soviet equipment, and is
likely to look elsewhere when the MiGs are due for
replacement.

With very few friends in the world, and surrounded by hostile Black African states on three sides,
the air force of the newly-named Zimbabwe-
Rhodesia has been on an operational footing for
several years. Supply of aircraft for daily anti-
guerrilla patrols and spares for the jet fighter-
bombers used in attacks against training camps in
neighbouring Zambia have presented a constant
headache for the ZRAF.

Following the dissolution of the Federation of
Rhodesia and Nyasaland, dating from secession
rights granted in March 1963, the largely British-
equipped Rhodesian Air Force was divided between

267

Above: No 2 Squadron, SAAF operates Mirage IIIs in the reconnaissance and fighter-bomber roles from its base at Waterkloof. The SAAF has acquired a total of 59 Mirage IIIs of various marks, earlier deliveries being from France with licence production following.

Above right and below: the SAAF has 12 Mirage IIID-2Z two-seat trainers, purchased from France in 1972. They are operated by No 85 Advanced Flying School at Pietersberg

SUB-SAHARAN AFRICA

Zambia and Rhodesia, the latter receiving the majority of aircraft. Financial constraints resulted in the disposal or storage of some aircraft, but the position was radically altered with Ian Smith's Unilateral Declaration of Independence on 11 November 1965. United Nations sanctions removed all possibility of obtaining armaments legally, while the black guerrilla campaign which began in mid-1972 forced the RhAF onto the defensive.

Sources and numbers of most aircraft obtained by devious means have been discovered, but there is some interchange of equipment with South Africa, resulting in reports of SAAF types of aircraft in Rhodesian service. ZRAF aircraft are camouflaged and fly without national insignia or serial numbers, rendering assessment of quantities in service impossible in some cases. Several Aérospatiale Alouette IIIs have reportedly been transferred from South Africa, but estimates of the number currently on ZRAF charge vary between 30 and 66.

Conjecture also surrounds recent reports of Dassault Mirages in ZRAF service, these supposedly comprising six IIIE and 16 IIIR-2Z aircraft, although South Africa only received eight Mirage IIIRs of all types. However, Rhodesian pilots are known to have undergone training in South Africa on Mirages, and some aircraft may be on temporary attachment.

Late in 1978, Indonesia was claimed to have supplied Rhodesia with four of its 16 North American OV-10F Broncos for COIN operations. At present their motives for risking world criticism are unknown, and apparently the US security services seem disinclined to accept the report that some of their aircraft have been illegally passed on

to the Smith Régime, as no formal disclaimer or warning to Indonesia has been forthcoming. Five North American T-28 Trojans or Fennecs are also stated to be in Rhodesian service, but their use is more difficult to comment upon, as many examples of the type have served in South American air forces and could have been shipped to a South African port with false documentation.

The SAAF is, however, known to have assisted in the maintenance of Rhodesian Vampires by the transfer of spares for these aircraft when they were withdrawn from South African service, and similarly, the joint operation of Alouette IIIs would lend itself to mutual support. Both countries may be experiencing difficulties with the maintenance of their BAC Canberras, although these and the Hawker Hunters seem to be regularly flown despite a 14-year spares embargo.

Other attempted purchases have been less effective. Projects which have failed to come to fruition have included 28 ex-Luftwaffe F-86K Sabre all-weather interceptors from Venezuela, 31 Hunters, plus 16 spare engines and ammunition from Jordan, both of which were blocked by international pressure, and 14 CT-4 basic trainers, vetoed by the New Zealand government.

Despite these problems, the ZRAF now has considerably more operational aircraft than at the time of UDI, when overall strength comprised six squadrons, including two in the training role, manned by 1,000 personnel. Present strength is 70–80 strike aircraft and a similar number of helicopters and second-line types, and 1,300 officers and men. Annual defence budget for 1978–79 is about £120 million. The majority of aircraft are based at New Sarum (Salisbury) and Thornhill

(Gwelo), but there are considerable numbers of forward airstrips, particularly near the Mozambique border to combat infiltration by ZANLA guerrilla forces.

Principal tactical fighter squadron is No 1 at Thornhill, flying about nine of the 12 Hunter FGA Mk 9s delivered in 1963. Hunters have been particularly active in attacks on guerilla camps in Zambia. Also at Thornhill is No 2 Squadron with some eight Vampire FB Mk 9 strike fighters remaining from the 15 received in 1953–54, plus a training flight with a further eight Vampire T Mk 11s out of the 15 delivered new in 1955, possibly augmented by a few ex-SAAF examples. Light fixed-wing attack aircraft are operated by No 4 Squadron with most of the 10 Aermacchi AL 60F-5s delivered from Italy, Cessna 337s and SIAI Marchetti SF 260 Warriors.

The origins of the last two types are particularly interesting as both arrived in Rhodesia via devious routes. In November 1975, 18 Reims-Cessna 337 Miliroles (known now as the Lynx in ZRAF service) were registered as French civil aircraft for delivery to a Spanish fishing company, based in Spain under Danish supervision for 'fishery spotting'. They were delivered southwards in the following months, finally arriving in Mozambique and Comores from where they disappeared without trace. Their French civil registrations–which had never been officially notified to the international authorities–were cancelled, and ultimately re-allocated to other genuine civil machines. It was not until 1977 that South African television showed Cessna 337s in combat with the RhAF and thus revealed the fate of the aircraft. A further four were also believed to have been delivered, although their origin and movements are

not known at the time of writing.

In 1978, the ZRAF received 22 SF 260 Warriors in a similar manner. The aircraft were bought by a Belgian dealer, supposedly for the Comores Islands, which already had three of the type in service. Again, it was South African television which first depicted the use of these aircraft in Rhodesia. Both the Lynx and Warriors are equipped with SNEB rockets for ground attack.

No 4 Squadron is also reported to be operating Islanders from New Sarum, although the often quoted figure of 14 is more likely to be in the region of about 10. Two aircraft were supplied to civilian operators before UDI, and a further two damaged examples were made into one airworthy aircraft. The remainder have most probably come from those left in Mozambique after the Portuguese withdrawal. At least two Islanders are used for VIP transport, while the others are employed on reconnaissance and light transport duties. None are known to have been converted to carry arms.

Three other squadrons at New Sarum operate in the bomber, transport and helicopter-attack roles, the latter usually dispersed to forward airstrips. No 5 Squadron has about eight Canberra B Mk 2s fitted with under-fuselage rocket racks, and two T Mk 4 trainers, survivors of the 15 B Mk 2s and three T Mk 4s supplied from RAF stocks in 1958. Transport tasks are allocated to No 3 Squadron which has a dozen Douglas C-47s and a Beech Baron, and may also use the three Cessna 310s and a single Cessna 182 also known to be in the inventory, together with a recently acquired Douglas DC-7 (and a second airframe for spares).

Bearing the brunt of anti-guerrilla roles is No 7 Squadron operating gunship versions of the Alouette III and UH-1 Iroquois. A dozen of the latter were revealed to be in service by late 1978, and are almost certainly Agusta-built AB 205As originally supplied to Israel. Considerable numbers of Portuguese Alouette IIIs were abandoned in Mozambique and have yet to be accounted for; it is probable that these have contributed to the figure of 66 claimed to be in use by the ZRAF.

Heavily involved in the 'search and destroy' missions mounted by the bi-service 'Fire Force', No 7 Squadron has suffered heavy casualties. Both types of helicopter are armed with 20mm or 0.303 machine guns and used as troop transports. No 3 Squadron also contributes to 'Fire Force', providing C-47s for paratroop operations. Remaining units at New Sarum are the Photographic Establishment, Air Movements Section, Aircrew Selection Centre, Apprentice Training School and a Parachute Training Section operated on behalf of the Army.

Pilot training is initiated on about a dozen Percival Provost T Mk 52s of No 6 Squadron from 16 aircraft delivered new in 1955, and continues in South Africa with jet-conversion to the Atlas Impala (Aermacchi MB 326G built under licence). A small number of Impalas is believed to be owned by Rhodesia, although these have remained in South Africa for purely training purposes.

Although threatened by neighbouring Black African states and subject to a UN embargo, South Africa is more fortunate than its Rhodesian ally in having facilities for the production of its own armaments. The South African Air Force is equipped with sophisticated aircraft such as the Mirage F1 as well as light strike jets, and is thus capable of defending itself against any large-scale invasion by regular forces.

Expansion of South African defences began during the 1960s in response to threats from newly-independent Black states, and has more recently been accelerated to combat armed insurgents into northern and eastern Transvaal and eastern Natal. Arms procurement has been made more difficult by world criticism of the Apartheid doctrine, and unimpressed by token gestures to Black self-determination, the UN reinforced arms embargoes late in 1977. France, until recently a major source of fighters and helicopters, now ostensibly provides only naval equipment, although Italy – surprisingly, in view of a strong left wing in parliament – continues to ignore the boycott.

A long-term effort to provide indigenous arms manufacturing capability has resulted in South Africa now producing about 75 per cent of its requirements, while further seeking to increase this proportion to 80 per cent and above. A major break-through occurred in 1971 with the granting of a licence to produce the Mirage F1, its Atar 9K-50 engine and avionics, as a follow-on to the Aermacchi MB 326 trainer and strike aircraft produced in quantity by Atlas Aircraft Corporation.

France has additionally been a prime source of missiles, South Africa having contributed to the development of the Crotale SAM and obtained a large number for air defence. It has also bought SS 11 and ENTAC anti-tank missiles as well as submarines and armoured cars. A French embargo on military equipment, including spares, was announced by President Giscard d'Estaing early in 1977, and according to some reports was having an adverse effect on serviceability of the Mirage F1s by early 1979.

The statement was categorically denied by a senior SAAF spokesman, who also supplied a local newspaper with photographic proof that the Mirage F1 fleet was still airworthy. The assertion may have been more credible had the photograph not been a well-worn print of Mirage IIIs (a substantially different aircraft) in formation, taken more than ten years ago. The Mirage III may also be suffering from spares-starvation but supplies could be coming from Israel, which operates the type and additionally is manufacturing the similar IAI Kfir.

One of the latest products of the South African arms industry is the V3 air-to-air missile which was acknowledged in May 1979 and is claimed to be ready for operational use on SAAF interceptors. Hitherto, Mirage F1s have been flown with the Matra R 550 Magic AAMs on wingtip attachments (and four pylon-mounted Matra 155 rocket pods), and externally, the V3 appears very similar to the French round. Jointly developed and produced by the Council for Scientific and Industrial Research, the Armaments Corporation and Atlas Aircraft, the V3 is supposedly an improvement on the Magic, although it may prove to be no more than a licence-produced example incorporating some local components. South Africa was known to be developing an AAM code-named 'Whiplash'

Above right: the Macchi MB 326M trainer is built under licence in South Africa by Atlas Aircraft Corporation as the Impala Mark 1. The type serves with the SAAF Flying Training School and Air Operational School at Langebaanweg. Additionally, five squadrons of the Active Citizen Force are equipped with the Impala, each unit's establishment being 15 aircraft. Right: No 19 Squadron operates the Puma assault helicopter from Zwartkop and Durban, 20 Pumas having been supplied by France to supplement earlier deliveries of Super Frelons. Below: long-range maritime reconnaissance is the role of No 35 Squadron, which flies from D. F. Malan airport, Cape Town. The Squadron operates seven Shackleton MR Mark 3s, which were supplied by Britain in 1957

as far back as 1966, claiming a successful trials launch against a Mach 3 target in 1971. Little has subsequently been heard of the missile, and it presumably fell below specification and development was abandoned.

Military expenditure will again increase in the coming financial year to the record level of R2 billion (£1,140 million), a 25 per cent increase over 1978, and will include provision for the formation of a parachute brigade and overhaul of the air defence system. Proportionally, however, the SAAF will have 18 per cent less financial allocations, at R74 million, although work will begin on modernisation of static air defence radars and expansion of the mobile systems. It was also revealed in the April 1979 estimates that helicopter spares were being produced in the country.

The SAAF underwent a change in command structure in 1970 with the disbandment of the previous Group formations in favour of the establishment of six functional Commands – Strike, Maritime, Air Transport, Light Aircraft, Training and

Top: reports have suggested that certain South African Air Force aircraft and helicopters have seen service with the Zimbabwe Rhodesia Air Force, devoid of national and unit markings. The Aérospatiale Alouette III (pictured) has been mentioned in this connection.
Above: the English Electric Canberra is flown by No 12 Squadron SAAF from its Waterkloof base

Maintenance. These control some 700 aircraft and helicopters and about 10,000 personnel, of which almost half are two-year conscripts. Nominal squadron establishment is 12 aircraft with 3–4 reserves, but actual numbers vary considerably.

Strike Command, headquartered at Waterkloof, controls fighter, bomber and Active Citizen Force (reserve) units. No 1 Squadron at Waterkloof operated CL-13B Sabre 6s until 1976, when it became the first unit to receive Mirage F1AZ ground-attack fighters from 32 on order for the SAAF. All 16 Mirage F1CZ interceptors, delivered from France in 1974 and equipped with Cyrano IV radar/fire-control systems and Magic AAMs are allocated to No 3 Squadron at the same base. Hopes of producing 50 or more extra Mirage F1s at the Atlas plant may be adversely affected by recent French arms restrictions.

Earlier Mirage IIIs are also stationed at Waterkloof, the SAAF having acquired 59 aircraft of various marks. A first order for 16 IIICZ interceptors armed with Matra R 530 AAMs was issued to No 2 Squadron and was augmented by three IIIBZ trainers and four IIIRZ tactical reconnaissance fighters. In 1965–66, 16 (later 17) IIIEZ strike/interceptors were delivered with three IIIDZ trainers and a supply of AS 20 and AS 30 air-to-surface missiles.

A further 16 aircraft order was placed in 1972 for Mirage IIIs fitted with the uprated Atar 9K-50 power plant, similar to that in the Mirage F1. While the designation should more correctly be Mirage 50, the batch comprised 12 IIID-2Z trainers and four IIIR-2Z reconnaissance variants. Mirage IIIs are operated by No 2 Squadron at Waterkloof and No 85 Advanced Flying School at Pietersberg. No 2 Squadron has a present complement of 16 IIICZs, four IIIRZs, four IIIR-2Zs and three IIIBZs.

Two medium bomber squadrons fly British-supplied equipment from their base at Waterkloof. No 12 Squadron has six Canberra B(I) Mk 12s ordered in 1962, together with three Canberra T Mk 4 trainers supplied from surplus RAF stocks. Aircraft are used for strike and reconnaissance roles, and the sixth SAAF Canberra was, in fact, the last of the type to be produced by English Electric.

Maritime strike is allocated to No 24 Squadron, with Buccaneer S Mk 50s supplied under the Simonstown Naval Agreement for defence of vital shipping lanes around the Cape. Delivery of the first of 16 Buccaneers began in November 1965, after crews had trained on their aircraft at Lossiemouth under Royal Navy supervision, although one was lost *en route*. A requirement for additional Buccaneers went unfulfilled following cancellation

of the Agreement by Britain.

SAAF Buccaneers are equipped with long-range underwing tanks and have a retractable Bristol Siddeley BS605 rocket boost motor for high-temperature take-off located in the rear fuselage, but are otherwise similar to the RAF's S Mk 2s. By early 1979, No 24 Squadron was reportedly reduced to only six aircraft.

While the Buccaneers were officially the last operational aircraft supplied by Britain, South Africa obtained 555 Short Tigercat SAMs and 162 practice rounds from Jordan in mid-1974 for point defence, supplementing the longer-range Crotale.

Earlier maritime equipment from the UK, in the form of seven Avro Shackleton MR Mk 3s, continues to serve in Maritime Command headquartered at Cape Town, for patrol, reconnaissance, ASW and naval co-operation. Eight new Shackletons were ordered for No 35 Squadron at DF Malan, Cape Town in 1957, one of which was lost in an accident in 1963. Purchase of Nimrods was ruled out by international embargoes, and the SAAF was forced to embark on an extensive refurbishment programme as aircraft neared the end of their fatigue limitation. The first aircraft of the programme, approaching the 4,300 hour wing-spar fatigue life, had the component removed and shipped to Britain in March 1973 for overhaul by Hawker Siddeley, while the remainder of the airframe received attention from SAAF technicians. After being grounded for three years, the Shackleton was returned to No 35 Squadron and followed by a second, similarly treated, aircraft. Others will probably receive refurbishment, although the SAAF may eventually be compelled to obtain a less sophisticated 'coastguard' type as a replacement.

Short-range surveillance and coastal patrol is performed by No 27 Squadron at Ysterplaat with radar-equipped Piaggio P166S Albatross aircraft, of which 20 were obtained from Italy in 1973–74. At the same base is No 25 Squadron flying six C-47s for transport and support roles, including target towing, and the Westland Wasp HAS Mk 1s of No 22 Flight attached to the navy's three frigates. Having lost half the original 10 Wasps, South Africa succeeded in placing an order for an additional seven, of which six were delivered in 1973–74, before a change in British government brought a veto on the last. Ironically, South Africa had shortly before given up one of its delivery positions to allow the Dutch Navy to obtain an urgently-needed attrition replacement for its Wasp complement. One or two Alouette IIs are used by No 22 Flight for helicopter conversion training.

Major component of Transport Command at Waterkloof is No 28 Squadron with seven Lock-

Top: in addition to the tactical transport element of C-130 Hercules and Transalls, the SAAF continue to operate the piston-engined Douglas C-47 and C-54 (pictured).
Above: the Hawker Siddeley Buccaneer was supplied under the terms of the Simonstown Naval Agreement in the mid-1960s. By 1979, however, only half a dozen aircraft remained serviceable

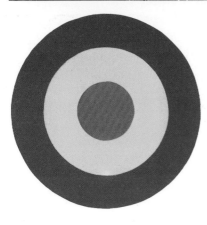

Top: having flown the Hunting (BAC) Jet Provost since 1962, the Sudan Air Force placed a repeat order seven years later, this time for five examples of the pressurised T Mark 5 equivalent. One of these T Mark 55 aircraft is illustrated. Above: the DHC-5 Buffalo has proved popular in the service of African air forces with its useful short take-off capability. Sudan took delivery of four Buffaloes in 1978

heed C-130B Hercules for long-range operations and nine C160Z Transall tactical freighters. Two other squadrons are based at Zwartkop, No 44 with 10 C-47s and five C-54s, and No 21 with a VIP fleet of one Viscount 781, four HS125s and five Swearingen Merlin IVAs, including one fitted out as an air ambulance. Three of the original HS125s were involved in a disastrous accident when they flew into Table Mountain shortly after delivery, although replacements were obtained soon afterwards. Nos 27 and 44 Squadrons are manned by both regular and reserve personnel.

Four helicopter units are attached to Air Transport Command for utility and support roles. No 15 Squadron at Zwartkop and Bloemspruit retains 14 of the 16 Aérospatiale Super Frelons delivered from 1967 for heavy-lift duties, while No 19 Squadron has two flights, each of 10 Pumas, at Zwartkop and Durban. Alouette IIIs are flown by No 16 Squadron with 10 aircraft at Ysterplaat and a further 10 at Port Elizabeth–and No 17 Squadron with 10 at Zwartkop.

Light Aircraft Command, headquartered at Zwartkop has four liaison squadrons with permanent and Citizen forces' personnel, the SAAF's 40 Atlas-Aermacchi AM 3C Bosboks equipping No 41 Squadron at Lanseria and No 42 Squadron at Potchefstroom. A few Bosbok developments, the C4M Kudu with similar wing and tailplane, are attached to No 41 from the 40 aircraft so far produced. Also at Potchefstroom, No 11 Squadron has about 20 Cessna 185s, a further number serving No 43 Squadron at Durban.

Active Citizen Force squadrons, recently con-

verted from Harvards to 15 Atlas MB 326M Impala Mk 1s for COIN duties, are now beginning to receive the first MB 326K Impala Mk 2s from an initial batch of 50 being built by Atlas. A further 13 optimistically-named Air Commando Squadrons exist in cadre form to be manned in a national emergency by private pilots and their aircraft.

Flying training begins at the FTS, Dunnottar with 125 hours' basic instruction on 45 of the 100 remaining Harvards, followed by 120 hours' advanced flying with 80 of the 151 Impala Mk 1s assembled or built by Atlas with the FTS's jet branch at Langebaanweg. Training Command also controls three Advanced Flying Schools for operational, multi-engine and specialised instruction.

At Pietersburg, No 85 AFS has recently replaced its last Sabres with 16 Mirage IIIEZs, 10 IIID-2Zs and three IIIDZs. No 86 AFS, Pietersburg flies six Dakotas for twin-conversion, while No 87 AFS, now transferred from Ysterplaat to Bloemspruit, has 10 Alouette IIIs. The Air Operational School at Langebaanweg also operates the Impala for weapons training, and has recently received some Mk 2s.

Sudan's air force, Silakh Al Jawwiya As Sudaniya, was formed with British equipment in the months prior to independence, this being granted on 1 January 1956. Since 1919, Britain, together with Egypt, had been responsible for the administration of internal security in the Sudan, and the RAF shouldered much of the burden of policing, supply and communications duties in the 1920s and 1930s.

The embryo SAF followed in the same footsteps, receiving four new Hunting Provost T Mk 53s for initial training duties in 1957, followed by a further four ex-RAF aircraft refurbished by the manufacturer in 1960. The first twin-engined transport also originated from Luton with the loan of a Hunting President in September 1958. Two more aircraft were delivered in March 1960, having originally been intended for a Spanish airline, whereupon the first aircraft was returned to Britain. In May 1960, two Pembroke C Mk 54s were transferred from the West German Luftwaffe, the type being used for light transport and twin-conversion.

Having firmly established their links with the Sudan, Hunting supplied 12 Jet Provost T Mk 52s in 1962 to replace the Provosts in training and secondary close-support roles. Sudan's initial jet operations made an inauspicious start: the first two arrivals were written-off within three weeks of each other in May and June 1962, while a third suffered a heavy landing at Fort Lamy in April of the following year, requiring its return to Hunting for repair.

Transport capability was increased with the arrival of two Douglas C-47s and a 1964 order for four F-27M Troopships, although two of these were transferred to Sudan Airways in 1971. Light utility aircraft came from Switzerland in the form of eight Turbo Porters, ordered in 1967.

In 1969 the SAF expanded its Jet Provost fleet with five T Mk 55s, built on the new assembly line at Warton. These represented the sole export order for the BAC 145 and were equivalents of the RAF's Jet Provost T Mk 5s rather than BAC 167 Strike-masters. Their operation was to be short-lived in view of a sudden switch in purchasing policy which accompanied the 1969 coup.

With Soviet and Chinese assistance, the SAF was re-organised and re-equipped with jet fighters, heavy transports and other equipment. China supplied a squadron of about 16 Shenyang F-4 (MiG-17F) fighter-bombers and one or two MiG-15UTI trainers while the Soviet Union later contributed a squadron of MiG-21s together with facilities for pilot and technician training. A small number of Antonov An-12 and An-24 transports and a few Mil Mi-4 and Mi-8 helicopters was received, but in 1971, relations with the Communist Bloc were cooled after an attempted coup which was allegedly Soviet-backed.

Sudan gradually improved its ties with the West and Egypt, and in 1976 approached the United States on the subject of weapons supplies. Having received a cautious, but favourable reply, 90 Soviet military advisors and half the Russian diplomatic staff were expelled from Sudan in May 1977, when Arab money was promised for re-equipment of the armed forces.

First US arrivals were six Lockheed C-130H Hercules from February 1978, but a request for Northrop F-5s was at that time turned down, and President Nemery travelled to Paris in April 1977 to negotiate the sale of 14 Mirage 5SOs, two 5SOD trainers, plus options on a further 10, and 10 Puma helicopters. Later in the year, four DHC-5D Buffaloes were delivered to the transport unit.

With a later improvement in US attitudes

Below: a Russian-built Antonov An-24 transport of the Sudan Air Force. Soviet and Chinese influence in the late 1960s and early 1970s was short-lived, but provided MiG fighters and Mil helicopters in addition to Antonov types. Bottom: representing Sudan's swing towards Western sources of supply, the Lockheed C-130 Hercules first appeared in Sudanese service in early 1978. Orders for the Northrop F-5 were confirmed shortly afterwards

towards Sudan, the F-5 request was reconsidered, and in April 1978, Congress received notification of a moderate defence package for Sudan, comprising 10 F-5Es and two F-5F trainers and six ground radars. The Mirage order was apparently cancelled at about the same time, but Sudan is also believed to be in the process of obtaining 20 MBB Bö 105 helicopters from Germany for communications and SAR work. Present personnel strength of the SAF is some 1,500, annual defence expenditure being in the region of £120 million.

Tanzania's moderate government, adopting a middle course between the power blocs of East and West, has recently gained world acclaim for its support of Ugandan exiles in overthrowing the reviled President Amin. Operations by the Tanzanian People's Defence Force Air Wing (Jeshi la Wanachi la Tanzania) were, however, restricted to a MiG-21 attack on Entebbe airport on 1 April and an unconfirmed strike against Kampala on

10 April. At least one Libyan bombing raid on Tanzania was made (in which six bombs were dropped with little effect), but there were apparently no air-to-air combats during the entire operation, which was primarily an army campaign.

Deriving its name from the 1964 federation of Tanganyika and the island of Zanzibar, Tanzania received aid from West Germany in the establishment of its Air Wing, accepting eight ex-Luftwaffe Piaggio P149Ds and a similar number of Dornier Do-28s for training and light transport/communications work. Plans were also made for the transfer of six Nord Noratlas transports, while Israel trained 120 paratroops and Canada supplied 30 officers to staff a military academy near Dar-es-Salaam.

Tanzanian recognition of East Germany in February 1965 provoked the withdrawal of Federal German assistance, and Canada then agreed a five-year plan for the further development of the

flegdling Wing with the Tanzanian government. About half the initial complement of 400 personnel was trained in Canada, de Havilland supplying six Beavers, eight Otters and five Caribou.

Canadian aid was restricted to non-offensive equipment, but at the conclusion of the five-year period, a growing Chinese economic influence was extended to cover the military field; in late 1970 the Tanzanian government announced plans for an expansion of the Air Wing. About 250 personnel were trained to fly or service the Shenyang F-4 (MiG-17) in China, and a new airfield constructed at Mikumi, near Dar-es-Salaam, for MiG operations.

In mid-1973, a dozen F-4s and two F-2s (MiG-15UTI trainers) were delivered, followed the next year by a few F-6s (MiG-19SFs) for a second combat squadron, and 16 F-7s (MiG-21MFs), one of which was lost in an accident soon after arrival. Initially the aircraft were not fully utilised, as Tanzania was short of both pilots and groundcrew. The supply of F-7s may not have been the generous act it at first appeared, for the MiG-21 copy was rejected by the Chinese air force after less than 100 had been built, in favour of renewed production of the earlier F-6.

Despite Chinese predominance, the Nyerere Government continues limited procurement of Western aircraft, mostly of a second-line nature. Otters were traded back to Canada in 1971 to obtain a further eight Caribou, and six Buffaloes were in the process of delivery in 1979, four having arrived in 1978. Other transport equipment includes three HS748 delivered between November 1977 and January 1978.

Following the arrival of two Bell 47Gs and two Agusta-Bell 206s in 1973, a further pair of the latter were received in 1978 to complete the present helicopter complement. Primary training has now transferred from the P149s to five Piper Cherokee

Left: the main transport currently operated by the Tanzanian Air Wing (insignia below far left) is the DHC-4 Caribou. A total of 13 was delivered, of which 12 remain in service. Since 1970 Tanzania has re-equipped with Chinese front-line types, but still procures second-line aircraft from the West.
Above: the Zairian air force (insignia below left) has three DHC-5 Buffalo STOL tactical transports

140s, while a Cherokee Six is used for communications together with six Cessna 310s. Annual defence expenditure is about £80 million and the Air Wing has some 1,000 personnel.

Known as Congo-Kinshasa until 1972, Zaire gained independence from Belgium in 1960, forming an air force (Force Aérienne Congolese) the following year with a handful of light aircraft manned by mercenary personnel in response to the seizure of most Belgian equipment by the break-away province of Katanga. Later *rapprochement* between the Central Government and Katanga brought a fusion of the FAC and Aviation de la Force République, and most foreigners were replaced as a training programme gathered momentum.

Continued rebel activity prompted a request for Italian aid, resulting in the supply of 12 North American Harvards for counter-insurgency roles to augment the three Douglas C-47s, single C-54 and remnants of Katangese forces then constituting the Force Aérienne Congolese. An Italian Air Force colonel was appointed assistant to the army C-in-C, while some FAeC personnel underwent training in Italy.

Despite the 1964 rebellion, the Italians re-organised and rebuilt the FAeC, which was then able to contribute materially to the success of Government forces in controlling the uprising. Aid from the United States included a detachment

of USAF personnel from Air Commando units and a supply of re-manufactured Douglas B-26K Invaders for light bombing. Operational requirements dictated the employment of civilian pilots to fly the Invaders, T-28 Trojans and further C-47s which came from the USA, several of these being exiled Cubans.

As late as 1970, 130 Belgians and 50 Italians were employed as military advisers and further Europeans were occupying key positions in the forces. A renewed training contract with Italy brought 24 SIAI-Marchetti SF 260MZ primary trainers and 18 Aermacchi MB 326GBs for advanced instruction and secondary light-strike duties, while in 1973, General Mobutu announced a requirement for three squadrons of Mirage 5s to re-equip the Force Aérienne Zairoise combat element.

A chronic shortage of funds reduced the order to 14 Mirage 5M fighter-bombers and three 5DM trainers, delivered from 1975, of which five have reportedly been written-off. The FAZ suffered further losses in May 1978 at the hands of Shaba rebels from north Angola who destroyed between five and eight MB 326s, an Alouette III, a Puma and several light aircraft on the ground at Kolwezi.

Current FAZ organisation comprises the support units of 1er Groupement Aérien and operational elements of 2eme Groupement Aérien Tactique, subordinate to the Chief of the Air Staff and C-in-C Defence Forces. The tactical group contains two wings, one combat and one support, each with three squadrons.

Tactical forces of 2eme GAT at Kamina comprise the Mirages, MB 326s and Cessna 337 Miliroles of 21 Wing and transports and helicopters of 22 Wing.

The original MB 326GBs have been augmented by a further two aircraft, plus six single-seat MB 326Ks, while 20 Reims-Cessna 337 Miliroles delivered in 1977–78 replaced some of the ageing T-6s and T-28s. Tactical transports are the three Buffaloes, two Caribou, and two C-46 Commandos, together with seven Bell 47 light helicopters for observation and liaison duties.

Units of 1er Groupement Aérien include transport, training and some close-support elements. No 11 Wing de Transport has 10 C-47s, four C-54s and two DC-6s, while 12eme Wing Tactique Appui Rapproche has two squadrons (120 and 121 Escadrilles) with MB 326s and about eight T-6s, plus 122 Escadrille de Sauvetage with Aérospatiale Pumas and Alouettes for SAR; some helicopters, however, are more likely to be operated in 22 Wing of the tactical group. Medium transports are flown by 19 Wing which has seven Hercules based at Kinshasa. Two Mitsubishi MU-2s were delivered for VIP use in 1974 as part-replacements for the de Havilland Doves, Herons and Beech 18s of the former 112 Escadrille/11 Wing, while a single Aérospatiale Super Frelon helicopter is reserved for presidential use.

Training is administered by 13eme Wing d'Entrainement comprising 131 Escadrille d'Ecolage Elementaire with SF 260s and 132eme Escadrille d'Ecolage Avance with MB 326GBs and T-6Gs. In mid-1976, 15 Cessna 150 Aerobats were delivered to Kinshasa for basic instruction, followed by 15 Cessna 310s for twin-conversion and liaison.

Annual defence expenditure is in the region of £100 million, and the FAZ has 3,000 personnel of all ranks including conscripts.

No 22 Wing, 2eme Groupement Aérien Tactique, Force Aérienne Zairoise, operates three DHC-5D Buffaloes (one of which is illustrated) and two DHC-4 Caribou in the tactical transport role. The Wing is deployed in support of 2eme Groupement's 21 Wing which is Zaire's main combat wing, operating Mirage fighter-bombers, ground-attack MB 326s and Cessna 337 Milirole COIN aircraft. Repeatedly in conflict with rebels from neighbouring Shaba province, Zaire has procured numbers of tactical combat aircraft since 1974. The 1e Groupement Aérien is mainly concerned with logistic transport, training and search and rescue

CHAPTER SIXTEEN
South America
Argentina/Brazil/Chile/Cuba
Ecuador/Peru/Venezuela

The British Aerospace Canberra B Mark 62 serves with the Argentine air force's I Grupo de Bombardeo. The unit's ten bombers, together with two Canberra T Mark 62 conversion trainers, were refurbished aircraft supplied by Britain in 1970–71

Increasing wealth has enabled many South American countries to relinquish their dependence on the United States and seek advanced combat aircraft in world markets. In this connection, Brazil has progressed most and established a thriving aircraft manufacturing plant capable of meeting the majority of its needs for strike aircraft, built under licence or indigenously designed, together with trainers and light transports. Argentina, too, is moving towards self-sufficiency, although in both of these countries United States' supplies have been curtailed because of President Carter's human rights policies.

Chile has also been the subject of an arms embargo for similar reasons, while Peru's fraternisation with the Soviet Union has resulted in the reduction of United States' supplies to a trickle of spares for its existing complement of MAP-funded aircraft. Ecuador's apprehensions over new Peruvian military power have prompted more Dassault Mirage sales and a willingness from the United States to supply some high-technology equipment

of a purely defensive nature.

Oil-rich Venezuela, though maintaining relations with the United States, has been able to pursue a moderate policy of independent purchasing for many years, but it is much closer to North American shores that the USA's main concern over arms proliferation lies. A few minutes' flying time from Florida, the island of Cuba is–in US eyes at least–the 16th Republic of the USSR. Since Castro seized power in 1958, Cuba has become the training ground and mentor for revolutionary and guerilla forces in Latin America and much of Africa. Despite the Kennedy-Khrushchev agreement over the issue of nuclear weapons to Cuba, Castro's island is carefully watched for signs of hostile activity.

As the second largest air force in the South American subcontinent, the Fuerza Aerea Argentina, has in recent years turned away from its traditional source of supply in North America, diversifying arms purchases in the face of United

States' opposition to further rearmament. Financial limitations have militated against ambitious self-sufficiency plans and, while deliveries of the indigenous Pucara counter-insurgency (COIN) aircraft are now well behind schedule, projected licence-production of the Dassault Mirage III has been abandoned in favour of off-the-shelf purchase of cheaper second-hand versions from Israel.

Argentina now spends about £900 million per year on defence, representing the highest proportion of gross national product in any of the major Latin American countries. Dissatisfaction in Washington over the Argentine record in human rights resulted in a sharp drop in military assistance during 1977 from 47 million to 15 million dollars, and restrictions on the supply of spares for aircraft and equipment already in FAA service. This was extended to a full embargo in October 1978, although even before that date, Argentina had experienced considerable difficulty in obtaining authorisation for second-line aircraft such as the Beech T-34C Turbo-Mentor trainer and Swearingen Merlin transport. Other early casualties of the embargo included Bell AH-1G/J Cobra helicopter gunships and AIM-9J Sidewinder air-to-air missiles.

The United States' policies have now accelerated the 'Plan Europa' formulated in the early 1970s, under which Argentina sought to widen the scope of its purchases through agreements with European manufacturers. This envisaged initial small-scale procurement of weapons as a preliminary to licence-production, yet little progress has subsequently been made. Consideration was also given to dealings with the Soviet Union, although once more, no final plans have been announced.

Most affected by the collapse of licence-production plans has been the Dassault Mirage III, up to 80 of which were to have equipped the FAA front-line by the late 1970s. Following delivery of 12 IIIEAs and two IIIDA two-seat trainers in 1972 to equip one squadron, the Mirage programme was allowed to laspe in favour of purchasing refurbished Douglas A-4 Skyhawks from the United States. Seven more examples of the Mirage IIIEA were ordered in mid-1977 as attrition replacements,

thoughts then turning to a cheaper alternative in the form of surplus Israeli aircraft.

In Autumn 1978, Argentina announced the purchase of 26 IAI Daggers, delivery of which began in January of the following year, for the relatively modest price of 185 million dollars, including associated weaponry and spares/support facilities. One of the interim stages between the French-built Mirage IIICJ and the Kfir-C2, the Dagger incorporates the basic Mirage airframe and engine married to Israeli avionics, although it is possible that some–if not all–the aircraft originated in Dassault's Bordeaux factory during the early 1960s and have been progressively modified during their service with the IDF/AF.

While Argentina would doubtless have preferred the Kfir, sales of its General Electric J79 engine are strictly controlled by the United States. The Dagger order was apparently placed with some speed in view of deteriorating relations with Chile over three small islands in the Beagle Channel, south of Tierra del Fuego, as well as border claims

Top: the McDonnell Douglas A-4 is flown by both the Argentine air force and the navy, an A-4Q of the naval 3ra Esquadrilla de Attaque being illustrated. Air force A-4P Skyhawks operate with IV and V Brigade Aerea.

Above: although intended for licence production in Argentina, in the event only French-built Dassault Mirage IIIEA interceptors entered service with the Fuerza Aerea Argentina. They fly with I Grupo de Caza from Moron air base near Buenos Aires.

Top right: a Canberra B Mark 62 pictured prior to delivery.

Above right: the Argentine navy's helicopter wing is 4 Esquadrilla, whose inventory includes the Aérospatiale Alouette III pictured.

Right: the navy flies about 20 North American T-28S Fennecs in the training and light strike roles

in Antarctica and development of natural resources including fish and oil. A deadline for negotiations in the long-standing dispute was fixed for 2 November 1978, prior to which more than 500,000 Argentinian reservists were called up and concentrated close to the Chilean border, while the FAA was put on alert. The seemingly inevitable war was averted at the eleventh hour by an interim agreement for joint exploration and development of the seas around the Beagle Islands and a unified claim for Antarctic territories presently under British jurisdiction.

Production of the FMA Pucara continues to progress slowly, despite repeated attempts to accelerate deliveries. Target figures of four aircraft per month during 1979 remain unrealised, the present figure being only 1.7, with delivery of the 28th production example effected in September 1979. FAA requirements are for three units of 15 aircraft each, a further 15 having been set aside against anticipated production for export. Orders for about six aircraft for Mauritania were cancelled as a result of financial problems, but Dominica is considering up to 24 aircraft and Venezuela is reportedly in the process of agreeing a licence-production arrangement.

Although a jet Pucara development has now been abandoned, FMA have recently produced a prototype of the IA 58B with heavier 20 mm cannon armament in a revised nose. The IA 58B is primararily aimed at the export market, but the FAA may be persuaded to convert its outstanding 20 or so aircraft to this standard prior to delivery, and to take up some of the 55 aircraft now on option. Joint service requirements also exist for 255 helicopters, including 120 observation types and 100 or more 7-10 seat examples, all of which are intended for licence production. Some 35 imported medium-lift helicopters are needed, of which only six Boeing Vertol CH-47C Chinooks have so far been ordered. Three army Chinooks are presently stored at the Boeing production facility as a result of the US arms embargo, although work continues on a further three FAA aircraft.

Under central control of the Defence Ministry, the FAA comprises four operational commands: Air Operations, Air Regions, Material and Personnel. Air Operations Command has overall responsibility for all military activity involving FAA aircraft, mainly through eight Air Brigades (*brigada aereas*), or wings. Each Brigade comprises two or three *grupos*, with varying numbers of aircraft operating from a single Military Air Base (*base aerea militar*). Additional independent bases accommodate special, non-operational units.

As currently organised, the FAA has a combat strength of about 150 fighter and strike aircraft, although some have only limited COIN capabilities. II Brigada Aerea, headquartered at General Urquiza BA, Parana, Entre Rios, fulfils several roles, its most important component being I Grupo de Bombardeo operating ten refurbished BAC Canberra B Mark 62s and two T Mark 64 trainers delivered in 1970–71. Unit establishment is completed by a photographic group, I Grupo de Fotografico with some 20 IA 50 Guarani II twin-turboprop transports converted for photographic survey work.

*Right: the Argentine navy's six
Grumman S-2A Trackers were retired
from service aboard the carrier 25 de
Mayo following delivery of four ex-US
Navy S-2Es in late 1978. They are now
flown as trainers and based at
Comandante Espora Naval Air Base.
Below: designed and built in Argentina,
the twin-turboprop Turbomeca Astazou
powered FMA IA 58 Pucará COIN
aircraft is intended to equip two
Argentine air force squadrons. The first
eight aircraft became operational in
October 1977 with II Escuadron de
Exploration y Ataque, based at
Reconquista. The unit currently
operates 12 Pucarás*

Previously a component of II BA, Strike Trials
Squadron, II Escuadron de Exploracion y Ataque
was the first recipient of the IA 58A Pucara and
transferred to Reconquista in 1977 to provide the
nucleus of III Brigada Aerea. Two *grupos*, each
with 12 aircraft plus three reserves, are now on
strength. A likely candidate for further Pucara
deliveries is IX Brigada Aerea, formed at Como-
doro Rivadavia BAM in late 1977 but without any
allocation of aircraft.

At El Plumerillo BAM, Mendoza, IV Brigada
Aerea has three fighter-bomber squadrons equipped
with Douglas A-4 Skyhawks and Morane-Saulnier
MS 760 Paris COIN aircraft. Three batches of
Skyhawks, each of 25 aircraft, were delivered to the
FAA in 1966, 1970 and 1975. Their correct designation, A-4P, is seldom used in Argentine service
and they are referred to under their former mark
numbers, A-4B and A-4C. Proposals exist for the
conversion of a few A-4Bs to two-seat trainers
similar to the TA-4S aircraft produced by Lockheed
Aircraft Service for Singapore.

I Grupo de Caza Bombardeo has some 20
Skyhawks, while the Paris aircraft are allocated
to II and III Grupos. As last operator of the
North American F-86F Sabre, II Grupo discarded
its final aircraft in late 1976, and 12 of the 14
remaining in the FAA inventory were prepared for

sale to Uruguay. Intervention by the US State
Department prevented completion of the transaction, and the aircraft are apparently still in
storage pending a decision on their future.

Designed as a fast communications aircraft,
the Paris adopts a more warlike role in the FAA.
During the late 1950s, 48 were assembled by FMA,
of which about 30 remain in operational use. In
the next few years, all will be uprated to Paris II
standard by the installation of Turboméca Marboré
VI turbojets, due for delivery by February 1980 for
local installation.

Main operating wing for the Skyhawk is V
Brigada Aerea at General Pringles BAM, Villa
Reynolds, San Louis, comprising IV and V Grupos
de Caza Bombardeo, each with some 20 A-4Bs
from the first two batches delivered. Skyhawks have
been updated with Ferranti Isis weapon sights and
are allocated the secondary role of air defence over
Buenos Aires and other principal cities. The Air
Surveillance and Instruction Group (Grupo de
Instruccion y Vigilancia Aerea) headquartered at
Merlo in Buenos Aires Province is responsible for
co-ordination of air defence, its radar network and
seven underground posts also covering Santa Fe
Province and Uruguay.

Primary interception tasks are allocated to I
Grupo de Caza with the Mirage III fleet, recently

transferred from VII BA to VIII Brigada Aerea at José C. Paz BAM, near Buenos Aires. Although no official announcement has been made, VIII BA is the likely operator of the newly-delivered IAI Daggers in a further two Grupos de Caza.

At Moron BAM, Buenos Aires, VII Brigada Aerea has a nominal COIN role, but also undertakes support and training duties with varied equipment in three *grupos*, including one with armed Beech T-34 Mentors. I Grupo de Exploracion y Ataque has 14 Hughes 500M, six UH-1H Iroquois, six Aérospatiale SA 315B Lama, two Sikorsky S-58T and three S-61N/R helicopters, while Grupo de Busqueda y Salvamento operates three Grumman HU-16 Albatross amphibians for search and rescue duties. Also housed at Moron is a calibration unit with a Douglas C-47 and two indigenous Dinfia Guarani transports.

Like many Latin American air forces, the FAA has large civil transport and communications commitments, which are the responsibility of I Brigada Aerea at El Palomar BAM. This comprises four *grupos de transporte* as operational elements of the Servicos de Transportes Aereos Militares, or Military Air Transport Service. The Lockheed Hercules of I Grupo comprise three C-130Es and four C-130Hs delivered between 1968 and 1972, a fifth C-130H having been lost in 1975. US approval

for the supply of two KC-130H tankers was granted in May 1978, deliveries being completed a year later.

I Grupo also administers LADE (Lineas Aereas del Estado) for communications with undeveloped areas of Argentina, to which other transport units contribute a variety of aircraft. Mainstay of LADE is the Fokker F-27 and F-28 fleet operated by VI Grupo, comprising 11 F-27 Troopships/Friendships and five F-28 freighters. II and III Grupos contribute half a dozen C-47s, 14 transport versions of the Guarani (from a total of 41 produced), five Twin Otters and two Swearingen Merlin IVA air ambulance transports.

The transport wing also includes the Presidential Flight with single examples of the Boeing 707, F-28, British Aerospace HS748, North American Sabreliner 75A and Guarani for VIP and government use. Servicios Aereas del Estado Nacional (SADEN), or National State Air Service, created in 1967, now flies 14 Shrike Commanders, six Cessna 182s and other light aircraft. Services to and in the Antarctic are handled by I Grupo Antarctica at Rio Gallegos, Santa Cruz with three DHC Beavers, three Otters, an S-61R and a ski-equipped LC-47 Dakota.

All FAA training establishments are administered by Grupo Aereo Escuela at Cordoba which controls the Escuela de Aviacion Militar with about 40 of the original 90 T-34 Mentors built under licence by FMA, together with 12 MS 760s. Approximately 40 pilots are trained each year, including a number of personnel from other Latin American countries.

Plans for future training aircraft have now been revised in the light of the Paris re-engine programme, and a requirement for the FMA IA 62 indigenous T-34A replacement as a lead-in to a new strike trainer in the Hawk/Alpha Jet class has been abandoned. Instead, FMA are concentrating on the IA 63, an ultra-light jet aircraft capable of handling basic, and some aspects of advanced training before students progress to the Paris and then dual control versions of operational types within the Air Brigades. T-34s are also used at the Curso de Aspirantes a Officiales de Reserva, José C. Paz BAM, for the training of a small number of conscripts and reserve officers. Liaison duties within

Training are performed by I Escuadron de Comunicaciones Escuela with various types, including MS 760s, T-34s, four Aero Commander 560s and one or two Huanqueros.

Comando de Aviacion Naval Argentina (CANA) is an autonomous naval command operating with 4,000 personnel from the aircraft carrier *25 de Mayo* (previously HMS *Venerable* and HrMS *Karel Doorman*) with additional shore-based reconnaissance and training units. The main combat element is 3ra Esquadrilla de Ataque operating the remaining 11 of 16 refurbished A-4Q Skyhawks ordered early in 1971, armed with Sidewinder AAMs and Bullpup ASMs. When not deployed on *25 de Mayo*, the unit is based at Puerta Belgrano and also operates 20 Sud T-28S Fennecs from the 59 aircraft obtained from France in 1966. Fennecs are modified with strengthened nosewheels and arrestor hooks for deck landing practice, and allocated secondary strike duties in support of the 7,000-strong Argentine Marine force.

Until 1977, Fennecs formed a component of the disbanded 2da Esquadrilla at Comandante Espora alongside ten Lockheed P-2 Neptunes and a similar number of Grumman S-2 Trackers, including four S-2E variants. Assuming sufficient funds are made available, the unit will re-form as a carrier-borne strike squadron with an aircraft in the Phantom or Etendard class.

Also at Comandante Espora is the CANA helicopter wing, 4 Esquadrilla, with a miscellany of types for first and second-line roles, spearheaded by four Westland Sea Kings for anti-submarine warfare (ASW) duties aboard the aircraft carrier. Two Westland Lynx helicopters were delivered in 1978 for deployment aboard frigates, some of the nine Alouette IIIs flying from the icebreaker *General San Martin*. The unit additionally operates two Beech King Airs delivered in early 1979, although the majority of transports are allocated to 5 Esquadrilla at Ezeiza which has three Electras, eight C-47s, five DC-4/C-54s, an HS125 and several other liaison types. Other aircraft are detached as required to Esquadrilla Aeronaval Antarctico, including Twin Otters, Beavers and Pilatus Turbo Porter lightplanes.

Above: the Argentine navy currently operates ten refurbished ex-US Navy Lockheed SP-2H Neptune maritime patrol aircraft, based at Comandante Espora Naval Air Base.
Right: the Argentine navy acquired two Aérospatiale-Westland Lynx HAS Mark 23 in 1978 to equip frigates.
Inset below right: the Argentine navy has four Sikorsky Sea Kings deployed in the anti-submarine role aboard the carrier 25 de Mayo and operated by the helicopter wing, 4 Esquadrilla

The Naval Aviation School or Escuela de Aviacion Navale at Punta de Indio includes 1 Esquadrilla with eight MB 326s, transferred from 3 Esq, and 16 T-34C Turbo Mentors delivered as T-6 and T-28 replacements in 1978. A few Beech AT-11s are retained for twin conversion.

Although Argentine ground forces have operated aircraft for many years, it was not until 3 November 1959 that the CAE came into existence, in the face of considerable opposition from the air force. In the early stages, however, Army Aviation was confined to liaison and VIP flying, with three Douglas C-47s operated for the support of IV Airborne Infantry Brigade until replaced by FAA Hercules in 1973. Transport capability returned in March 1977 following delivery of the first of three Aeritalia G 222s.

Lightplanes and helicopters of FAE, principally some 15 Cessna U-17s and 20 UH-1H Iroquois respectively, form part of 601 Aviation Battalion at Campo de Mayo on the outskirts of Buenos Aires, but are in practice permanently detached to the Aviation Sections of each army corps for liaison and transport. Technical support is provided by the FAA at Moron.

During the mid-1970s, the FAE underwent considerable expansion, taking delivery of five Beech Queen Airs, three Twin Otters, five Turbo

Commanders, five Cessna Turbo Skywagons and five Cessna T-41 trainers. At the same time, the rotorcraft fleet was expanded with the arrival of two Bell 212s, and in 1977, six more FMA-built Cessna 182Js were acquired.

Most recent orders have been for a camera-equipped Cessna Citation and three Turbo Porters, but requirements for a helicopter gunship have yet to be translated into hardware. A long-standing order for Agusta A.109 Hirundos is believed to cover nine aircraft, but Bell AH-1 Cobras are subject to a US embargo, as are three Boeing Vertol CH-47C Chinooks.

Several paramilitary organisations operate aircraft on nominal army charge. The Gendarmerie Nacional border patrol force has some Fairchild Hiller FH 1100 helicopters, a dozen Cessna A 182Js, two Super Skymasters and Bell Jet Rangers, while a few Turbo- and Shrike Commanders are flown by military factories. The photographic Citation and a similarly-equipped Queen Air are used by the Military Geographic Institute.

One of the largest and most modern air arms in South America, the Forca Aerea Brasileira recently completed a re-equipment programme which disposed of the majority of antiquated types for which Latin American air forces are renowned. Backed by the design and manufacturing resources of Empresa

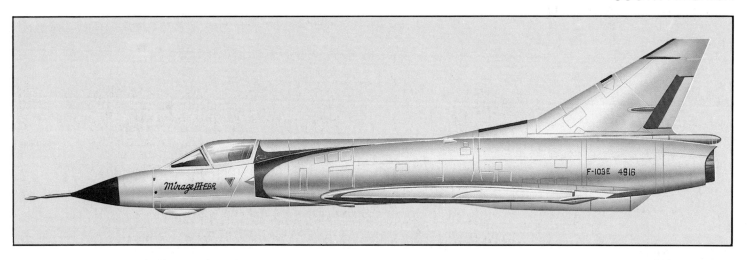

Brasileira de Aeronautica SA (Embraer), the FAB's inventory includes a sizeable proportion of indigenously-built aircraft.

Defence agreements with the United States through which the FAB previously received much of its equipment were rescinded in 1977 in view of increased self-sufficiency and US human rights conditions being attached to any further aid. As a measure of protection for Embraer, all aircraft imports have been banned until at least 1982, except in a few special cases. Defence expenditure was averaging some £1,000 million per year in the late 1970s.

FAB organisation comprises a series of specialised Commands (Air Defence, Tactical, Maritime, Transport and Training), with individual units subordinated to the HQs of six regional commands (COMARs). The latter comprise 1^O COMAR at Belem, 2^O at Recife, 3° at Rio de Janeiro, 4° at Sao Paulo, 5° at Porto Alegre and 6° at Brasilia.

Following delivery of 13 Dassault Mirage IIIEBRs (local designation F-103E) and five IIIDBR (F-103D) trainers ordered in 1970, a new air defence organisation known as 1^a ALADA (ala de defesa aerea or air defence wing) was formed at Anapolis in early 1972. Supported by ground radars, these aircraft were intended for the defence of the triangle formed by the cities of Rio de

Janeiro, Sao Paulo and Brasilia. This area was extended to cover the entire south of Brazil from 1977, reinforced by army-operated mobile Euromissile Roland SAM launch units.

Remaining FAB combat elements comprise 13 light strike and COIN units, almost half of which operate the AT-26 (EMB 326GC Xavante). At Santa Cruz, 1^O Grupo de Aviacao de Caca (GAvCa or fighter group) replaced its AT-33 armed trainers with Xavantes in 1971, but these were themselves replaced in the Group's two squadrons (1^O and 2^O Esquadrao) by 12 F-5Es and two F-5B trainers for strike interdiction roles, from 36 and six of each type ordered in 1973. A requirement exists for two more F-5 squadrons when finance permits, but import restrictions now provide a second, if self-induced obstacle.

Two further fighter groups are equipped with the Xavante, and a recent follow-up order for a further 40 brings total FAB procurement to 172, for delivery by 1981. IV GAvCa at Fortaleza AFB, Ceara State comprises 1^O and 2^O Esquadrao, while at Canoas AFB, near Porto Alegre, XIV GAvCa has only one component squadron. A successor to the AT-26 has been under consideration since 1975 in the AX programme for joint development with Aermacchi, but the proposed MB 339 derivative has apparently been abandoned in favour of

Above: the Dassault-Breguet Mirage IIIEBR interceptors of 1^a Ala de Defesa Aerea are an important part of Brazil's French-installed computerised air defence system.
Below: the two squadrons of the 1^O Grupo Aviacao de Caca, Fuerza Aerea Brasileira operate Northrop F-5B trainers and F-5E fighters.
Opposite below: the Embraer AT-26 (EMB 326GC Xavante) fulfills several roles in the Fuerza Aerea Brasileira. They equip three fighter squadrons and are also used for training.
Far left: the badge of the 1st Fighter Group's 1st Squadron.
Centre far left: Brazilian unit badge.
Above left: the Neiva T-25 Universal is Brazil's standard basic trainer. Some 150 equip the Air Force Academy flying school, Pirassununga and the reserve pilots' school and tactical weapons training unit based at Natal

extension of AT-26 service lives and possible replacement by a wholly indigenous type at a later date yet to be decided.

The GAvCas of Comando Aerotatico are augmented by counter-insurgency, reconnaissance and attack squadrons, Esquadraoes Mistos de Reconhecimento e Ataque. Of five EMRAs, 3° at Galeao AFB, Rio de Janeiro, 4° at Cumbica AFB, Sao Paulo, and 5° at Canoas AFB, have converted from T-6s to Xavantes, while two further units, 1^O EMRA, Belem, Para State and 2^O EMRA, Recife, Parnambuco have armed versions of the NEIVA T-25 Universal II trainer. The T-25 has recently been returned to production against an order for 20 more aircraft to complement the original order for 150, and in anticipation of sales abroad.

Additional COIN roles are undertaken by Esquadraoes de Ligacao e Observacao (liaison and observation squadrons or ELOs) with light-planes and helicopters. Two units, 1^O ELO at Campo dos Afoncos, Rio, and 3° ELO, Porto Alegre, operate the majority of the 40 NEIVA L-42 Regente light liaison aircraft produced for the FAB, plus a dozen JetRangers, a few Cessna O-1 Bird Dog lightplanes and most of the 24 UH-1H Iroquois. At Sao Pedro de Aldea, 2^O ELO has T-25s, fitted with weapons aiming sights and arma-ment selector systems, in common with those of 1^O and 2^O EMRAs.

Since a Presidential decree of 1965, all fixed-wing naval aircraft, including those on the air-craft carrier *Minas Gerais* (ex HMS *Vengeance*), have been under control of the FAB. I Grupo de Aviacao Embarcada has eight Grumman S-2E Trackers at Santa Cruz, replacing earlier S-2As and CS2F-1s, six of which form the normal carrier complement. In April 1978, VII Grupo de Aviacao at Salvador, Bahia State, began conversion of its single squadron from Lockheed P-2 Neptunes to 12 EMB 111A(A) maritime reconnaissance versions of the Bandeirante, further deliveries being planned for a further four squadrons to conduct fishery patrol, shipping surveillance and sonar search missions. EMB 111s will also go to VI Grupo de Aviacao at Recife, Pernambuco, to join three RC-130E Hercules in long-range SAR and photo-graphic survey roles. Main SAR tasks, however, are undertaken by Servico de Busca e Salvamento which administers the two squadrons of X Grupo de Aviacao at Florionapolis: 2^O Esquadrao (HU-16A Albatross) and 3° Esquadrao (UH-1D Iroquois and Bell 47G).

FAB transport elements have received priority in re-equipment, the modernised fleet being organi-

Above: a Northrop F-5E of the 1^O Grupo de Aviacao de Caca (1st Fighter Group) at Santa Cruz air base. The breech and ammunition feed of the port 20mm M-39 cannon can be seen. Below: F-5B trainers (nearest the camera) and F-5E fighters lined up at Santa Cruz, where they serve with the FAB's 1^O Esquadrao

sed in four *grupos de transporte* for logistic and support tasks including operation of Correio Aero Nacional, the National Airmail Service. Heavy logistic support is delegated to 1^O GT at Galeao AFB, Rio de Janeiro with seven Lockheed C-130Es, three C-130Hs and two KC-130H tankers (for in-flight refuelling of Northrop F-5s) in its single 1^O Esquadrao. At the same base, II Grupo comprises 1^O Esquadrao operating six British Aerospace HS748s and a few Bandeirantes and 2^O Esquadrao with six newer HS748 Srs 2Cs with large freight doors. The FAB has received 86 EMB 110 Bandeirantes, comprising 58 C-95 transports, 20 C-95A freighters, six R-95 photo-survey aircraft and two EC-95 navaid calibration aircraft.

Support for the army's 1st Parachute Division is provided by 2^O Esquadrao of the Grupo de Transporte de Tropas, based at Campo dos Afoncos with most of the 24 DHC-5 Buffaloes delivered from 1968 onwards, the few surviving Fairchild C-119s of 1^O Esquadrao having been withdrawn. Official communications aircraft are flown by Grupo de Transporte Especial at Brasilia, comprising the Presidential fleet of two Boeing 737s and three Bell Jet Rangers in 1^O Esquadrao, and VIP aircraft of 2^O Esquadrao: five Xingus delivered in 1978, nine HS125s and a Vickers Viscount turboprop airliner.

Independent units in Comando de Transportes Aereos include four Esquadraoes, 1^O at Belem, 2^O at Recife, 3° at Galeao and 4° at Cumbica, with C-95 Bandeirantes, plus 5° and 6° Esquadraoes de Transporte de Tropas at Campo Grande and Manaus, equipped with Buffaloes. At Belem, II Grupo de Aviacao still operates six Consolidated Catalinas, recently refurbished by the installation of new avionics and power plants from retired Douglas DC-6s. Each Command and major base has one or two Bandeirantes for utility transport roles, augmented by a recent delivery of 12 EMB 810Cs (FAB type U-7), a locally-built version of the Piper Seneca.

Flying training establishments include the Air Force Academy at Pirassununga, Sao Paulo, where flying tuition for regular officers comprises some 350 hours to wings standard, including 50 hours on the Aerotec T-23 Uirapuru primary trainer. Delivery of 100 T-23s was completed in 1975, 24 of which have recently been modified by minor airframe changes to improve performance, although a new Mark 2 version is to be flown shortly. Cessna T-37s were withdrawn from the training programme in 1977, students now progressing from the T-23 to 100–150 hours on T-25 Universals and a further 100 hours on Xavantes.

A similar syllabus is followed by reserve pilots at Centro de Formacao de Pilotes Militares at Natal, Rio Grande do Norte, while T-25s and AT-26s are used by the tactical weapons training unit, CATRE, also at Natal. Transport pilots are converted to multi-engined aircraft via the Bandeirante. T-23s will ultimately be replaced by the EMB 312 (FAB T-27) turboprop trainer, two prototypes of which are on order, with first flight scheduled for September 1980.

Helicopter operations are centred on Santos, Sao Paulo, where Ala 435 has six UH-1H Iroquois, four JetRangers and four Hughes OH-6As, princi-

pally for COIN missions, supported by Centro de Instrucao de Helicopteros with some 30 Bell H-13s (Model 47s). A helicopter production plant under construction at Itajuba by Helibras is likely to provide licence-built types to satisfy an FAB requirement for additional helicopters.

Restricted to helicopter operations since 1965, the Aeronaval maintains a rotary-wing ASW force, Grupo de Caca do Destruicao, for deployment aboard NAel *Minas Gerais* and several frigates. Five Westland Sea Kings of 1^O Esquadrao de Helicopteros Anti-Submarine are shore-based at Sao Pedro de Aldeida when not detached to the carrier, while six *Niteroi*-class frigates are equipped with some of the nine Westland Lynx helicopters delivered in 1978.

Also at Sao Pedro, 1^O Esquadrao de Helicopteros de Emprego General operates on support duties with three Westland Whirlwinds, five Wasps (from two batches of three delivered ex-Royal Navy in 1977–78), 18 Bell JetRangers and eight Brazilian-assembled Aérospatiale AS 350 Ecureuils which replaced Fairchild Hiller FH 1100s attached to survey vessels from mid-1979. A few Hughes 300s provide training facilities in 1^O Esquadrao de Instrucao at Sao Pedro.

Complications arising from the 1973 right-wing

Top: responsible for heavy logistic transport and in-flight refuelling duties, 1^O Esquadrao, 1^O Grupo de Transporte, operates 12 Lockheed C-130 Hercules of varying marks. A C-130E is pictured. Three RC-130Es are additionally flown by the 6th Grupo de Aviacao on long-range SAR duties.
Above: deliveries of nine Westland Lynx HAS Mk 21 ordered by the Brazilian navy began in 1978. They will equip six Niteroi-class frigates

Top: based at Salvador, 1º Esquadrao, 7º Grupo de Aviacao operates 12 examples of the Embraer EMB 111A(A) Bandeirante. Designated P-95 in Brazilian air force service, it is a maritime patrol version of the EMB 110 (C-95) transport. The pictured aircraft appeared at the 1978 Farnborough Air Show, carrying eight five-inch rockets. Above: six Consolidated PBY-5A Catalinas, designated CA-10 in Brazilian service, are flown by II Grupo de Aviacao of Comando de Transportes Aereos, based at Belem. They have recently been refurbished

20 airworthy Hunter F Mk 71s and T Mk 77s out of 39 refurbished aircraft delivered from 1967 onwards. Ala 2 at Quintero AFB, Valparaiso has both first and second-line responsibilities with mixed equipment including a dozen Grumman HU-16B Albatross amphibians operated by Grupo 2 on ASW and search and rescue duties, together with nine Beech 99s for light transport and training with Grupo 11, also known as the Escuelo de Vuelo por Instrumentos y Navagacion Aerea (Instrument and Air Navigation Flying School).

At Chucumata AFB, Iquique/Los Condores, Ala 4's single squadron, Grupo 1, previously operated all 34 Cessna A-37s for COIN roles. Half of these have now been transferred to Ala 5 at El Temual AFB, Punta Montt, replacing the Lockheed F-80C and T-33 in Grupo 12 and serving alongside ten DIIC Twin Otters of Grupo 5.

Autonomous squadrons include Grupo de Helicopteros 3 at Maquehue AFB, Temuco, equipped mainly with a dozen Bell UH-1H Iroquois, plus seven Helitec-Sikorsky S-55Ts and two Hiller UH-12Es for liaison and SAR. A single Aérospatiale Puma is operated on VIP missions by the main FAC transport unit, Grupo 10 at Los Cerriloos, Santiago, together with two C-130H Hercules, three DC-6As, a King Air and about eight C-47s. Grupo 6 flies from Bahia Catalina AFB, Punta Arenas with a small number of DHC Twin Otters and Beech Twin Bonanzas, while Servicio de Aerofotogrametrico has camera-equipped pairs of Twin Otters and Learjet 35s and a single Beech King Air.

Pilot training is centred on the Escuela de Aviacion 'Capitan Avalos' at El Bosque AFB, Santiago, commencing with primary instruction on eight Cessna T-41Ds. Students then transfer to about 50 survivors of 66 Beech T-34As delivered from 1953 onwards, and thence to some 25 T-37Bs for the advanced stage. Gliding instruction is also given at Escuela de Vuela Sin Motor at Lòs Condes with two O-1A Bird Dogs fitted as tugs in addition to a complement of various sailplanes. Search and rescue (SAR) activities throughout Chile are co-ordinated by Servicio de Busqueda y Salvamento from headquarters at Los Cerrillos.

From an original establishment of two or three Bell 47 and JetRanger helicopters, the Servicio de Aviacion de la Armada de Chile (SAAC) has been expanded to include a small fixed-wing element of both combat and support types, manned by 500 personnel and operating from the main base of El Belloto, near Valparaiso. For maritime surveillance, Grumman HU-16B amphibians were replaced by six EMB-111A Bandeirantes in 1977, following a delivery of three EMB-110 transports the previous year as C-47 substitutes. Additional transport capacity was added in 1978 with the receipt of four CASA C.212 Aviocars.

Short-range ASW duties are performed by four specially-equipped Bell 206AS JetRangers, while eight Bell 47s and a few Alouette IIIs operate in training and SAR roles. Executive transport is provided by a single Piper Navajo, while six Mentors are used for basic flying-training. Seaborne forces have Exocet ship-to-ship missiles, and Short Seacat missiles for air-defence.

Air elements of the Chilean Army are equipped with nine Pumas, two JetRangers, three UH-1H

coup which deposed the world's first democratically-elected Marxist administration have compelled the Chilean Air Force to change its sources of supply twice in recent years. An embargo on Hawker Hunter spares imposed by Britain prompted a 1974 order for 15 Northrop F-5Es and three F-5F trainers plus 34 Cessna A-37B strike aircraft from the United States. However, these too became the subject of an arms embargo as a result of US dissatisfaction with Chilean human rights. As a counter to Argentina's purchase of IAI Daggers, Chile recently placed an order for 16 Mirage 50s from France and is believed to be in the process of negotiating with Brazil for AT-26 Xavante counter-insurgency aircraft.

The Fuerza Aerea Chilena comprises some 12 Grupos Aeros, each consisting of a single *esquadrilla*, within four combat and support wings (*alas*), plus associated units with varying strengths and roles. Present complement is about 70 first-line and 145 second-line types, manned by 11,000 personnel and drawing from a total defence expenditure of some £400 million.

As FAC operational spearhead, Ala 1 operates three combat units from Cerro Moreno AFB, Antofagasta, comprising Grupo 7 with all 18 F-5E/Fs, and Grupos 8 and 9 sharing the remaining

Iroquois and an SAR component of six Alouette III/Lamas. In 1978, a sizeable transport component was added with delivery of six C.212 Aviocars to join the existing liaison and utility fleet of four Cessna O-1 Bird Dogs, two Piper Cherokee Sixes and four Navajos.

Basic training is conducted on 18 Cessna R172K Hawk XPs delivered in 1978, followed by a course on 10 NEIVA T-25 Universals obtained from Brazil in mid-1975. The latter are equipped with armament for secondary light-strike roles, a requirement for a further eight being, as yet, unfulfilled.

Since the 1958 revolution installed a pro-Soviet administration on its very doorstep, the United States has suffered many anxious moments over subsequent events in Cuba. The 'crisis' of 1962 resulted in an agreement whereby the US pledged not to invade Cuba in return for Soviet restraint in offensive weapons deliveries, but periodic surveillance by satellite and some SR-71 overflights have continued to monitor developments.

Recent discoveries of MiG-23s and a Soviet brigade of some 2,000 men stationed on the island have rung alarm bells in Washington, seemingly out of all proportion to the threat – real or imagined – which they pose. It is, however, in Africa that Cuba chooses to deploy a large proportion of its forces, and ground elements have been heavily involved in Angola, Mozambique and Ethiopia.

Initial re-equipment of Castro's Fuerzas Aerea Revolucionaria took the form of about 30 MiG-17 and 20 MiG-19 interceptors for three squadrons, 12 Ilyushin Il-14s for an *escuadron de transport*, 20 Antonov An-2 utility aircraft and 24 Mi-4

helicopters. Pressure from the United States resulted in the removal of 33 Il-28 jet bombers and six squadrons of Sandal medium-range ballistic missiles in late 1962, but Guideline, Frog, and Snapper missiles remained.

Early deliveries of 42 MiG-21Fs, originally flown by Soviet pilots from Santa Clara, Camaguey and San Antonio, have been augmented by some 50 MiG-21MFs and two further interceptor squadrons with a similar number of MiG-19s. Four fighter-bomber squadrons have a total of about 80 MiG-17s, alongside an operational conversion unit with a dozen MiG-15UTIs. From July 1978, about 15 MiG-23/27s were shipped to Cuba for two squadrons at San Julian and Guines. These include Flogger-B interceptors, Flogger-D strike aircraft and a few MiG-23U trainers, although the US is satisfied that none are, at present, equipped with nuclear weapons.

Transport forces have recently been strengthened

Top: six of 1⁰ Grupo de Aviacao's Grumman S-2E Trackers are normally deployed aboard the Brazilian navy's carrier Minas Gerais. The older S-2As have been relegated to training and transport duties. The Grupo, which comprises two squadrons, is shore-based at Santa Cruz.
Above: the Cessna T-37 was supplied by the US for basic jet training, but examples were withdrawn in 1977 in in favour of the Neiva T-25 and the EMB AT-26 Xavante

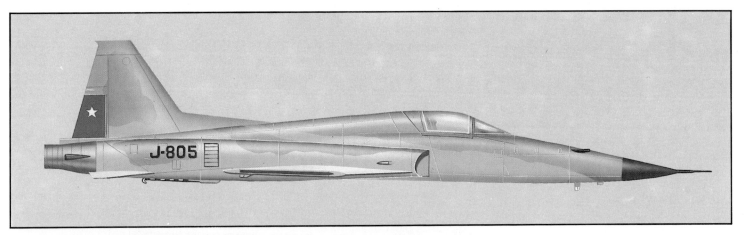

Top: Grupo 7 operates 15 Northrop F-5E fighter-bombers and three F-5F two-seat trainers delivered in 1978 from the US. It is based at Cerro Moreno AFB, Antofagasta with the Hawker Hunter-equipped Grupos 8 and 9, and these three Grupos are the main front-line units of the Fuerza Aerea de Chile. Above: Grupo 5, Fuerza Aerea de Chile operates some 14 DHC-6 Twin Otters on communications and light utility duties

by deliveries of no fewer than 20 Antonov An-24s in 1978–79, but a further batch of Il-14s received in early 1979 is apparently destined for the civil aviation training school. Unconfirmed reports mention some ten Mi-8 helicopters and a few Su-7 variable geometry fighter-bombers in service.

Developed mainly with United States' assistance since the Rio Mutual Defence Pact of 1947, the Fuerza Aerea Ecuatoriana has in recent years both expanded and diversified its sources of supply, currently sharing in an annual defence budget of some 60 million dollars. Concerned at the build-up of Soviet equipment in neighbouring Peru, to which one-third of its territory was lost in the war of 1941, Ecuador has recently attempted to obtain Douglas A-4 Skyhawks and Northrop F-5Es from the United States without success. While withholding more sophisticated weapons, the US shares misgivings over Peru's intentions and is proposing to supply anti-aircraft equipment including Chaparral SAMs and 20mm Vulcan cannon for defensive purposes.

Ecuador was one of the first two customers for the Jaguar International, taking delivery of 12, including two trainers, from the British production line in 1977. Attempts to obtain 24 Kfir-C2s were blocked by US objections to the sale by Israel of their J79 engines, and despite an offer of ATAR-engined Neshers, the FAE ordered 16 Mirage F1JA fighter-bombers and two F1JB trainers in 1977, first arrivals being in February 1979.

Jaguar delivery brought about the retirement of a squadron of Lockheed F-80 Shooting Stars in 1977, but the FAE continues to carry out photo-

reconnaissance with seven of the 12 Meteor FR Mk 9s received in 1954. Other British equipment includes three Canberra B Mk 6s from six delivered and 14 Strikemasters for basic training and secondary light-strike roles from purchases of eight, and two supplementary batches of four in 1974–76. A second COIN esquadrilla was established in 1976 with 12 Cessna A-37Bs provided by the United States of America.

FAE's Transport Group, including Transportes Aereos Militares Ecuatorianos (TAME), has also been substantially re-equipped in recent years, to supplement the original establishment of a dozen or so Douglas C-47s, four DC-6Bs and six Beech C-45 Expeditors. The FAE operates scheduled services with four ex-airline Lockheed L-188 Electras obtained in 1976, together with one of the two C-130H Hercules delivered in 1977, the second having crashed.

In 1974 a further two British Aerospace HS748s with large freight doors were ordered to supplement the two remaining earlier aircraft, while Canada also supplied two Buffaloes and three Twin Otters. There are additionally six IAI Aravas (four more are operated by the army and navy), a Piper Navajo and two Presidential Learjets. A single Consolidated PBY-5A amphibian is still operated for re-supply flights to the Galapagos Islands and on coastal patrol duties. The present TAME fleet of two HS748s, four Electras and two Twin Otters and the five Boeing 707/720s of Ecuatoriana have dual civil/military registrations, despite the flamboyance of the latter's colour schemes.

The original rotary-wing component, comprising three Bell 47s and six Alouette IIIs has been increased by two armed SA 315 Lamas and a similar number of Pumas.

Modernisation of the training programme has resulted in the delivery of 12 SIAI-Marchetti SF 260C basic trainers in 1977 and 20 Beech T-34C Turbo Mentors for the advanced stage, replacing T-28 Trojans at Guayaquil and the Air Academy, Maya Cesma Renella. Primary instruction continues to be undertaken on Cessna 150s and T-41s. Jet conversion is via the British Aerospace Strikemaster, assisted by a few Lockheed T-33s.

Aviacion Naval Ecuatoriana has a small naval air element with two IAI Arava transports, two Aérospatiale Alouette IIIs and an assortment of Cessnas. Somewhat larger is the Aviacion del Ejercito Ecuatoriana, or Army Aviation, which operates two Short Skyvans, two IAI Aravas, a

Above: the insignia of the Cuban Fuerza Areea Revolucionaria which is entirely equipped with Soviet Aircraft.
Below: the Fuerza Aerea Revolucionaria has four fighter-bomber squadrons currently operating some 80 Mig-19Fs.
Bottom: two Cuban interceptor squadrons are equipped with MiG-17MF Frescoes, a total of 40 having been received from the Soviet Union. MiG-17s and MiG-19s were the first Soviet front-line aircraft received by Cuba. However, the force also operates some 70 MiG-21s and has recently received 15 MiG-23s from the Soviet Union

photo-survey Learjet and several other light aircraft for liaison and communications.

Peru's revolutionary military administration, which seized power in 1968, pursues a procurement policy from both East and West and has armed forces relatively well-equipped with modern technology. Shortly after the revolution, Peru became the first Latin American country to obtain Mach 2 fighters, in the form of 12 Dassault Mirage 5P strike-interceptors and two 5DP trainers, followed by a further eight 5Ps in 1974 and a later order for an additional four.

Most recent, and potent, additions to the armoury have been 36 Sukhoi Su-22 Fitter-C fighter-bombers supplied on favourable credit terms by the USSR, after offers of Douglas Skyhawks and Northrop F-5Es from the United States had been turned down. Peruvian forces have also received large numbers of Mil Mi-8 medium-lift helicopters and SA-3 and SA-7 surface-to-air missiles from Russia in addition to comparatively small-scale military aid from the USA mostly taken up in spares and support facilities for existing American equipment.

The Fuerza Aerea Peruana currently comprises some 140 combat aircraft, plus transport and support types, manned by 10,000 personnel. National defence spending averages 125 million dollars per year–equivalent to the purchase price of the Su-22s alone–and Peru has been forced to re-negotiate debts to the Soviet Union to lessen the burden of re-equipment.

The FAP is currently organised within about eight main *grupos* or wings, each with two or three squadrons. Sukhoi Su-22s are believed to have replaced 14 North American F-86F Sabres and 20 AT-33 armed trainers in Grupo de Caza 12 at Limatambo AB, but the unit apparently retains ten of the 16 Hunter F Mk 52s and a single T Mk 62 purchased in 1956. Pilots for the Su-22s were trained in Cuba, the FAP also receiving 12 MiG-21s from President Castro to assist conversion. Continuation training is undertaken on the four two-seat Su-22s included in the batch under the direction of 100 Cuban and 75 Soviet advisers. The latter are additionally involved in the establishment of a radar and communications network associated with deliveries of SA-2, SA-3 and SA-7 SAMs to cover the border with Chile.

Second strike/interceptor wing is Grupo 13 at Chiclayo with 26 Mirages in two squadrons, plus an additional *escuadron* with half the 24 Cessna A-37B light-attack aircraft ordered in late-1973. The Mirages have been retro-fitted with inertial navigation systems and AS 30 air-to-surface missiles obtained from France.

Twelve remaining A-37s are flown by the third *escuadron* of Grupo de Bombardeo 21, which has as main equipment, two squadrons of Canberras at Jorge Chavez International Airport, Lima. Canberra purchases began with nine new B(I) Mk 8s in 1956, followed by a single replacement in 1960, and six refurbished B Mk 2s and two T Mk 4s in 1966. A further six B Mk 56s were obtained in

1969, joined by batches of eight and three B(I) Mk 68s overhauled by Marshall of Cambridge in 1974–77, bringing total procurement to 35, of which all except two apparently remain in service. Maritime reconnaissance is practised in conjunction with the Servicio Aeronavale by four Grumman HU-16B Albatross amphibians in Escuadron de Reconocimiento of Grupo 31 at Lima-Callao.

Extensive FAP transport commitments are managed through two major organisations formed in 1960, comprising Servicio Aerea de Transportes Comerciales (SATCO) for medium-lift operations, and Transportes Aereas Nacionales la Selva (TA-NS), the light aircraft division. Both organisations comprised several *grupos* and *escuadrons*, although in mid-1974 SATCO's Fokker F-27 and F-28 routes were taken over by the new national airline, AeroPeru. As mainstay of the transport force, Grupo 41 operates a squadron of two Lockheed L-100 and four C-130E Hercules from Jorge Chavez AB, while a second squadron in the wing has 16 DHC Buffaloes. Older equipment, two Curtiss C-46 Commandos, four C-54s and five DC-6s, are flown by a third *escuadron* of Grupo 41.

At Iquitos, Grupo 42 provides the main element of TANS with a dozen DHC Twin Otters, including several on floats, 12 Pilatus Turbo Porters and a number of C-47s. Several of FAP's older piston-engined transports have been replaced by deliveries of 16 Antonov An-26 twin-turbo prop tactical freighters which began in late-1977. Additional transport units based at Lima-Callao include Grupo 8 with 18 Beech Queen Airs, three King Airs and two Beech 99s in two Escuadrons, and the Presidential Flight, comprising a single Fokker

F-28 Fellowship twin-jet airliner.

Helicopters are concentrated in Grupo 3 at Callao, which maintains detachments to several bases, deliveries having included 20 or more Bell 47Gs, 17 Bell 212s, 12 Alouette IIIs, six UH-1Ds, five Mil Mi-6s and eight Mi-8s. The first two Mi-8s were presented by the Soviet Union in 1970, and the Peruvian Government then ordered a further six for the FAP and more for the army.

Second-line units also include Direccion General de Aerofotografia (DGAF) and its operating unit, Servicio Aerofotografino Nacional (SAN), provides aerial survey facilities for various civil engineering projects. Based at Las Palmas, SAN originally operated two specially-equipped Queen Airs, but in 1974 received two Learjets complete with cameras and inertial navigation systems.

First Latin American air arm to start *ab initio* jet training, the FAP obtained 20 Cessna T-37Bs for the Las Palmas Air Academy in 1960, to which were later added 12 T-37Cs. This was later found to

Top: an Ecuadorean British Aerospace HS748 displays its port-side loading door, which may be employed for paratrooping.
Middle: three English Electric Canberra (BAC) medium bombers remained in 1979 of the six originally delivered.

Above left: Peru has pursued a relatively wide-ranging programme of equipment, seeking sources of supply from Europe, the United States and the Soviet bloc. In common with many Latin American countries, its interceptor force includes the Dassault Mirage. Initial deliveries of the type in 1968 – the first to any air arm on the continent – were followed by further orders for the Mirage 5P.

Left: Grupo 41 is responsible for Peru's military transport capability. Largest aircraft operated is the Lockheed C-130, one of which is pictured, flown from Jorge Chavez AB. Other types employed in the medium-lift roles include the DHC Buffalo, the Douglas C-54 and DC-6, together with a pair of Curtiss C-46s

result in excessive student wastage and basic training on T-41s was then introduced. Advanced instruction now follows on eight T-33s, followed by operational conversion in first-line units. A few North American T-6s, Beech T-34As and Cessna 150s are used for miscellaneous duties.

Having started in the 1960s as a small helicopter force with eight Bell 47s, the Servicio Aeronavale de Marina Peruana (Peruvian Naval Aviation Service) quickly expanded by procurement of further rotorcraft and increasing numbers of fixed-wing aircraft, operating mainly from Callao on ASW patrol, reconnaissance, fleet support and transport duties.

Mainstay of the SAMP is a fleet of nine Grumman S-2E Trackers, delivered in 1976 to replace earlier S-2As, together with two F-27MPA Friendships received in 1977–78, representing the first order for the maritime Friendship. Transfers from the FAP have added seven C-47s and two T-34A Mentors to the naval inventory.

Original helicopter equipment has been augmented by two Aérospatiale Alouette IIIs, ten Bell JetRangers and six UH-1D/H Iroquois. Six Agusta-Bell 212AS helicopters are being delivered for deployment aboard four frigates built in Italy, armament of the latter including OTOMAT long-range ship-to-ship missiles and Selena Aspide SAMs for air defence. MM 38 Exocets are fitted to six fast patrol boats ordered from France, while

Italy is also supplying four SH-3D Sea Kings for ASW. In early 1978, the Netherlands frigate *de Zeven Provincen* was transferred to Peruvian Navy control, taking delivery of the first AB 212AS while still in Holland. More ambitious plans for the purchase of HMS *Bulwark* and a small number of Sea Harriers expressed in 1976, have now been abandoned. Six Beech T-34C Turbo Mentors were delivered to SAMP in March 1978 for basic training duties.

Peruvian army aviation received a massive infusion of equipment in the early 1970s following delivery of 48 Mil Mi-8 medium helicopters to supplement the previous small force of eight Bell 47s and six Alouette IIIs. Although made available by the Soviet Union at bargain prices, Mil supply greatly exceeded demand, 36 remaining in storage to the present day.

Fixed-wing aircraft are used for liaison and communications, comprising five Helio Couriers, two of which are float-equipped, and five Cessna 185s. Interest was expressed in two GAF Nomads for medium transport tasks, but the requirement has now apparently been withdrawn.

As a Rio Pact signatory, Venezuela has received considerable United States military assistance from 1947 onwards, supplemented by procurements from Britain, Canada and France, in an annual defence budget of some £300 million. Present strength of the

Above: the Mil Mi-8 formed a major part of a Soviet arms package to Peru, but does not appear to play a significant role in the country's order of battle in the late 1970s. A number of the medium-lift helicopters have been utilised for oil exploration duties, while the majority of the army's Mi-8s is in storage.
Below: two examples of the Fokker F-27MPA maritime Friendship were delivered to Peru's naval air arm in 1977–78, being the first such aircraft to enter military service anywhere in the world. The Friendship purchase also represents the Navy's movement from rotorcraft to fixed-wing equipment, as exemplified by the nine Grumman S-2E Trackers delivered in 1976

Fuerzas Aereas Venezolanas is about 240 aircraft and 8,000 personnel.

Current organisation comprises three major commands: Comando Aereo de Combate, controlling all first-line and transport unit, Comando Aereo de Instruccion, responsible for training activities and Comando Aereo Logistico for supply. Aircraft units are administered within five *grupos*, two of which form the FAV combat element of six *escuadrones*.

Fighter elements are concentrated in Grupo de Caza 12 at Base Aerea El Libertador, Palo Negro with three squadrons of varying complements. Escuadron de Caza 34, previously one of two units equipped with 47 North American F-86K Sabres bought from Germany in 1966, converted to 16 Canadair CF-5A fighter-bombers and two CF-5B trainers transferred from CAF stocks in 1970,

and has since received a further two twin-seat CF-5Ds. Sabres are still retained by Escuadron de Caza 35, surplus aircraft having gone to other Latin American air forces. Third fighter squadron, Escuadron de Caza 36, re-equipped with nine Dassault Mirage IIIEV strike-interceptors, four Mirage 5V fighter-bombers and two 5DV trainers from its previous complement of 20 F-86F Sabres and has subsequently received one IIIEV attrition replacement from the French manufacturers.

As remaining combat element, Grupo de Bombardeo 13 operates from Teniente Vicente Landaeta BA, Barquisimento (Lara State) with three squadrons. Two of these, Esquadrones de Bombardeo 38 and 39 operate 23 of the 30 Canberras bought from 1952 onwards. Between 1975 and 1979 all surviving aircraft were re-worked by British Aerospace at Salmesbury for continued service. The third squadron of Grupo 13, Escuadron de Bombardeo 40, exchanged its North American B-25J Mitchells for 16 OV-10E Broncos in 1972, operating three *esquadrilla* (flights) on border patrol and counter-insurgency tasks.

Transport is the responsibility of Grupo de Transporte 6 at Francisco de Miranda BA, La Carlota, Caracas with three attached units. For VIP and governmental transport, Escuadrilla Presidencial has a single Boeing 737, delivered in early 1976, a DC-9, a Cessna Citation and two Bell UH-1N twin-turbine helicopters. Heavy work is allocated to the six Hercules (a seventh crashed near the Azores in 1976) of Escuadron de Transporte 1, which also has some 20 Douglas C-47s and two British Aerospace HS748s. Escuadron de Transporte 2 still has about 12 of the 16 Fairchild C-123 Providers ordered in 1956.

Additional transport capability, together with tactical reconnaissance, forward air control, liaison

Above: the Fuerza Aerea Venezolanas has long been an operator of the English Electric (BAC) Canberra, taking delivery of its first example in 1952. An extensive refurbishing programme is planned to extend still further the airframe lives of these pioneer jet bombers in FAV service.

Below: a Venezuelan Dassault Mirage 5V (nearest camera) formates with an Argentine Mirage IIIE. The type succeeded the North American F-86F

and communications is undertaken by Grupo Mixto de Enlace y Reconocimiento from Teniente Luis del Valle Garcia AB, Barcelona. This has some six Beech Queen Airs, a dozen Cessna 182Ns and 30 helicopters including most of the FAV's 20 UH-1 Iroquois and several Alouette IIIs, from 20 delivered. Six JetRangers and one LongRanger were received in 1977, some of which have been passed on to the army.

Grupo de Entrenamiento Aereo provides all flying tuition up to advanced standard under Comando de Instruccion, headquartered at Mariscal Sucre BA, Boca de Rio, Maracay. Initial training is begun at Escuela de Aviacion Militar on 25 of the 34 Beech T-34A Mentors obtained in 1958–59, followed by twin conversion for transport pilots on Queen Airs and C-47s before posting to Grupo de Transporte for advanced training. Future combat pilots move from Mentors to the advanced school at Palo Negro, where Jet Provost T Mk 52s were replaced from 1973 onwards by 12 Rockwell

T-2D Buckeyes for completion of 120-hour wings courses. T-2D complement was doubled in 1976 and the type allocated a secondary light-strike role.

Venezuela's army (Ejercito Venezolana) has an air element, equipped mainly with some 20 Aérospatiale Alouette IIIs and Bell 47Gs for liaison and observation. At least one King Air is used for staff transport, ground forces additionally employing SS 11 anti-tank missiles. Two Swearingen Merlins were received in early-1979.

A small shore-based air arm operates on behalf of the Marina Venezolana with six S-2E Trackers received in 1974–75 for ASW patrol. Other naval aircraft include four HU-16A Albatross amphibians used as SAR aircraft, an HS748 transferred from the national postal service in 1977, a King Air, Piper Aztec and two Bell 47J Rangers. These are to be supplemented by ten Agusta-Bell 212AS helicopters for deployment on six frigates on order from Italy, the vessels also being armed with Albatross anti-aircraft missiles and ship-to-ship missiles.

Index

INDEX

INDEX